T0408237

MANAGEMENT OF INSECT PESTS IN VEGETABLE CROPS

Concepts and Approaches

Innovations in Horticultural Science

MANAGEMENT OF INSECT PESTS IN VEGETABLE CROPS

Concepts and Approaches

Edited by

Ramanuj Vishwakarma, PhD
Ranjeet Kumar, PhD

APPLE ACADEMIC PRESS

Apple Academic Press Inc.
4164 Lakeshore Road
Burlington ON L7L 1A4, Canada

Apple Academic Press Inc.
1265 Goldenrod Circle NE
Palm Bay, Florida 32905, USA

© 2020 by Apple Academic Press, Inc.

Exclusive worldwide distribution by CRC Press, a member of Taylor & Francis Group

No claim to original U.S. Government works

International Standard Book Number-13: 978-1-77188-859-2 (Hardcover)
International Standard Book Number-13: 978-0-42932-884-8 (eBook)

All rights reserved. No part of this work may be reprinted or reproduced or utilized in any form or by any electric, mechanical or other means, now known or hereafter invented, including photocopying and recording, or in any information storage or retrieval system, without permission in writing from the publisher or its distributor, except in the case of brief excerpts or quotations for use in reviews or critical articles.

This book contains information obtained from authentic and highly regarded sources. Reprinted material is quoted with permission and sources are indicated. Copyright for individual articles remains with the authors as indicated. A wide variety of references are listed. Reasonable efforts have been made to publish reliable data and information, but the authors, editors, and the publisher cannot assume responsibility for the validity of all materials or the consequences of their use. The authors, editors, and the publisher have attempted to trace the copyright holders of all material reproduced in this publication and apologize to copyright holders if permission to publish in this form has not been obtained. If any copyright material has not been acknowledged, please write and let us know so we may rectify in any future reprint.

Trademark Notice: Registered trademark of products or corporate names are used only for explanation and identification without intent to infringe.

Library and Archives Canada Cataloguing in Publication

Title: Management of insect pests in vegetable crops : concepts and approaches / edited by Ramanuj Vishwakarma, PhD, Ranjeet Kumar, PhD.

Names: Vishwakarma, Ramanuj, 1984- editor. | Kumar, Ranjeet, 1982- editor.

Description: Includes bibliographical references and index.

Identifiers: Canadiana (print) 20200177079 | Canadiana (ebook) 20200177133 | ISBN 9781771888592 (hardcover) | ISBN 9780429328848 (ebook)

Subjects: LCSH: Vegetables—Diseases and pests—Control. | LCSH: Insect pests—Control. | LCSH: Vegetables—Diseases and pests—Control—Environmental aspects.

Classification: LCC SB608.V4 M36 2020 | 635/.0497—dc23

Library of Congress Cataloging-in-Publication Data

Names: Vishwakarma, Ramanuj, 1984- editor. | Kumar, Ranjeet, 1982- editor.

Title: Management of insect pests in vegetable crops : concepts and approaches / edited by Ramanuj Vishwakarma, Ranjeet Kumar.

Other titles: Innovations in horticultural science.

Description: Palm Bay, Florida, USA : Apple Academic Press, [2020] | Series: Innovations in horticultural science | Includes bibliographical references and index. | Summary: "This new book on the sustainable management of insect pests in important vegetables offers valuable management strategies in detail. It focuses on eco-friendly technology and approaches to mitigating the damage causes by insect pests with special reference to newer insecticides. Chapters in Management of Insect Pests in Vegetable Crops: Concepts and Approaches provide an introduction to vegetable entomology and go on to present a plethora of research on sustainable eco-friendly pest management strategies for root vegetables, spice crops, tuber crops, and more. It looks at the damage done by snails, slugs, phytophaous mites of vegetable crops, and more and presents management strategies. Vegetable crops that are infested by several insect pests right from the nursery to the harvesting stage cause enormous crop losses. Given that it is estimated that up to 40 percent of global crops are lost to agricultural pests each year, new research on effective management strategies is vital. The valuable information provided in this book will be very helpful for faculty and advanced-level students, scientists and researchers, policymakers, and others involved in pest management for vegetable crops"-- Provided by publisher.

Identifiers: LCCN 2020003666 (print) | LCCN 2020003667 (ebook) | ISBN 9781771888592 (hardcover) | ISBN 9780429328848 (ebook)

Subjects: LCSH: Vegetables--Diseases and pests--Control. | Insect pests--Control. | Vegetables--Diseases and pests--Control--Environmental aspects.

Classification: LCC SB608.V4 M356 2020 (print) | LCC SB608.V4 (ebook) | DDC 632/.7--dc23

LC record available at https://lccn.loc.gov/2020003666

LC ebook record available at https://lccn.loc.gov/2020003667

Apple Academic Press also publishes its books in a variety of electronic formats. Some content that appears in print may not be available in electronic format. For information about Apple Academic Press products, visit our website at **www.appleacademicpress.com** and the CRC Press website at **www.crcpress.com**

DEDICATION

Dedicated to our wives with love

INNOVATIONS IN HORTICULTURAL SCIENCE

Editor-in-Chief:

Dr. Mohammed Wasim Siddiqui Assistant Professor-cum-Scientist
Bihar Agricultural University | www.bausabour.ac.in
Department of Food Science and Post-Harvest Technology
Sabour | Bhagalpur | Bihar | P. O. Box 813210 | INDIA
Contacts: (91) 9835502897
Email: wasim_serene@yahoo.com | wasim@appleacademicpress.com

The horticulture sector is considered as the most dynamic and sustainable segment of agriculture all over the world. It covers pre- and postharvest management of a wide spectrum of crops, including fruits and nuts, vegetables (including potatoes), flowering and aromatic plants, tuber crops, mushrooms, spices, plantation crops, edible bamboos etc. Shifting food pattern in wake of increasing income and health awareness of the populace has transformed horticulture into a vibrant commercial venture for the farming community all over the world.

It is a well-established fact that horticulture is one of the best options for improving the productivity of land, ensuring nutritional security for mankind and for sustaining the livelihood of the farming community worldwide. The world's populace is projected to be 9 billion by the year 2030, and the largest increase will be confined to the developing countries, where chronic food shortages and malnutrition already persist. This projected increase of population will certainly reduce the per capita availability of natural resources and may hinder the equilibrium and sustainability of agricultural systems due to overexploitation of natural resources, which will ultimately lead to more poverty, starvation, malnutrition, and higher food prices. The judicious utilization of natural resources is thus needed and must be addressed immediately.

Climate change is emerging as a major threat to the agriculture throughout the world as well. Surface temperatures of the earth have risen significantly over the past century, and the impact is most significant on agriculture. The rise in temperature enhances the rate of respiration, reduces cropping periods, advances ripening, and hastens crop maturity, which adversely affects crop productivity. Several climatic extremes such as droughts, floods, tropical cyclones, heavy precipitation events, hot extremes, and heat waves cause a negative impact on agriculture and are mainly caused and triggered by climate change.

In order to optimize the use of resources, hi-tech interventions like precision farming, which comprises temporal and spatial management of resources in horticulture, is essentially required. Infusion of technology for an efficient utilization of resources is intended for deriving higher crop productivity per unit of inputs. This would be possible only through deployment of modern hi-tech applications and precision farming methods. For improvement in crop production and returns to farmers, these technologies have to be widely spread and adopted. Considering the above-mentioned challenges of horticulturist and their expected role in ensuring food and nutritional security to mankind, a compilation of hi-tech cultivation techniques and postharvest management of horticultural crops is needed.

This book series, Innovations in Horticultural Science, is designed to address the need for advance knowledge for horticulture researchers and students. Moreover, the major advancements and developments in this subject area to be covered in this series would be beneficial to mankind.

Topics of interest include:
1. Importance of horticultural crops for livelihood
2. Dynamics in sustainable horticulture production
3. Precision horticulture for sustainability
4. Protected horticulture for sustainability
5. Classification of fruit, vegetables, flowers, and other horticultural crops
6. Nursery and orchard management
7. Propagation of horticultural crops
8. Rootstocks in fruit and vegetable production
9. Growth and development of horticultural crops
10. Horticultural plant physiology
11. Role of plant growth regulator in horticultural production
12. Nutrient and irrigation management
13. Fertigation in fruit and vegetables crops
14. High-density planting of fruit crops
15. Training and pruning of plants
16. Pollination management in horticultural crops
17. Organic crop production
18. Pest management dynamics for sustainable horticulture
19. Physiological disorders and their management
20. Biotic and abiotic stress management of fruit crops
21. Postharvest management of horticultural crops
22. Marketing strategies for horticultural crops
23. Climate change and sustainable horticulture
24. Molecular markers in horticultural science
25. Conventional and modern breeding approaches for quality improvement
26. Mushroom, bamboo, spices, medicinal, and plantation crop production

BOOKS IN THE SERIES

- **Spices: Agrotechniques for Quality Produce**
  Amit Baran Sharangi, PhD, S. Datta, PhD, and Prahlad Deb, PhD

- **Sustainable Horticulture, Volume 1: Diversity, Production, and Crop Improvement**
  Editors: Debashis Mandal, PhD, Amritesh C. Shukla, PhD, and Mohammed Wasim Siddiqui, PhD

- **Sustainable Horticulture, Volume 2: Food, Health, and Nutrition**
  Editors: Debashis Mandal, PhD, Amritesh C. Shukla, PhD, and Mohammed Wasim Siddiqui, PhD

- **Underexploited Spice Crops: Present Status, Agrotechnology, and Future Research Directions**
  Amit Baran Sharangi, PhD, Pemba H. Bhutia, Akkabathula Chandini Raj, and Majjiga Sreenivas

- **The Vegetable Pathosystem: Ecology, Disease Mechanism, and Management**
  Editors: Mohammad Ansar, PhD, and Abhijeet Ghatak, PhD

- **Advances in Pest Management in Commercial Flowers**
  Editors: Suprakash Pal, PhD, and Akshay Kumar Chakravarthy, PhD

- **Diseases of Fruits and Vegetable Crops: Recent Management Approaches**
  Editors: Gireesh Chand, PhD, Md. Nadeem Akhtar, and Santosh Kumar

- **Management of Insect Pests in Vegetable Crops: Concepts and Approaches**
  Editors: Ramanuj Vishwakarma, PhD, and Ranjeet Kumar, PhD

ABOUT THE EDITORS

Ramanuj Vishwakarma, PhD

Dr. Ramanuj Vishwakarma is presently working as an assistant professor-cum-junior scientist in the Department of Entomology, Bihar Agricultural College, Bihar Agricultural University, Sabour, Bhagalpur, Bihar, India. He is also the Nodal Officer of beekeeping-cum-honey production unit of the university.

Dr. Vishwakarma has significantly contributed to the field of agricultural entomology. During his entire tenure, he has specialized in integrated pest management in broccoli, bottle gourd, brinjal, chili, cabbage, mustard, and sunflower by using a number of plant products and entomopathogenic fungi. He did research on the pollinating efficiency of insect pollinators, including honeybees, for enhancing crop productivity and for conserving the ecosystem. He also worked on various aspects of insect enemies of honeybees with special reference to the greater wax moth.

He has contributed as organizing secretary and successfully organized a two-day national seminar on Quality Honey Production for Livelihood Security, which was held on 5 and 6 August, 2016 at Bihar Agricultural University, Sabour, India

He has received many prestigious awards, namely, Outstanding Scientist in Agriculture Award, Bharat Shiksha Ratan Award, Junior Scientist of the Year Award, and Best Young Teacher Award; and he was elected as a Life Fellow of the Entomological Society of India and International Society for Research and Development, India.

He has published about 13 research articles in reputed national and international journals. He has written/edited and published five books, 10 book chapters, one bulletin, three manuals, three souvenirs, three Kisan Diaries, four popular articles, two folders and teaching materials, and also handled/handling three research projects funded by ICAR, New Delhi, and BAU, Sabour, India. He has successfully guided three MSc (Ag.) students for their degree programs and is currently guiding three students for the same.

He did his MSc (Ag.) in agricultural entomology from Narendra Deva University of Agriculture and Technology, Kumarganj, Faizabad, Uttar Pradesh, India and PhD from Bidhan Chandra Krishi Viswavidyalaya, Mohanpur, Nadia, West Bengal.

Ranjeet Kumar, PhD

Dr. Ranjeet Kumar is presently working as an assistant professor-cum-junior scientist in the Post Graduate Department of Entomology, Bihar Agriculture University, Sabour, Bhagalpur, Bihar, India.

Dr. Ranjeet Kumar has worked on stored product entomology for the sustainable and herbal management of stored grain and seed insect pests. During the course of investigation Dr. Kumar developed several herbal fumigants for sustainable management of stored grain and seed insect.

Dr. Kumar has significantly contributed to the field of agricultural entomology as well as stored product entomology. Dr. Kumar published two books from International Book Distributing Company, Lucknow, India; one manual from Danteshwari College of Horticulture, Indira Gandhi Krishi Vishwavidyalya, Raipur, India; one book from AAP–CRC Press, USA; one book in Hindi from Agricultural Technology Management Agency, Gaya, Bihar, India. In addition, Dr. Kumar has published 28 research papers in journals of international and national reputes and 19 popular articles in magazines and periodicals. Dr. Kumar is a life member fellow of several scientific committees and societies.

Dr. Kumar received an award from the International Conference on Entomology at Punjabi University, Punjab, India. Dr. Kumar has vast experience in agricultural entomology as well as stored product entomology for teaching, research, and extension.

Dr. Kumar did his PhD (stored product entomology) in the year 2010 from G.B. Pant University of Agriculture and Technology, Pantnagar, Uttarakhand, India.

CONTENTS

CONTRIBUTORS

Rahul Kumar Chandel
Division of Entomology, Indian Agriculture Research Institute, New Delhi, India

P. Chand
Department of Agricultural Entomology, Bidhan Chandra Krishi Vishwavidyalya,
Mohanpur, Nadia, West Bengal, India

Pritam Ganguli
Department of Soil Science and Agricultural Chemistry, Bihar Agricultural University,
Sabour, Bhagalpur 813210, Bihar, India

Geetanjly
Division of Entomology, Indian Agriculture Research Institute, New Delhi, India

Dinesh Prasad Gond
Department of Endocrinology, Institute of Medical Science, Banaras Hindu University,
Varanasi 221005, Uttar Pradesh, India

Hari Chand Ingle
Department of Entomolgy, Dr. Rajendra Prasad Central Agricultural University, Pusa,
Samastipur, Bihar, India

Jaba Jagdish
International Crops Research Institute for the Semi-Arid Tropics, Patncheru, Hyderabad,
Telangana, India

Ranjeet Kumar
Department of Entomology, Bihar Agricultural University, Sabour, Bhagalpur 813210, Bihar, India

Mrinalini Kumari
Department of Entomology, Bihar Agricultural University, Sabour, Bhagalpur 813210, Bihar, India

Kalmesh Managanvi
Department of Entomology, Bihar Agricultural University, Sabour, Bhagalpur 813210, Bihar, India

Vijay Kumar Mishra
Department of Entomology, Institute of Agricultural Sciences, B. H. U. Varanasi,
Uttar Pradesh, India

Navendu Nair
Department of Entomology, Central University of Tripura, Lembucherra 799210, West Tripura, India

Atanu Seni
Department of Entomology, Regional Research and Technology Transfer Station,
Odisha University of Agriculture and Technology, Chiplima 768025, Sambalpur, Odisha, India

Preeti Sharma
Development Center, International Crops Research Institute for the Semi-Arid Tropics,
Ptancheru, Hyderabad, India

Dipak Shyamrao
Department of Entomology, Institute of Agricultural Sciences, B. H. U. Varanasi,
Uttar Pradesh, India

Amit Singh
Department of Agricultural Entomology, Bidhan Chandra Krishi Vishwavidyalya, Mohanpur,
Nadia, West Bengal, India

Pushpa Singh
Department of Entomolgy, Dr. Rajendra Prasad Central Agricultural University, Pusa,
Samastipur, Bihar, India

R. N. Singh
Department of Entomology, Institute of Agricultural Sciences, B. H. U. Varanasi,
Uttar Pradesh, India

C. P. Srivastva
Department of Entomology, Institute of Agricultural Sciences, B. H. U. Varanasi,
Uttar Pradesh, India

Budhachandra Thangjam
Department of Entomology, Central University of Tripura, Lembucherra 799210,
West Tripura, India

Ramanuj Vishwakarma
Department of Entomology, Bihar Agricultural University, Sabour, Bhagalpur 813210,
Bihar, India

ABBREVIATIONS

ACS   American Chemical Society
ADI   acceptable daily intake
APC   Agricultural Price Commission
APEDA   Agricultural and Processed Food Products Export
            Development Authority
CA   controlled atmosphere
CACP   Commission on Agricultural Costs and Prices
CIB   Central Insecticides Board
CNS   central nervous system
CT   carbon tetrachloride
CWC   Central Warehousing Corporation
DBM   diamondback moth
DIPA   destructive insect pest act
DPPQS   Directorate Of Plant Protection and Storage
EC   Economic Community
EC   emulsifiable concentrate
ECD   electron capture detector
EDB   ethylene dibromide
EPA   Environmental Protection Agency
FAO   Food and Agriculture Organization
FCI   Food Corporation of India
FID   flame ionization detector
FIFRA   Federal Insecticide, Fungicide, and Rodenticide Act
FPD   flame photometric detector
FQPA   Food Quality Protection Act
FSSA   Food Safety and Standards Act
FSSAI   Food Safety and Standards Authority of India
GAP   Good Agricultural Practice
GC   gas chromatography
GCMS   gas chromatography and mass spectra
GC–MS   gas chromatography–mass spectrometry
GIT   gastrointestinal tract effects
GLC   gas liquid chromatography

HPLC	high performance liquid chromatography
IGSI	Indian Grain Storage Institute
IGSRTI	Indian Grain Storage Research and Training Institute
IOBC	International Organization on Biological Control
IPM	integrated pest management
IPPC	International Plant Protection Convention
IR	infrared
LC	lethal concentration
LD	lethal dose
LOQ	level of quantification
MB	methyl bromide
MRLS	maximum residual limits
MS	mass spectrometer
NBARD	National Bank for Agriculture and Rural Development
NCI	National Cancer Institute
NCIPM	National Center for Integrated Pest Management
NE	natural enemies
NIR	near infrared
NRI	Natural Resource Institute
OC	organochlorine
PDA	photo diode array
PFA	Prevention of Food Adulteration Act
PFAA	Prevention of Food Adulteration Act
PPO	Plant Protection Organization
PTM	potato tuber moth
RBD	randomized block design
RC	Registration Committee
RCA	Royal Commission on Agriculture
REACH	registration, evaluation, authorisation and restriction of chemical substances
SGC	save grain campaign
SGRL	stored grain research laboratory
SPE	solid phase extraction
SWC	State Warehousing Corporation
TC	toxic concentration
TCD	thermal conductivity detector
TD	toxic dose
TLC	thin layer chromatography

USDA	United Nation Department of Agriculture
UV–VIS	ultraviolet–visible
WCA	Warehousing Corporation Act
WDRA	Warehouse Development and Regulation Act
WFD	World Food Day
WHO	World Health Organization
WP	wettable powder

ACKNOWLEDGMENT

Due to blessing of god **HANUMAN JI**, we are able to complete this book. We express our heartiest gratitude to god **HANUMAN JI** for his continuous blessing to write this book.

We express our sincere thanks to Bihar Agricultural University, Sabour, Bhagalpur, India, for providing an opportunity to execute this work.

We feel elated to offer our thanks to the founder and former honourable Vice Chancellor Dr. M. L Chaudhary and former Dean Agriculture Dr. D. Roy Bihar Agricultural University for his constant encouragement and guidance.

We are very much grateful to Dr. Wasim Siddiqui, Assistant Professor cum Junior Scientist, Department of Food Science and Technology, Bihar Agricultural University, Sabour, India for his inspiration and consideration of this project work in very short time.

We are grateful to Mr. Ashish Kumar, President, Apple Academic Press to complete our dream in form of book. We are also grateful to Ms. Sandra Jones Sickels and Mr. Rakesh Kumar of Apple Academic Press for their constant support and inspiration to publish this book.

We feel immense pleasure to express our gratitude to our adored parents and family members for their continuous inspiration and blessings in our life.

We express our special thanks to our better halves and heartiest benevolent for silent prayer, moral support, and encouragement for completion of this book.

PREFACE

Fruits and vegetables play an important role in the balanced diet of human beings by providing energy-rich nutrients and minerals. Vegetables are a rich and comparatively cheap source of vitamins and minerals. Their consumption in sufficient quantities provides taste, palatability, and increased appetite, and provides a fair amount of fiber. Vegetables are currently reckoned as an important adjunct for maintenance of good health and are beneficial in protecting against some degenerative diseases. Vegetables also play a key role in neutralizing the acids produced during digestion of proteinous and fatty foods and also provide valuable roughage, which promotes digestion and helps in preventing constipation. Some vegetable are good sources of carbohydrates, proteins, and vitamins A and C.

India is fortunate enough to have a varied agro-climatic zone found across the country, which enables it to produce several kinds of vegetables. India contributes a sizable share in terms of area and production under important vegetable crops. Vegetables are an integral part of our daily nutritional requirement. It is a well-known fact that vegetables are important sources of income as well as nutritional security. In the last five decade, there has been a tremendous increase in vegetable production, and India is now the second largest vegetable producer, next to China, with about 75 million tons. The existing area under vegetable cultivation in India is around 4.5 million ha. The majority of Indians are vegetarian, with a per capita consumption of 135 g per day as against the recommended 300 g per day, which is still very less than the recommended diet level. In near future, there is a need of around 5–6 million tons of food to feed our 1.3 billion Indian population expected by the year 2020.

The vegetable crops become infested by several insect pests right from the nursery to harvesting stage, which cause enormous losses. Vegetable crops from sowing till harvesting get infested by several insect pests in India; most of the time farmers use many kinds of insecticide to combat the pest population.

The present book elaborates on the insect pests of important vegetable crops, like root vegetables, turmeric and ginger, cumin, cucurbits, spice crops, tomato, and tubers. In this book, we discuss non-insect pests like mites infesting vegetable and spice crops as well as pesticide residues in vegetables and snails and slugs infesting vegetable crops. This book focuses on major insect pests and their effective management strategies.

This book covers the aspects of insect pest and pest management strategies in detail. It consists of ten chapters. Chapter 1 describes eco-friendly pest management strategies of root vegetable crops. The insect pests of ginger and turmeric and their management are discussed in Chapter 2. Chapter 3 deals with the insect pests of cumin and their management. The insect pests of cucurbitaceous vegetable and their management are elaborated on in Chapter 4. The phytophagous mites of vegetable crops and their management are presented in Chapter 5. Chapter 6 deals with mite problems in spice crops and their management. Chapter 7 discusses pesticide residues in vegetables. Chapter 8 deals with insect pests of tomato and their management. Chapter 9 deals with insect pests of tuber crops and their management. Chapter 10 discusses the snails and slugs that infest vegetable crops.

In all, this book provides comprehensive information in the field of vegetable entomology in order to facilitate the sustainable management of insects and other non-insect pests.

The editors would appreciate receiving comments and suggestions from readers that will be helpful in subsequent editions.

—Ramanuj Vishwakrma
Ranjeet Kumar

CHAPTER 1

ECO-FRIENDLY PEST MANAGEMENT STRATEGIES OF ROOT VEGETABLE CROPS

KALMESH MANAGANVI[1*], RAMANUJ VISHWAKARMA[1], and JABA JAGDISH[2]

[1]Department of Entomology, Bihar Agricultural University, Sabour, Bhagalpur 813210, Bihar, India

[2]International Crops Research Institute for the Semi-Arid Tropics, Patncheru, Hyderabad, Telangana, India

*Corresponding author. E-mail: kalmesh.managanvi@gmail.com

ABSTRACT

Root vegetable crops are second only in importance to cereals as a global source of carbohydrates. They provide many minerals and essential vitamins, although a proportion of the minerals and vitamins may be lost during processing. When it comes to replacing grains in your diet with root vegetables, there are many benefits. Root vegetables are truly natural, unadulterated sources of complex carbohydrates, antioxidants and important nutrients and they cause less digestive or inflammatory issues than many grains do. The quality and quantity of the protein in starchy staples are variable and relatively low on a fresh weight basis. Root vegetables have been a staple in many counties diets for thousands of years. In fact, records show that certain root veggies like sweet potatoes were an important ingredient in folk medicine over thousands of years ago, and they've supported undernourished populations around the world ever since. Today, there exists strong evidence that some of the vital nutrients found in many root vegetables including vitamin C, vitamin A, magnesium potassium and dietary fibre can help fight cancer, diabetes, obesity and inflammatory based disorders like heart disease and arthritis.

1.1 INTRODUCTION

Root vegetables have been a staple in many counties' diets for thousands of years. In fact, records show that certain root veggies like sweet potatoes were an important ingredient in folk medicine over thousands of years ago, and they have supported undernourished populations around the world ever since. Today, there exists strong evidence that some of the vital nutrients found in many root vegetables including vitamin C, vitamin A, magnesium, potassium, and dietary fiber can help fight cancer, diabetes, obesity, and inflammatory-based disorders like heart disease and arthritis.

Root vegetable crops are second only in importance to cereals as a global source of carbohydrates. These provide many minerals and essential vitamins, although a proportion of the minerals and vitamins may be lost during processing. When it comes to replacing grains in your diet with root vegetables, there are many benefits. Root vegetables are truly natural, unadulterated sources of complex carbohydrates, antioxidants, and important nutrients and they cause less digestive or inflammatory issues than many grains do. The quality and quantity of the protein in starchy staples are variable and relatively low on a fresh weight basis.

1.2 PESTS OF NATIONAL SIGNIFICANCE OF ROOT VEGETABLES

Common name	Scientific name	Family	Order
1. Potato			
Tuber moth	*Phthorimaea operculella*	Gelechiidae	Lepidoptera
Colorado potato beetle	*Leptinotarsa decemlineata*	Chrysomelidae	Coleoptera
Green peach aphid, potato aphid	*Myzus persicae* *Macrosiphum euphorbiae*	Aphididae	Hemiptera
Cutworm	*Agrotis ipsilon*	Noctuidae	Lepidoptera
Hadda beetle	*Henosepilachna vigintioctopunctata* (*Epilachna vigintioctopunctata*)	Coccinellidae	Coleoptera

Common name	Scientific name	Family	Order
2. Carrot			
Aster leafhopper	*Macrosteles quadrilineatus*	Cicadellidae	Hemiptera
Flea beetle	*Systena blanda*	Chrysomelidae	Coleoptera
Willow carrot aphid	*Cavariella aegopodii*	Aphididae	Hemiptera
Green peach aphid	*Myzus persicae*		
Carrot Weevil	*Listronotus oregonensis*	Curculionidae	Coleoptera
Carrot rust fly	*Psila rosae*	Psilidae	Diptera
Cutworm	*Agrotis* spp.	Noctuidae	Lepidoptera
3. Radish			
Diamondback moth	*Plutella xylostella*	Yponomeutidae	Lepidoptera
Aphids	*Myzus persicae*	Aphididae	Hemiptera
Root maggots	*Delia radicum*	Anthomyiidae	Diptera
Black cutworm	*Agrotis ipsilon*	Noctuidae	Lepidoptera
Granulate cutworm	*Feltia subterranea*		
4. Yam			
Scale insect	*Aspidiella hartii*	Diaspididae	Hemiptera
San Jose scale	*Quadraspidiotus perniciosus*	Diaspididae	Hemiptera
Mealy bugs	*Geococcus coffeae,* *Planococcus citri,* *Phenococcus gossypii*	Pseudococcidae	Hemiptera
Vine borers	*Apomecyna saltator*	Cerambycidae	Coleoptera
Wood boring beetle	*Clytocera chinospila*	Cerambycidae	Coleoptera
Leaf eating beetles	*Galerucida bicolor,* *Lema lacorelairei,* *Crioceris impressa*	Galerucidae	Coleoptera
Yam tuber beetle	*Heteroligus meles,* *H. appius*	Dynastidae	Coleoptera
White grub	*Leucopholis coneophora*	Scarabaeidae	Coleoptera
Coffee bean weevil	*Araecerus fasciculatus,* *A. laevigatus*	Anthribidae	Coleoptera
Termite	*Odontotermes escherichi*	Termitidae	Isoptera
Tussock caterpillar	*Dasychira mendosa*	Lymantriidae	Lepidoptera
Leaf feeder	*Ansioarthra coerulea*	Tenthredinidae	Hymenoptera
Yam nematode	*Hoplolaimus* spp., *Scutellonema* *Bradys*	Hoplolaimidae	Secernentea
Root-knot nematode	*Meloidogyne* spp.	Heteroderidae	Secernentea

Common name	Scientific name	Family	Order
5. Sweet potato			
Sweet potato weevil	*Cylas formicarius*	Curculionidae	Coleoptera
West Indian sweet potato weevil	*Euscepes postfasciatus*	Curculionidae	Coleoptera
Sweet potato stem borer	*Omphisia anastomasalis*	Pyralidae	Lepidoptera
Sweet potato butterfly	*Acraea acerata*	Nympalidae	Lepidoptera
Tortoiseshell beetle	*Aspidomorpha* spp	Chrysomelidae	Coleoptera
Sweet potato hornworm	*Agrius convolvuli*	Spingidae	Lepidoptera
Armyworms	*Spodoptera eridania* *S. exigua,* *S. litura*	Noctuidae	Lepidoptera
6. Cassava			
Scale insect	*Aonidomytilus albus*	Diaspididae	Hemiptera
Cassava green spider mite	*Mononychellus tanajoa*	Tetranychidae	Acari
Cassava mealy bug	*Phenacoccus manihoti*	Pseudococcidae	Hemiptera
Cassava lace bug	*Vatiga illudens*	Tingidae	Hemiptera
Cassava whitefly	*Bemisia tabaci,* *Aleurodicus disperses*	Hemiptera	Aleurodidae
7. Sugar beet			
Sugar beet root Aphid	*Pemphigus populivenae*	Aphididae	Hemiptera
Beet webworm	*Spoladea recurvalis*	Crambidae	Lepidoptera
Beet leafhopper	*Circulifer tenellus*	Cicadellidae	Homoptera
8. Elephant foot yam			
Mealy bug	*Rhizoecus amorphophalli,* *Pseudococcus longispinus,* *P. citriculus*	Pseudococcidae	Hemiptera
Aphids	*Lipaphis erysimi,* *Aphis gossypii*	Aphididae	Hemiptera
Red cotton bug	*Dysdercus cingulatus*	Pyrrhocoridae	Hemiptera
Leaf eating beetle	*Galerucida bicolor*	Gallerucidae	Coleoptera
Leaf eating caterpillars	*Pericallia ricini*	Arctiidae	Lepidoptera
	Spodoptera littoralis	Noctuidae	Lepidoptera
	Theretra gnoma	Sphingidae	Lepidoptera

The economically important pests that cause severe damages in the field are being discussed here as follows:

1.2.1 POTATO TUBER MOTH (PTM): PHTHORIMAEA OPERCULELLA (GELECHIIDAE: LEPIDOPTERA)

Distribution: Cosmopolitan in distribution, especially in warm temperate and tropical regions where host plants are grown. Native of South America and was introduced to India in 1906 with seed potatoes imported from Italy. During recent years, the species has been inadvertently introduced into Georgia (Markosyan, 1992) and Ukraine and there is a threat of its spreading to neighboring states (Sikura and Shendaraskaya, 1983). It has also been newly recorded from the Arabian Peninsula (Povolny, 1991; Kroschel and Koch, 1994) and more widely in East Africa (Parker and Hunt, 1989).

Hosts: Potato is the principal host for *P. operculella* but has also been reported from other Solanaceae crops viz. tomato, tobacco, chili, aubergine, sugar beet, and cape gooseberry. It may also bore in above-ground stems. Foliar damage to potatoes is usually not economically important (Rondon, 2010; Trivedi and Rajagopal, 1992). They also attack many weeds and wild plants (Das and Raman, 1994: Weber, 2013).

Biology and description of pest: Female moth can lay 150–200 eggs. The eggs are laid singly or in batches on the leaves of the host plants or on exposed tubers near the eye buds and have incubation period of 3–6 days (Langford and Cory, 1932). Eggs are oval, smooth and yellowish, iridescent, and measure less than 1 mm in diameter. Newly emerged larvae are gray yellowish-white with brown head. The fully-grown, *P. operculella* larvae are about 15 mm in length. Head dark brown; prothoracic plate sometimes pinkish; body greyish-white or pale greenish-grey. Pinacula small, dark brown, or black. Anal plate brown. Pupa yellowish or reddish-brown, eighth abdominal segment with spiracles on slightly raised, backward-pointing spiracles; cremaster with a median, dorsal, thorn-like spine, and eight slender hooks. Full-grown caterpillars come out of the tubers and pupate in silken cocoons either in dried leaves, soils, over the stored *tubers* or in cracks and crevices in the store. Pupal period lasts for 5–9 days. Adult moth measuring about 1 cm in length when at rest, colored pale brown with darker marbling. Wingspan 15–17 mm. Head and thorax pale brown, palpi curved, ascending, and terminal segment about as long as second. The moth breeds continuously where conditions permit; up to 13 generations a year have been recorded in India (Mukherjee, 1948; Trivedi et al., 1994).

Nature of damage: The pest breeds throughout the year under favorable conditions. Tuber moth infestation exists both during vegetative stage and postharvest stage. On growing plants, the larvae mine the leaves, petiole, and terminal shoots causing wilting. After tuberization, the larvae enter into the tubers and feed on them. Bore the tubers in stores also. Larvae tunnel into the pulp, which ultimately becomes unfit for use as seed or for human consumption. The infested tubers are further exposed to microbial infection, which leads to rotting. The extent of damage to stored tubers varies from 20–85 percent.

Management:

1. Follow proper earthing up operation, as infested seed tubers are the main cause of reinfestation, the use of healthy tubers will reduce levels of field infestation (Lal, 1991).
2. Destroying of self-grown potato plants and harvesting the crop as soon as possible is recommended to restrict damage (Herman, 1999).
3. Harvested potatoes should be lifted to cold stores immediately, if cold store facilities are not available, only healthy tubers should be stored.
4. Light irrigation, every 4 days, and mulching with neem leaves during the last 4 weeks before harvest were the most effective treatments (Ali, 1993).
5. The larval infestations of *P. operculella* on potatoes were consistently reduced when potatoes were grown with chilies, onions, or peas compared with potato alone (Lakshman Lal, 1991; Afifi et al., 1990).
6. The parasitoids, *Chelonus curvimaculatus*, *Bracon gelechiae*, *Apanteles subandinus*, *Melanis* sp. *Copidosoma koehleri*, and *Diadegma molliplum* have become established in a number of countries and found to be successful biological control agents (Sankara and Girling, 1980; Horne, 1990; Herman, 2008).

7. The nematodes *Steinernema feltiae, S. carpocapsae*, and *Heterorhabditis heliothidis* were used to control the tuber moth (Ivanova et al., 1994).

8. *Bacillus thuringiensis* has also been reported to suppress this pest. Pheromone traps are used both for monitoring and control in the field and in storage.

9. The sex pheromone of *P. operculella* was identified as a mixture of trans-4, cis-7-tridecadienyl acetate (PTM1), and trans-4, cis-7, cis-10 tridecatrienyl acetate (PTM2) (Persoons et al., 1976). Under field conditions, more than 20 traps/ha are required.

10. Cover the stored tubers with 2.5-cm layer of chopped dry leaves of Lantana or Eucalyptus or Eupatorium below and above the potato.

11. Spray of crop with chlorfenvinphos (0.4 Kg a.i./ha) or quinalphos (0.375 Kg a.i./ha) or acephate (0.5 Kg a.i./ha). In stores, dusting the tubers with 5% malathion or 1.55 quinalphos dust @ 125 g dust/100 Kg of potatoes. Alternatively, dipping of tubers before storage with 0.0028% deltamethrin.

1.2.2 HADDA BEETLE: HENOSEPILACHNA VIGINTIOCTOPUNCTATA (COCCINELLIDAE: COLEOPTERA)

Distribution: Hadda beetles are universally distributed. It originated in east of Russia. Presently they occur mainly in tropical and semitropical parts of the world. Their existence also well noticed throughout India, Pakistan, China, Japan, South East Asia, Australia (CABI, 2010), New Zealand (MAFBNZ, 2010), Brazil, Argentina, and Oceania (Richards, 1983; Schroder et al., 1993; Folcia et al., 1996). In India, the beetle is present in higher hills and in plains of Jammu and Kashmir, Punjab, Himachal Pradesh, Uttar Pradesh, Karnataka, and Bengal and in the plains (Shankar et al., 2010).

Hosts: In addition to potato, the phytophagous coccinellids that feed on foliage of the Solanaceae, Cucurbitaceae, Fabaceae, and Asteraceae (Richards, 1983; Webber, 2013).

Biology and description of pest: Two types of hadda beetles are commonly found in India-*Epilachna vigintioctopunctata* and *Epilachna dodecastigma*. The *E. dodecastigma* is 12-spotted and *E. vigintictopunctata* is 28-spotted beetles. The black dark spots are present on the elytron. These two species can interbreed among themselves. Adult beetles are about 8 mm in length and 5–6 mm in breadth. *E. dodecastigma* is copper-colored while the *E. vigintioctopunctata* is deep red-colored. The body is hemispherical and smooth. Adults are good fliers and move from plant to plant.

After mating, the female starts laying eggs in the month of March–April. A female lay about 120–180 eggs. Eggs are laid in the cluster (Deshmukh et al., 2012). The eggs are cigar-shaped, yellowish in color, and are arranged side to side on the surface of the leave in an erect position. The larva hatches in 3–4 days in summer months and in 4–9 days in winter. Newly hatched 1st instar is yellowish in color and has six rows of long branched spines (Tayde and Simon, 2013). The grubs are oval, fleshy, and yellow in color bearing hairs and spines on the body surface. The grubs restrict their feeding to the epidermis of the leaves. A fully-grown larva measures about 8 nm in length. The larval passes through four different stars. The larva changes into pupa. The pupation takes place on the leaf surface or on stem or at the base of the plants. Pupa is oval and dark in color. The pupal period lasts for 3–6 days. Life cycle is completed in 17–18 days in summer but, in winter, it may prolong up to 50 days. The pest completes 7–8 generations in year (Hossain et al., 2009).

Nature and symptoms of damage: Both larvae and adults are destructive. Adult and grubs scrap the lower epidermis on the green tissues of leaves in characteristic manner leaving behind stripes of uneaten areas. The leaves give a stifled appearance. In severe infestation, all leaves may be eaten off leaving only the veins intact and which ultimately dries up.

Management:

1. The beetle's larvae, pupae, and eggs can be hand collected and destroyed.
2. Shake plants to dislodge grubs, pupae, and adults in kerosenated water early in the morning or collect them mechanically and destroy.
3. Thorough irrigation of infested crop can minimize the increase in pest population.
4. The spray of carbaryl 50% WP 2 kg + wettable sulfur 2 kg or malathion 50 EC 1.5 L or Azadirachtin 0.03% 2.5–5.0 L in 500–750 L of water is recommended for the control of pest.

5. Use of 1 L of Neem oil with 60 g of soap dissolved in ½ L of water, dilute emulsion by adding 20 L of water, then mix about 400 g of well-crushed garlic and spray.
6. In India and Pakistan, hadda beetle is attacked by three predatory bugs: *Rhynocoris fuscipes* (Reduviidae), *Cantheconidea furcellat* (Pentatomidae), and *Geocoris tricolor* (Lygaeidae) (Patalappa and Basavanna, 1979; Schaefer, 1983; CABI, 2010).

1.2.3   POTATO APHID: MYZUS PERSICAE, MACROSIPHUM EUPHORBIAE (APHIDIDAE: HEMIPTERA)

Several species of aphids feed on potatoes throughout the world; the most important are green peach aphid *Myzus persicae*, potato aphid *Macrosiphum euphorbiae*, Buckthorn aphid *Aphis nasturtii*, and Foxglove aphid *Aulacorthum solani*; they may vector several damaging viruses in a persistent or nonpersistent manner (Ragsdale et al., 1994; Flint, 2006). Virus transmission is of highest concern for growing seed potatoes, which are used to plant subsequent crops (Nakata, 1995; Flint, 2006).

Distribution: The potato aphid is worldwide in distribution. *Macrosiphum euphorbiae* originated in North America but it has spread to the temperate parts of Europe and Asia and is found in all areas in which potatoes are grown.

Hosts: The potato aphid attacks over 200 plants including vegetable and ornamental crops as well as weeds. Cultivated food hosts include apple, bean, broccoli, cabbage, corn, eggplant, ground cherry, lettuce, mustard cabbage, papaya, pea, pepper, etc.

Biology and description of pest: Green peach aphid arrives on potatoes in the spring from weeds and various crops where it has overwintered as nymphs and adults, or from peaches and related trees where it overwinters as eggs. The life cycle varies considerably, depending on the presence of cold winters (van Emden et al., 1969). Development can be rapid, often 10–12 days for a complete generation and with over 20 annual generations reported in mild climates. The eggs measure about 0.6-mm long and 0.3-mm wide and are elliptical in shape. Eggs initially are yellow or green but soon turn black. Mortality in the egg stage sometimes is quite high. Nymphs initially are greenish but soon turn yellowish, greatly resembling viviparous adults. There are four instars in aphids and the average length of life was about 23 days (Horsfall, 1924). The adults vary in appearance, occurring in a green or pink form. Winged adults have a black head and thorax and a yellowish-green abdomen with a large dark patch in the middle of the abdomen as viewed from above. They measure about 2 mm in length. Wingless adults are yellowish, greenish, or reddish. The cornicles are long and colored similar to the body.

Nature of damage: Green peach aphids can build large populations on a variety of crops. On young plants, they can cause wilting and stunting. Colonies of nymphs and adults are seen on the ventral surface of leaves and shoots and suck sap therefrom. Infested leaves become yellowish, wrinkled, and cupped, whereas tender shoots turn yellowish and die away. They also excrete honeydew on which sooty mold develops covering affected parts with a thin superficial black coating that hinders photosynthetic activity of leaves resulting in stunted growth of plants. Aphids feed by sucking sap from their host plants. When aphid populations are large, feeding can cause plants to become deformed and the leaves curled and shriveled (Metcalf, 1962). Wilting of the plants and honeydew on the leaves may be an indication of high numbers of potato aphids on the crop. The major damage caused by aphids is through transmission of plant viruses. Nymphs and adults are equally capable of virus transmission (Namba and Sylvester, 1981) but adults by virtue of being so mobile, probably have greater opportunity for transmission. Both persistent viruses, which move through the feeding secretions of the aphid, and nonpersistent viruses, which are only temporary contaminants of aphid mouthparts, are effectively transmitted. Kennedy et al. (1962) listed over 100 viruses transmitted by this species. The potato aphid transmits potato virus Y on tobacco but almost never on potato. It is also considered a poor vector of potato virus A and potato leaf roll.

Management:

1. The most important source of virus in a potato field is infected plants already in that field. Therefore, purchasing certified seed, with low or no virus infection, is the best first step in controlling aphid related damage to potatoes.

2. Green peach aphids are generally controlled with application of insecticides; however, insecticide resistance has been widely documented in this species.

3. Potato aphids are often controlled by the natural occurrence of predators, such as coccinellid larvae, predatory bugs in genera *Orius*, *Nabis*, and *Geocoris*, lacewings, spiders, syrphid fly larvae, flower flies, predatory gall midge larvae, entomopathogenic fungi (Hautier et al., 2006; Straub and Snyder 2006; Alyokhin et al., 2005, 2011; van Emden et al., 1969). Aphids may occasionally be cannibalistic or predatory (Banks et al., 1968).

1.2.4 *POTATO CUTWORM:* AGROTIS IPSILON (*NOCTUIDAE: LEPIDOPTERA*)

Cutworms, (*Agrotis segetum, A. ipsilon, A. flammatra, A. Spinifera*) are major pests causing damage to potato in many parts of the world including northwest Himalayan region (Das and Ram, 1988; Lal, 1990; Mrowc-zynski et al., 2003; Kumar and Tiwary, 2009; Anonymous, 2011a).

Distribution: The origin cutworm is uncertain, though it is now found in many regions of the world, being absent principally from some tropical regions and cold areas. It is more widespread and damaging in the northern hemisphere than the southern hemisphere (Capinera, 2012). It is reported from China, India, North Europe, Canada, Japan, South America, and New Zealand. In India, cutworms are more serious in northern region than in south.

Hosts: Black cutworm has a wide host range. Nearly all vegetables can be consumed, and this species feeds on alfalfa, clover, cotton, rice, sorghum, strawberry, sugarbeet, tobacco, and sometimes grains and grasses.

Biology and description of pest: The adult moths fly from late May to early July, laying eggs on plants or on pieces of litter and debris in the soil. Eggs are globular in shape, 0.5 mm in diameter, ribbed and whitish in color. Each female lies on an average 300–450 egg in clusters of 30–50. Larvae smooth, stout, cylindrical, 40–50 mm long, blackish-brown dorsally and grayish-green laterally with dark stripes. They coil up at the slight touch. Tiny caterpillars feed gregariously on foliage for a few days, then segregate, and enter into the soil. The caterpillars are nocturnal. Pupae 18–22 m long and reddish-brown in color and pupation takes place in soil. Moths are medium-sized, stout, dark greenish-brown with a reddish tinge and have grayish-brown wavy lines and spots on the fore wings. Hind wings are hyaline having ark terminal fringe which is darker in females than in male. Wings expanse is 45–50 mm. Total life cycle is completed in 30–68 days depending on the climatic conditions.

Nature of damage: Caterpillars are damaging stages of cutworms and most of the feeding occurrs at soil level as they cut the seedling at ground level. However, larvae will feed above ground until about the fourth instar. Larvae can consume over 400 sq cm of foliage during their development but over 80% occurs during the terminal instar and about 10% in the instar immediately preceding the last. Thus, little foliage loss occurs during the early stages of development. Once the fourth instar is attained, larvae can do considerable damage by severing young plants, and a larva may cut several plants in a single night. Caterpillars also nibble the tubers of plants.

Management:

1. Deep plowing and stirring of soil, flooding of fields so that caterpillars are exposed to birds and other enemies.
2. Hand-picking and destruction of caterpillars found just under the damaged plant.
3. Poison baits containing wheat bran + carbaryl + molasses be spread on the ground to attract and kill larvae and mixing of insecticidal dust are some of the practices to control the cutworms.

1.2.5 COLORADO POTATO BEETLE: LEPTINOTARSA DECEMLINEATA (CHRYSOMELIDAE: COLEOPTERA)

It is a common and destructive pest of potato and brinjal. This insect was first discovered in the Rocky Mountains feeding on a common weed called buffalo bur (*Solanum rostratum*).

Distribution: The Colorado potato beetle, *Leptinotarsa decemlineata* is native to Mexico (Alyokhin, 2008) and was first reported in Florida in 1920 but it is not often a major pest. It also occurs in southern Canada and is a pest in Central America. The species has been introduced into Europe and parts of Asia (Capinera, 2001).

Hosts: Potatoes are the preferred host for the Colorado potato beetle, but it may feed and survive on a number of other plants in the family Solanaceae, including belladonna, common nightshade, eggplant, pepper, tobacco, ground cherry, henbane, horse-nettle, thorn apple, tomato, and its first recorded **host plants:** buffalo bur (Capinera, 2001).

Biology and description of pest: The eggs are bright orange in color, about 1.7–18 mm long and 0.8 mm wide. They are usually deposited in batches of about 30 on the underside of host leaves. Under field conditions, females can lay 200–500 eggs. After 4–15 days, the eggs hatch into reddish-brown larvae with humped backs and two rows of dark brown spots on either side. Larvae progress through four distinct growth stages. Larvae bear a terminal proleg at the tip of the abdomen as well as three pairs of thoracic legs (Capinera, 2001). Upon reaching full size, each fourth instar spends an additional several days as a nonfeeding prepupa, which can be recognized by its inactivity and lighter coloration. The prepupae drop to the soil, burrow the soil to a depth 2–5 cm, and after about 2 days begin to pupate. The adults are yellowish-orange with multiple black stripes down the back with five per elytron (Wilkerson et al., 2005). They are robust and oval when viewed from above. The head has a triangular black spot and the thorax has irregular dark markings (Capinera, 2001).

Nature of damage: Colorado potato beetle is one of the most important defoliators of potato. Both adults and larvae feed on leaves. Approximately 40 cm^2 of potato leaves are consumed by a single beetle during the larval stage (Ferro et al., 1985; Logan et al., 1985).

Management:

1. Colorado potato beetle can be reduced with crop rotation practices and can be easily implemented.
2. Manipulating planting time or early planting also practiced to reduce the populations.
3. Planting trap crops that attract beetles away from the main crop may be effective in intercepting overwintered beetles colonizing a field in the spring.
4. Mulching increases the time required by the beetles to find potatoes.
5. Larval populations of the beetle were significantly reduced in straw-mulched plots of potato (Stoner, 1993) and eggplant (Stoner, 1997).

6. Potatoes were the first successful transgenic crop plants (An et al., 1986). Genetically-modified potatoes with *Bacillus thuringiensis* delta-endotoxin that is toxic to the Colorado potato beetle were registered and sold in the U.S. from 1995–2000 (Alyokhin, 2008).
7. Digging plastic-lined trenches along a field border will intercept migrating Colorado potato beetles.
8. Predaceous stink bugs. *Perillus bioculatus* and *Podisus maculiventris* have been shown to have significantly controlled the Colorado potato beetle.
9. The lady beetle *Coleomegilla maculata* (Coleoptera: Coccinellidae) preys on eggs and larvae (Groden et al., 1990; Hazzard et al., 1991).

1.2.6 CARROT ASTER LEAFHOPPER: MACROSTELES QUADRILINEATUS (CICADELLIDAE: HEMIPTERA)

Distribution: Aster leafhopper is native to North America. It is the most common, in the central region of the continent and it overwinters poorly in cold areas (Capinera, 2008).

Host plants: It is a polyphagous species that uses cereals, vegetables, and weeds as host plants (Szendrei, 2012). Aster leafhopper feeding does not cause the damage to plants. Rather, the infection of aster yellows phytoplasma produces negative effects on the crop.

Biology and description of pests: Leafhopper has been found to overwinter on the Prairies as adults or else hatch from overwintering eggs. The adults usually arrive in early mid-May to mid-June. Adults are 3.5–4 mm long, light green or yellowish and have 6 dark spots arranged in 3 rows on the head. The adult are narrow and wedge-shaped, with a beak and tiny antennae. It has long hind legs fringed with hairs. Leafhoppers jump and fly readily at temperatures above 15°C. Rain and cool temperatures will slow migration, and in these conditions, the insects are more concentrated at field edges and in patches. Eggs are laid in plant tissue and nymphs emerge within a few days. Nymphs hatch from eggs and go through five nymphal stages before adulthood is reached. Nymphs are small, wingless versions of the adults about 0.6–3 mm in size (Seaman, 2014). They are generally completed 2–5 generations per year.

Nature of damage: Both nymphs and adults feed by inserting and sucking the plant to extract sap. If leafhoppers feed on an infested plant, it ingests the aster yellow pathogen. The insect is a transmitter of various 'viral' diseases, most notably Aster yellows. It has also been implicated in clover phyllody. Symptoms vary from crop to crop and across the spectrum of diseases and appear 10–40 days after infestation (Seaman, 2014).

Management:

1. Removal of weeds from the field edges act as reservoir for the pathogen.
2. Aster leafhopper may effectively be controlled by excluding them from the carrot planting with floating row covers.
3. Place yellow sticky card in the field early in the spring when plants are newly sprouted.
4. Use of nonorganic mulches can reduce both leafhopper populations and the incidence of Aster yellows disease.
5. Sticky traps and sweep nets can be used for early detection and monitoring of leafhoppers to give producers an early warning of potential problems.

1.2.7 CARROT FLEA BEETLE: SYSTENA BLANDA (CHRYSOMELIDAE: COLEOPTERA)

Distribution: Flea beetle is an occasional pest of carrot. Its larvae are delicate and thread-like with white bodies and brown head capsules. They have characteristically large hind legs, which makes them excellent jumpers (Delahaut and Newenhouse, 1998).

Hosts range: Flea beetle attack only one plant group or closely related groups. Other hosts include members of the brassica (mustard, broccoli, kale, cabbage, collards, etc.) and solanaceous (potatoes, tomatoes, eggplant, peppers, etc.)

Biology and description of pest: The adult flea beetle is a small (1.5–2.0 mm), oblong to oval-shaped shiny black beetle with reddish legs and antennae. The hind femora are enlarged and help the beetle to jump when disturbed. Adults overwinter in leaf litter or soil in protected places such as field margins, tree rows, or less frequently, within cultivated fields. Adults emerge in spring when soil temperatures warm to approximately 15°C and begin feeding on winter annual weeds. Adult beetles mate in the spring and summer and lay eggs on the soil surface plant, near a plant stem.

Nature of damage: Adult flea beetles chew small holes in leaves that can give them a "shot hole" appearance. The emerging larvae burrow down and feed on plant roots. Like the adults, the larvae are generally not considered major pests because they cause minimal superficial damage to roots. In addition to the large feeding cavities the larvae create, there are also many small superficial lesions present. The lesions are randomly scattered but may be more concentrated on the upper third of the root (CABI, 2008).

Management:

1. Early planting helps to avoid the population of flea beetles while the plants are small and vulnerable.
2. Enclosing seedbed with floating row cover to get protection from laying by adults.
3. Applications of diamotecoeus earth or neem oil for control.
4. Adults overwinter under soil clods and plant debris; therefore, good sanitation practices are important to reduce overwintering flea beetle populations.
5. Parasites and predators viz. lacewing (*Chrysopa* spp.), adult big-eyed bugs (*Geocoris* spp.), and damsel bugs (*Nabis* spp.) feed on adult flea beetles.

1.2.8 WILLOW CARROT APHID: CAVARIELLA AEGOPODII AND GREEN PEACH APHID MYZUS PERSICAE (APHIDIDAE: HEMIPTERA)

Distribution: This aphid is globally distributed in temperate and warm temperate regions of the world.

Host range: It is a polyphagous species that feeds on many vegetables viz. mustard, broccoli, kale, cabbage, collards, potatoes, tomatoes, eggplant, peppers, etc.

Biology and description of pest: Most species of aphids have similar life cycles. Aphid females give birth to live offspring all year without mating. When vegetable crops are not available, aphids live on a wide variety of weed hosts. In summer and fall, aphids may produce winged females and later winged males. They mate and produce eggs for overwintering, especially in colder climates. Otherwise, the adult aphids overwinter on crops, weeds, or trees. There may be as few as two generations or as many as 16 generations each year, depending on the species and climate. Aphids overwinter as eggs on a hardwood host, e.g., *C. aegopodii* (willow-carrot aphid) on *Salix alba* or *Salix fragilis*. In May and June, winged females migrate to carrot crops, where a new colony of the species is established. A series of wingless and winged generations take place here until autumn. In warmer climates, adult females may overwinter on carrots, which stay in the soil during the winter (OEPP/EPPO, 2000).

The presence of cornicles, which project backward from the body of the aphid and will generally not move very quickly when disturbed; willow carrot aphid also attacks turnip, parsley, and celery (CABI, 2008). The green peach aphid is slender, dark green to yellow, and has no waxy bloom. The wingless form of the green peach aphid is pale green. The winged form has a black head and thorax. It is primarily an early year pest and transmits virus diseases (Seaman, 2014; Delahaut and Newenhouse, 1998).

Nature of Damage: Aphids may cause direct injury, giving rise to plants with twisted and malformed foliage, which may be stunted or even killed. They also transmit carrot motley dwarf disease, causing stunting, reddening of the outer leaves and a fine chlorotic mottle on the inner leaves. Some aphids can cause leaf curling when the insect infests emerging leaves (Cranshaw, 2009).

Management:

1. On shrubs and garden plants, aphids can be managed by simply washing them off of plants with a forceful jet of water.
2. Use tolerant varieties, sanitation for curbing the spread of the viruses.
3. Plowing all crop residues under as soon as harvest.
4. Use of reflective mulches, such as silver-colored plastic, can deter aphids from feeding.
5. Use of predators, such as green lacewing larvae, lady beetles, and syrphid fly larvae prey on this aphid. Even several species of minute stingless wasps parasitize aphids.
6. Use of insecticidal soaps or oils, such as neem or canola oil, etc., are followed to control the aphids.

1.2.9 *CARROT WEEVIL:* LISTRONOTUS OREGONENSIS (CURCULIONIDAE: COLEOPTERA)

Distribution: The carrot weevil is native to northeast North America and was first described in Pennsylvania in 1842. The first damage reported on carrots was from Washington, DC in 1922.

Hosts plants: Other than carrot important hosts are celery, potato, onion, lettuce, etc.

Biology and description of pest: Carrot weevils overwinter as adults in and near fields where carrots were grown the previous year emerging in April and May. Weed species also serve as hosts of the carrot weevil. Although the weevils have wings, they seldom fly and instead, they walk to nearby carrot fields. The adults feed directly on the leaves of carrots. Eggs are deposited in small cavities that the females chew in the stems, crowns, or exposed roots of the plant. Larvae hatch out and crawl down to the carrot roots. Each female can lay up to 200 eggs. The economic damage is done by the larva that tunnels in the root of the plant. After entering the carrot, the larva blocks the tunnel entrance with its frass. The tunnels eventually become filled with frass and the epidermal cells over the tunnel die and become dark brown. Weevil completes one generation a year—mostly overwinters as adults in vegetation at the edge of fields (Boivin, 1999). Adult weevils spend overwinter in crop debris remaining in the ground; larvae feed for approximately 2 weeks before pupating in the soil; insect undergoes several generations each year. Snout-nosed beetles that are about 6-mm long. Larvae are white to pinkish-white C-shaped grubs with a yellow–brown head.

Nature of damage: Their grub, like larvae, either tunnel down into the root or leave the stalk and bore into the side of the root from beneath the soil. Larvae may kill young plants. Damage to older plants is typically observed in the upper one-third of the root. Feeding injury may allow entry by pathogens that will cause roots to rot.

Management:

1. Early control was achieved using baits of apple pomace and calcium arsenate.
2. Applications of 50 pounds of apple pomace plus calcium arsenate (95–5) reduced carrot weevil damage by 90% (Pepper, 1942).

3. Crop rotation is effective if a carrot field is set remotely from fields previously in carrots.
4. Use of Azadirachtin is quite effective against carrot weevil.
5. Proper timing of sprays is critical for good control pests and minimizing the cost involved.

1.2.10 CARROT RUST FLY: PSILA ROSAE (PSILIDAE: DIPTERA)

Distribution: The carrot rust fly is native to Europe where it was recognized as early as 1794. The insect was first noted in Canada in 1885 and in the US in 1893.

Hosts plants: The host range of the carrot rust fly extends to 107 different plant species, all in the same family as the carrot. Many of the host species are also grown for food, including celery, parsnips, celeriac, parsley, and dill (Degen et al., 1999).

Biology and description of pest: The adult female lays its eggs in the soil at the base of the carrot. Six to ten days later the larva hatches and feeds on the carrot, rendering the carrot impossible to market. The larvae feed for about a month. Carrot rust flies obtain their common name from the rust color frass that they deposit in the feeding tunnels on the carrot. 100% of the plants in a field can be damaged. Individual carrots can yield as many as 100 maggots each. The rust fly adult spends most of its time in the periphery of the fields, flying into the field to lay eggs at the base of the carrot, and then leaving the field. When the adult emerges from the pupal case, it flies to the periphery of the field. This behavioral pattern leaves only limited opportunities for control with insecticides (Gianessi, 2009).

Nature of Damage: Damaged plants are wilted or stunted.

Management:

1. Use row covers or to rotate the carrot crop every other year.
2. Application of crude naphthalene flakes, broadcast on the soil surface to deter the fly from laying eggs in the carrot field.
3. Delayed planting or harvest early is practiced to reduce the damage by pest.

1.2.11   *DIAMONDBACK MOTH:* **PLUTELLA XYLOSTELLA (YPONOMEUTIDAE: LEPIDOPTERA)**

Distribution: They may have originated in Europe but on the basis of the presence of its biocontrol agents and host plants speculated that it originated in South Africa. This pest is now present wherever its host plants exist and is considered to be most widely distributed of all Lepidoptera (Shelton, 2004, CIE, 1967). The diamondback moth (DBM) is found in Europe, Asia, Africa, the Americas, Australia, New Zealand, and the Hawaiian Islands. In India, DBM was first recorded in 1914 (Fletcher, 1914).

Hosts: DBM is an important pest of cruciferous crops, it infests cabbage, cauliflower, radish, knol khol, turnip, beetroot, mustard, and Amaranthus. Noncruciferous crop, like *Amaranthus virdis*, has also been reported to be the host of this species (Vishakantaiah and Visweswara Gowda, 1975). The mustard (*Brassica juncea*) is a second alternative host of DBM (Ahmad et al., 2008).

Biology and description of pest: Female moths lay eggs singly or in small clusters on stems and both sides of the leaves. A female moth may lay more than 150 eggs during her lifetime. The first and some second growth-stage caterpillars are leaf miners tunneling inside the leaf. Subsequent growth stages feed on the underside of leaves or tunnel into the plant. Caterpillars that are disturbed from feeding will wriggle backward rapidly across the leaf surface and may drop to the ground on silken threads. Pupation occurs in an open-mesh cocoon. The time taken to complete the life cycle depends on temperature. In 1 year, the moth will complete six to seven life cycles. Generations overlap throughout the warmer months of the year.

Nature of damage: Caterpillars form many small holes in the leaves of the host plants, often leaving the leaf epidermis intact, making a "feeding window." Most damage is caused by the caterpillars tunneling into the heads of plants, such as cabbage and Brussels sprouts.

Management:

1. Use of nonglossy collards as a trap crop for control of the DBM.
2. Remove and destroy all the remnants and stubbles after the harvest of the crop.
3. Need-based spray of Quinolphos or Fenvalerate.

4. Intercropping with nonhost plants is a very good method for reducing pests.
5. Dead-end trap cropping: A technique can be adapted to control the pest effectively (Shelton and Nault, 2004).
6. Parasitoids recorded on pests are *Trichogramma chilonis, Cotesia plutella, Microplitis* sp., *Oomyzus sokolowskii, Diadromus collaris, Itoplectis naranyae, Exochus* sp., and *Brachymeria excarinata* (Shu-Sheng Liu, 2000).
7. Entomopathogenic viruses, nematodes, and microsporidia also have potential as biopesticides for DBM (Sarfraz, et al., 2005).

1.2.12 *APHIDS:* MYZUS PERSICAE (APHIDIDAE: HEMIPTERA)

(REFER POTATO APHID)

1.2.13 *RADISH ROOT MAGGOTS:* Delia radicum (Anthomyiidae: Diptera)

Distribution: It is found commonly in Japan, India, Australia, northern Africa, and New Zealand.

Hosts: *Delia* sp. feeds on the roots of a broad range of hosts, including crucifers, onions and leeks, solanaceous crops, cucurbits, and cereal crops such as corn (Griffiths, G. C. D, 1993).

Biology and description of pest: Delia spp. overwinters in pupa 5–20 cm below the soil surface. Adults emerge from mid-May to mid-July; shortly after emergence flies mate and females lay eggs, singly or in batches of up to 25–30, near the base of host plants, usually in cracks or under a thin layer of soil. The eggs hatch in 3–10 days and the white legless maggots burrow into the soil to feed on root hairs and on secondary roots. Developing through three instars, older maggots tunnel into the taproot of the plant. Pupation lasts 2–3 weeks. When multiple generations occur, they usually overlap, so that adult flies can be present in crucifer fields from late spring to October (Juliana and Lloyd, 2011).

Nature of damage: The scars left by the feeding maggots leave the roots unmarketable.

Management:

1. Avoid close spacing while planting.
2. Follow crop rotation.
3. Field sanitation

1.2.14   BLACK   CUTWORM: AGROTIS IPSILON (NOCTUIDAE: LEPIDOPTERA)

Distribution: The black cutworm is a cosmopolitan pest that poses an economic threat to many agricultural plant species. The origin pest is uncertain, though it is now found in many regions of the world, being absent principally from some tropical regions and cold areas.

Hosts: The black cutworm has a wide host range, feeding on nearly all vegetables and many important grains.

Biology and description of pest: Newly hatched larvae are about a quarter-inch long and grow to be about two inches long when full-sized. Their color ranges from gray to nearly black. There is a pale rather indistinct narrow stripe along the center of the back. The texture of the skin is characteristic and distinguishes them from all other cutworms. The skin texture consists of convex, rounded, coarse granules with smaller granules interspaced between.

Nature of damage: Black cutworms exhibit two types of feeding patterns depending upon the amount of moisture in the soil and size of plants. Where soil moisture is adequate and plants are small, the larvae hide in the soil during the day and move to the soil surface at night where they cut off plants just above the soil surface (Tooker, 2009).

Management:

1. Adult populations can be monitored with both black light and sex pheromone traps.
2. Destruction of weeds hosts and field sanitation.
3. Broadcast pesticide or bait application may be used as a rescue treatment.
4. Transgenic control also provides very good control of black cutworm (Herculex® I and Herculex® XTRA, etc.)

1.2.15   INSECT PESTS OF YAM: SCALE INSECT: ASPIDIELLA HARTII (DIASPIDIDAE: HEMIPTERA)

Distribution: West Africa, India, Oceania. It is recorded from Fiji, Solomon Islands, and Tonga. In India, yams are infested by 12 insect pest species of which A. hartii is the most important (Pillai et al., 1993).
Hosts: Ginger, taro, and turmeric are minor hosts.

Biology and description of pest: This scale is called the armored scales. They covered themselves with the formation of a circular-shaped scale termed as puparium that protects the body of insects, composed of discarded skins and secreted matter. The eggs are elongated, with ends equally rounded and laid within the puparium. The early-stage nymphs are active crawlers for a short period before selecting a feeding site. Once they start feeding, the waxy protective scale starts to form. The second-stage nymphs emerge after the first molts in which the discarded skin. The last stage nymphs are sometimes referred to as pupae, which posses' rudimentary legs, antennae, wings, and stylus (mouth). Adults are moderately convex, about 0.05 inches in diameter, and brownish-gray in color with a slight purplish shade. Males are generally oval-shaped where females are circular-shaped.

Nature of damage: Wilting due to continuous removal of plant sap by scales. The infested tubers shrivel, so reducing their quality, viability, and marketability and in severe infestation, it kills the whole plant.

Management:

1. Check the surface of yams before they are stored for the presence of scales.
2. Inspection of yams in storage regularly and remove those that become infested with the scale.
3. Select scale-free planting material for planting.
4. Use white oil (made from vegetable oils), soap solution, or horticultural oil (made from petroleum) on yams infested with scale.
5. Seed treatments can be done by using Dichlorvos 76 WSC @ 0.7 mL/L or Phosalone 35 EC @ 1.5 mL/L (for 15 min) or Monocrotophos 36 WSC @ 1.5 mL/L or Quinalphos 25 EC @ 2 mL/L of water (for 5 min).

6. Apply well-rotten sheep manure @ 10 t/ha in two splits or poultry manure in 2 splits followed by drenching dimethoate 30 EC @ 2 mL or phosalone 35 EC @ 2 mL/L of water.

1.2.16   MEALY BUGS: GEOCOCCUS COFFEAE, PLANOCOCCUS CITRI, PHENOCOCCUS GOSSYPII (PSEUDOCOCCIDAE: HEMIPTERA)

Distribution: It is widely distributed in India, Bangladesh, Pakistan, Myanmar, Malaysia, Java, and New Guinea, Philippines.

Hosts: It has a very wide host range, including citrus, cacao, coffee, croton, cyperus, ferns, mango, oleander, palms, philodendron, pineapple, and syngonium.

Biology and description of pest: Their reproduction may be sexual or parthenogenetic. Eggs are laid in groups and covered by a layer of cotton-like wax or by an egg sac of crystalline wax filaments. A single female may deposit 300–600 eggs. Upon eclosion, the small nymphs start looking for an appropriate place to settle on the plant root; at the selected site, they insert their mouthparts and feed by suctioning the sap from the root. As they feed and develop, the nymphs and adults excrete their characteristic waxy cover and form compact colonies. The sugary substances also attract certain ant species, which live in a symbiotic association with the mealybugs. The lifecycle, from egg to adult, requires from 30 to 120 days, according to the species and the temperature.

Nature of damage: Both nymphs and adults suck the sap from roots, tender leaves, petioles, and fruit. They excrete honeydew on which sooty mold develops and ultimately affects photosynthetic activity.

1. **Management:** Removing mealy bugs by rubbing or picking from affected plants.
2. Pruning and destruction of affected plant parts.
3. It is attacked by numerous natural enemies, which usually keep them under control. Most common natural enemies include parasitic wasps, ladybird beetles, hover flies, and lacewings. The most effective natural enemies are parasitic wasp, *Apoanagyrus* (*Epidinocarsis*) *lopezi*). Use 1% hexane extract of neem seeds to repel mealy bug.

4. Spray soap and water solution (1–2%) to control mealy bugs.

1.2.17   YAM TUBER BEETLE: HETEROLIGUS MELES, H. APPIUS (DYNASTIDAE: COLEOPTERA)

Distribution: It is widely spread in tropical Africa, Asia, South America, South Pacific.

Hosts: It is a very serious insect pest of yam in riverine areas particularly in the forest zones, up to the Savana regions along the Benue–Niger rivers and tributaries.

Biology and description of pest: Adult *H. meles* is about 23–33 mm in length while H. appius is about 21–23 mm long. Both are brown to black, short-distance fliers, and usually oviposit in moist and damp sites (Taylor, 1964). The females lay eggs in the soil. Newly emerged larvae look like creamy white to gray color that feed on grassroots and other organic matter. They complete their life cycle in 22–24 weeks.

Nature of damage: The beetle feed and creates spherical and semispherical lesions of varying sizes on tubers in the range of 1–20 holes in a tuber. Adult *H. meles* feeds on yam tuber creating large semispherical holes of 1–2.5 cm that predisposes tubers to fungal and bacterial attacks on the field and during storage (Okoroafor et al., 2009, 2014).

Management:

1. Timing of planting is the most important preventive measure. Planting late in the season as possible can significantly reduce the damage caused by the beetles.
2. Yams should be planted near wet areas, along rivers, creeks, or tributaries where the beetles breed should be avoided.
3. Mulching with the leaves of lemongrass (*Cymbopogon citrates*) or mosquito plant (*Ocimum viride*) can increase yields and reduce the damage caused by the yam beetle.
4. Application of *Parkia biglobosa* husk extract, *Azadirachta indica* seed, and leaf extract @ 0.84 g/210g on yam sett and @ 2.5 g/500 g soil per cage is highly effective against this beetle under lab (Okoroafor et al., 2014).

1.2.18   WHITE GRUB: *LEUCOPHOLIS CONEOPHORA* (SCARABAEIDAE: COLEOPTERA)

Hosts: *Leucopholis coneophora* is a polyphagous pest of affecting yam, coconut, banana, colocasia, cassava, elephant foot yam, sweet potato, and fodder grasses, etc.

Biology and description of pest: The pest has an annual life cycle and adult emergence coincides with the onset of monsoon (Abraham, 1993). Brown colored beetle with striated wings not covering the abdomen fully.

Nature of damage: The damaged tubers are adversely affected and lose the quality and culinary taste.

Management:

1. Collecting and killing adults during peak emergence periods.
2. Set up light trap @ 1/ha or bonfire.
3. Soil application of Malathion 5 D 25 kg/ha at the time of planting.

1.2.19   *TERMITE:* ODONTOTERMES ESCHERICHI (TERMITIDAE: ISOPTERA)

Hosts: It is the occasional pest of yam and it is cosmopolitan and polyphagous.

Biology and description of pest: Adults are cream colored, tiny insects resembling ants with dark colored head.

Nature of damage: Wilting of seedlings observed in the initial stages.

Management: Locate termite mounds in or near the coconut nursery or garden and destroy.

1.2.20   *YAM NEMATODES:* SCUTELLONEMA BRADYS (HOPLOLAIMIDAE: SECERNENTEA) HOPLOLAIMUS SPP., (HOPLOLAIMIDAE: SECERNENTEA) AND MELOIDOGYNE SPP. (HETERODERIDAE: SECERNENTEA)

Distribution: These nematodes have worldwide distribution, namely, West Africa, Korea, South America, Caribbean, India, China, Pakistan, etc. (Park and Khan, 2007).

Hosts: Nematodes are polyphagous in nature and affect many crops and weeds. Cowpea, sesame, green gram, pigeon pea, okra, tomato, and melon are susceptible to *S. bradys.*

Biology and description of pest: Nematodes lay eggs mainly in plant tissues where they hatch and develop into adults. Nematodes feed intra-cellularly in yam tuber tissues resulting in rupture of cell walls, loss of cell contents, and the formation of cavities. They are mainly confined to the subdermal, peridermal, and underlying parenchymatous tissues in the outer 1–2 cm of tuber. *S. bradys* continues to feed and reproduce in yams stored after harvesting. (Adesiyan et al., 1975). *S. bradys* is a migratory endoparasite of roots and tubers and will also be present in soils around host plants.*Meloidogyne* spp. can also cause infield and postharvest losses to yam (Bridge et al., 2005).

Nature of damage: Due to nematodes infestation the quality, market value, and consumption qualities are lost. Sometimes even cracks are developed on tubers. *Meloidogyne* sp. causes malformation and reduc-tion in size of the tubers. It causes disease of tubers and is commonly known as dry rot.

Management:

1. Use of planting material which is free of nematodes
2. Treatment of seed material (tubers and setts with yams) prior to planting to reduce or eliminate nematodes from propagative material.
3. Treatment of tubers after harvesting to prevent storage losses
4. Crops susceptible to nematodes should be avoided. Crops, such as *Tagetes* spp. and groundnut (peanut), have been recommended for use to lower nematode populations and restore fertility for yam production.
5. Hot-water treatment of 50–55°C for up to 40 min can reduce or eliminate *S. bradys* from tubers.
6. Granular nematicides, like oxamyl, carbofuran, or isazophos, applied as postplant treatments in yam mounds 2 weeks after planting @ 2 kg a.i./ha reduce soil populations of nematodes.

1.2.21 INSECT PESTS OF SWEET POTATO: SWEET POTATO WEEVIL: CYLAS FORMICARIUS (CURCULIONIDAE: COLEOPTERA)

Distribution: It is a serious insect pest of sweet potato worldwide. It is supposed to be originated in India and later widely distributed including tropical and subtropical regions of the world and recently has been found in higher latitude areas as well (Korada et al., 2010).

Hosts: Sweet potato, morning glory, and other plants of the same family. Alternate hosts of sweet potato weevils are *Ipomoea* spp. weeds.

Biology and description of pest: The adult insect is ant-like and the basic color of the insect is red. After mating, the female lays eggs on the tuber or on the leaf. The eggs hatch in 5–14 days and the larvae live for 10–35 days. The pupal stage lasts for 7–28 days and pupation takes place within the larval tunnels. The adult weevil can survive up to 94 days.

Nature of damage: Larvae feed and develop on the stems of sweet potato vines causing thickening and malformation. Both the larvae and adult damage tubers by making tunnels filled with frass making them unsuitable for consumption.

Management:

1. Mixed cropping systems with sweet potato and other crops (ginger, okra, maize, colocasia, and yam) are practiced.
2. Removal of alternate host plants, like *Calystegia soldanella, C. hederacea* and *Ipomoea indica* reduced the pest infestation in Japan (Komi, 2000).
3. Many species of Braconidae are parasitoids on the pests viz. *Bracon* sp., *Bracon yasudai B. cylasovorus* and entomopathogenic nematodes viz, *Steinernema carpocapsae, Heterorhabditis bacteriophora* helps significantly in reducing the pests (Korada et al., 2010).
4. Ants, spiders, carabids, and earwigs are important generalist predators that attack weevils.
5. The species-specific pheromones traps are used as monitoring, training, and management tools. Many effective traps have been designed by farmers using locally available materials.

6. Storage in controlled atmospheres, principally low oxygen and high carbon dioxide, is very effective for destruction of weevils but requires good storage conditions.

1.2.22 WEST INDIAN SWEET POTATO WEEVIL: EUSCEPES POSTFASCIATUS (CURCULIONIDAE: COLEOPTERA)

Distribution: The pest has widespread distribution. Caribbean, Central, and South America and with a restricted distribution in the USA (California and Hawaii) and Asia (Japan).

Hosts: Sweet potato and wild relatives in the genus Ipomoea.

Biology and description of pest: Adult weevils enter sweet potato fields by crawling in or when tubers infested with weevils are dumped near the field (Yasuda 1997). The larvae or grubs do the most damage and this is similar to that caused by the sweet potato weevil. The eggs are round, yellowish, laid singly in pits in the stems or in storage roots. After laying the eggs the pits are sealed with a fecal plug. Storage roots are preferred to stems. The eggs hatch in about 10 days and the legless grubs molt four times over the next 20–30 days, reaching a length of about 8 mm. The pupae are white about 5 mm long. After another 10 days, the adults emerge. Adults are 3.5–4 mm long and 1.5 mm wide, reddish-brown to greyish black, covered with stiff hairs with two white patches on the wing. Hot and dry weather favors weevil's development.

Nature of damage: Larvae of pest bore into roots and make sweet potatoes taste bitter, drastically decreasing the market value of the crop. The weevil seriously damages sweet potatoes not only in the fields but also during storage.

Management:

1. Choose short-season varieties, i.e., those that produce a crop early.
2. Always use tip cuttings for planting, 25–30-cm long, taken from young shoots.
3. Use varieties that set their roots deep in the soil.
4. Remove alternate hosts in and around the field.
5. Remove and destroy infested vines during harvesting.

1.2.23   SWEET POTATO STEMBORER: OMPHISIA ANASTOMASALIS (PYRALIDAE: LEPIDOPTERA)

Distribution: Widely distributed in the Philippines, Indonesia, India, Sri Lanka, Malaysia, Taiwan, Hawaii, and Vietnam. It also occurs in China, Japan, Cambodia, Laos, Burma, and Thailand.

Hosts: Sweet potato stemborer also lives on another Convolvulaceae: *Ipomea pescaprae.*

Biology and description of pest: The eggs are laid singly along the underside of the leaves, along the leaf margins. The egg, larval, and pupal stages last an average of 55–65 days. There are six larval instars. A newly emerged larva has a brown head and a reddish or pinkish body. After a few days, it turns creamy with black markings. Full-grown larvae are 30-mm long. Infested plants usually have a pile of brownish frass around their base. Before pupating, the larva makes an exit hole that is covered with the epidermis of the stem. Pupation lasts about two weeks and takes place in a web covered cocoon within the tunnel. The adults emerge by breaking through the dry papery covering of the exit hole. They live 5–10 days and the females lay an average of 150–300 eggs. The moths are 15-mm long and have reddish brown heads and bodies and light brown wings. The total life cycle ranges from 22–30 days (Ames, et al., 1996).

Nature of damage: The larva bores into the main stem and sometimes penetrates the storage roots. Larval feeding produces large tunnels causing hollow cavities in the stem resulting in wilting and eventual death of infested plants.

Management:

1. Crop rotation with nonhost plants is a valuable means of control.
2. Hilling-up, often practiced to reduce damage from sweet potato weevil, also contributes to the containment of a stem borer infestation.
3. Earwigs and ants may attack the larvae developing within sweet potato vines.

1.2.24 SWEET POTATO BUTTERFLY ACRAEA ACERATA (NYMPALIDAE: LEPIDOPTERA)

Distribution: The sweet potato butterfly is a pest in East and Central Africa.

Biology and description of pest: Eggs are laid in batches of 100–400 on both surfaces of the leaves which are pale yellow in color. The caterpillars are greenish black and covered with branching spines. These larvae are concentrated in protective webbing during the first 2 weeks after hatching. They then become solitary and hide from the sunlight on the ground during the day. The pupae hang singly on the underside of leaves or on another support. The attractive adult butterfly has orange wings with brown margins. The life cycle takes 27–50 days depending on temperature.

Nature of damage: The larvae feed on the leaves. Complete defoliation in young plants results in plant establishment failure which in turn leads to reduced storage root yields (Ames et al., 1996; Smit et al., 1997). Outbreaks of *A. acerata* normally occur during the dry season (Lugojja et al., 2001).

Management:

1. Handpicking and destroying young caterpillars.
2. Use of contact insecticides.
3. Early planting and harvesting enable the crop to escape severe attacks.
4. Intercropping sweet potato with onion, *Allium cepa* and/or silverleaf desmodium, *Desmodium uncinatum*.
5. Natural control from parasitoids, predatory ants, and entomopathogenic fungi is also effective in controlling the pests (Azerefegne et al., 2001)

1.2.25 TORTOISESHELL BEETLE: ASPIDOMORPHA SPP (CHRYSOMELIDAE: COLEOPTERA)

Distribution: Tortoise shell beetles are widely distributed in Kenya, South East Asia, etc.

Hosts: Alternative host plants include members belonging to Convolvulaceae family and coffee, round potato, and various flowers.

Biology and description of pest: Eggs are laid on the underside of the sweet potato leaves or other Convolvulaceae. Larvae are characteristically flattened and spiny. In some species, the tail is held up over the back and the larva may carry excreta and previous cast skins. The pupa is less spiny compared with the larva and is fixed inert to the leaf. The adults are broadly oval and may be bright and patterned. Larvae, pupae, and adults are found on both sides of the foliage. Development from egg to adult takes 3–6 weeks depending on the species (Ames et al., 1996).

Nature of damage: The young larvae scrape the upper surface of the leaves, whereas older larvae and adults eat large round holes in the leaves. Severe attacks can sometimes skeletonize the leaves and peel the stems.

Management:

1. Planting away from alternative host plants can reduce populations.
2. Several natural enemies including egg and larval parasites (*Tetrastichus* sp., Eulophidae; Chalcidae) and predators (*Stalilia* sp., Mantidae) have been reported against the pest.

1.2.26 SWEET POTATO HORNWORM: AGRIUS CONVOLVULI (SPINGIDAE: LEPIDOPTERA)

Distribution: *A. convolvuli* occurs worldwide but is prevalent in Africa, Asia, Australia, the Pacific, and Southern Europe.

Hosts: The sweet potato hornworm larvae also attack eggplant, legumes, pepper, tomato, and taro.

Biology and description of pest: The larvae have a conspicuous posterior "horn." They vary in color from green to brown and are marked with distinct striped patterns. Hornworms are found mainly on young shoots. The large, reddish-brown pupa is characterized by a prominent proboscis, which is curved downward. Adults are large, gray hawk moths with black lines on the wings (Ames et al., 1997; Stathers et al., 2005). The small, shiny eggs are laid singly on any part of the plant. The last instar caterpillars reach 9.5 cm in length. The larval period lasts 3–4 weeks. Pupation takes place in the soil and takes 5–26 days, depending on the temperature.

Nature of damage: A large caterpillar can defoliate a plant and a large population of late instar larvae can defoliate a field overnight. The larvae feed on the leaf blades, causing irregular holes, and may eat the entire blade, leaving only the petiole.

Management:

1. Handpicking the larvae from the leaves, plowing the land between crops to expos the pupae.
2. Manual removal of small larvae can prevent the buildup of an outbreak population of the voracious late instar larvae.
3. Use of light traps to monitor the population of adults.

1.2.27   SUGAR BEET: SUGAR BEET ROOT APHIDS: PEMPHIGUS POPULIVENAE (APHIDIDAE: HEMIPTERA)

Hosts: The primary host plants of sugar beet root aphids are certain deciduous trees in the genus *Populus*, including narrow-leaved cotton-wood, balsam poplar, and black cottonwood. Secondary hosts include the roots of beets, lambs quarter, pigweed, foxtail, and dock (Foottit et al., 2010).

Biology and description of pest: Sugar beet root aphids are small oval-shaped with a body length ranging from 1.9–2.4 mm. In contrast to most other aphid species the *P. populivenae* have reduced appendages (siphun-culi, legs, antennae, and cauda) as an adaptation to their subterranean exis-tence. Individual aphids secrete a waxy material, giving their subterranean colonies a moldy appearance. The amount of wax present on a sugar beet-root is useful for rating sugar beet root aphid infestation levels in the field (Hein and Bradshaw, 2016).

Nature of damage: Heavy feeding will give the root a rubbery and flaccid appearance under conditions of extreme stress and inadequate moisture levels. The aphid damage is often most severe under drier conditions

Management: Integrated pest management is an important approach for sugar beet root aphid management and incorporates biological control, cultural control, and host plant resistance. Natural enemies (ladybird beetles, lacewings, wasps) and diseases generally keep aphids in check.

1.2.28 BEET WEBWORM: SPOLADEA RECURVALIS (CRAMBIDAE: LEPIDOPTERA)

Biology and description of pest: Larvae are slender, greenish-black, or pink. The adult beet webworm moth is brown with two incomplete white stripes on each forewing and a complete white stripe across each hind wing. The Caterpillar is brownish-green, with sparse white hairs and a brown head. It pupates in a silken web of leaves and debris on its food plant.

Nature of damage: Spin webs and feed on beet leaves, usually near the leaf base. Weedy fields are more favorable because females deposit eggs on some weed species.

Management: Many parasites and predators. Insecticides generally not needed.

1.2.29 BEET LEAFHOPPER: CIRCULIFER TENELLUS (CICADELLIDAE: HOMOPTERA)

Hosts: Sugar beet, radish, dry bean, potato, carrot, and tomato.

Biology and description of pest: The leafhopper is described as a small insect, 3–3.5 mm in length, which is often greenish-yellow in color. The leafhopper may have darker markings on its wings, pronotum, abdomen, and head if it has developed during colder temperatures. The general shape of the body has been described as "wedge-shaped" with the body tapering off at the posterior end of the insect (Hudson et al., 2010).

Nature of damage: Major damage is mostly to sugar beets. Sucking sap is of lesser importance than being a vector of the virus beet curly top virus (family Geminiviridae, genus Curtovirus) to several crops and is the only known vector of this destructive plant pathogen. (Munyaneza and Upton, 2005).

Management:

1. In areas where beet leafhopper and curly top virus are a problem, several cultural practices can reduce the potential for leafhopper build up and damage potential to the crop.

2. If pesticides are used early in overwintering areas, use of insec-
    ticides on a large-scale basis treating entire beet field can be
    avoided.

1.2.30   ELEPHANT FOOT YAM: MEALY BUG: RHIZOECUS AMORPHOPHALLI, PSEUDOCOCCUS LONGISPINUS, P. CITRICULUS (PSEUDOCOCCIDAE: HEMIPTERA)

Distribution: It is widespread in Federated States of Micronesia,
Hawaiian Islands, Indonesia, Indonesia, India, Malaysia, Malaysia, Phil-
ippines, and Thailand. In India, it is distributed in Maharashtra, Goa, and
Kerala.

Hosts: *Amorphophallus* spp. *Colocasia esculenta, Cordyline terminalis,
Dioscorea elephantipes, Curcuma longa, Curcuma domestica, Kaemp-
feria galangal, Zingiber officinale.*

Biology and description of pest: Its multiplication is rapid during the
summer season. Eggs are ovoid shape, pale white in color and 62–74 eggs
lay in clusters inside the egg sac. Eggs hatch in 7–9 day and turned into
light brown color. There are three nymphal instars in its life cycle. The
pupal period for female and male requires about 27 days and 22 days
(Sreerag et al., 2014).

Nature and symptom of damage: Same as *Dioscorea* spp.

Management:

1. Application of salt (NaCl) solution (1000 ppm), cow urine, cow
    dung slurry (2 kg of cow dung in 1 L of water) and clay slurry
    (1 kg of clay in 1 L of water) treatments were effective against
    reduction in population of mealybug (Nedunchezhiyan et al.,
    2011).
2. Rubbing of infested corms with a dry cloth and forcefully washing
    the corms with water (Misra et al., 2003).
3. Application of cassava leaf-based fumigant has proved to be effec-
    tive and can be used for successful management (Sreerag et al.,
    2014).

KEYWORDS

- **potato**
- **carrot**
- **radish**
- **yam**
- cassava
- **sugarbeet**
- **pest management**
- **eco-friendly approaches**

REFERENCES

Abraham, V. A. Behaviour of Adult Beetles of the Coconut White Grub, *Leucopholis coneophora* Burm. *Indian Coconut J.* **1993,** *24* (1), 2–4.

Adesiyan, S. O.; Odihirin, R. A.; Adeniji, M. O. Economic Losses Caused by the Yam Nematode, *Scutellonema bradys*, in Nigeria. *Plant Dis. Rep.* **1975,** *59* (6), 477–480.

Afifi, F. M. L.; Haydar, M. F.; Omar, H. I. H. Effect of Different Intercropping Systems on Tomato Infestation with Major Insect Pests; *Bemisia tabaci* (Genn.) (Hemiptera: Aleyrodidae), *Myzus persicae* Sulzer (Homoptera: Aphididae) and *Phthorimpa operculella* Zeller (Lepidoptera: Gelechiidae). Bulletin of Faculty of Agriculture, University of Cairo, 1990, Vol. 41 (3, Suppl.1), pp 885–900.

Ahmad, T.; Ali, H.; Ansari, M. S. Biology of Diamondback Moth, *Plutella xylostella* (Linn.) on *Brassica juncea* cv. Pusa Bold. *Asian J. Bio. Sci.* **2008,** *3* (2), 260–262.

Ali, M. A. 1993. Effects of Cultural Practices on Reducing Field Infestation of Potato Tuber Moth (*Phthorimaea operculella*) and Greening of Tubers in Sudan. *J. Agr. Sci.* **1993,** *121* (2), 187–192.

Alyokhin, A. Colorado Potato Beetle Management on Potatoes: Current Challenges and Future Prospects. *Fruit Veg. Cer. Sci. Biotechnol.* **2008,** *3* (Special Issue 1), 10–19.

Alyokhin, A.; Drummond, F. A.; Sewell, G. Density Dependent Regulation in Populations of Potato-colonizing Aphids. *Popul. Ecol.* **2005,** *47*, 257–266.

Alyokhin, A.; Drummond, F. A.; Sewell, G.; Storch, R. H. Differential Effects of Weather and Natural Enemies on Coexisting Aphid Populations. *Environ. Entomol.* **2011,** *40*, 570–580.

Ames, T.; Smit, N. E. J. M.; Braun, A. R.; Sullivan, J. N. O.; Skoglund, L. G. Sweet potato: Major Pests, Diseases, and Nutritional Disorders. International Potato Center (CIP), Lima, Peru, 1997, 1–153.

Ames, T.; Smit, N. E. J. M.; Braun, A. R.; O'Sullivan, J. N.; Skoglund, L. G. Sweet potato: Major Pests, Diseases, and Nutritional Disorders. International Potato Center (CIP). Lima, Peru, 1996, p 152.

An, G.; Watson, B.; Chiang, C. Transformation of Tobacco, Tomato, Potato, and *Arabidopsis thaliana* Using a Binary *Ti* Vector System. *Plant Physiol.* **1986,** *81*, 301–305.

Anonymous. Annual Report. Department of Entomology, CSK Himachal Pradesh Krishi Vishvavidyalaya Palampur, Himachal Pradesh (India), 2011a, p 63.

Azerefegne, F.; Solbreck, C.; Ives, A. R. Environmental Forcing and High Amplitude Fluctuations in the Population Dynamics of the Tropical Butterfly *Acraea acerata* (Lepidoptera: Nymphalidae*). J. Anim. Ecol.* **2001,** *70* (6), 1032–1045.

Banks, C. J.; Macaulay, E. D. M.; Holman, J. Cannibalism and Predation by Aphids. *Nature* **1968,** *218*, 491.

Boivin, G. Integrated Management for Carrot Weevil, Integrated Pest Management Reviews, 1999, Vol. 4, pp 21–37.

Bridge, J.; Coyne, D.; Kwoseh, C. K. Nematode Parasites of Tropical Root and Tuber Crops. In *Plant Parasitic Nematodes in Subtropical and Tropical Agriculture*; Luc, M., Sikora, R. and Bridge, J., Eds.; 2nd ed; CAB International, Walingford, UK, 2005; pp 221–258.

CABI (Centre for Agriculture and Biosciences International). The Crop Protection Compendium, 2010 Edition, CABI Publishing, Wallingford, UK, 2010, www.cabi.org/cpc2010.

CABI Crop Protection Compendium. *Daucus carota* Datasheet, 2008. DOI: http://www.cabi.org/cpc/datasheet/18018(http://www.cabi.org/cpc/datasheet/18018).

Capinera, J. L. *Handbook of Vegetable Pests*; Academic Press: San Diego, 2001, p 729.

Capinera, J. L. Aster Leafhopper, *Macrosteles quadrilineatus* Forbes (Hemiptera: Cicadellidae). *Encycl. Entomol* **2008,** 320–323.

Capinera, J. L. Black Cutworm, *Agrotis ipsilon* (Hufnagel) (Insecta: Lepidoptera: Noctuidae), *EY395* **2012,** 1–5.

CIE. Distribution Maps of Pests-Map No. 32 (Revised), Commonwealth Institute of Entomology, London, UK, 1967.

Cranshaw, W. S. Aphids on Shade Trees and Ornamentals. Colorado State University Extension. 1/00. Revised 2/09. Fact Sheet No.5.511, 2009.

Das, B. B.; Ram, G. Incidence, Damage and Carryover of Cutworm (*Agrotis ipsilon*) Attacking Potato (*Solanum tuberosum*) Crop in Bihar. *Indian J. Agr. Sci.* **1988,** *58* (8), 650–651.

Das, G. P.; Raman, K. V. Alternate Hosts of the Potato Tuber Moth, *Phthorimaea operculella* (Zeller). *Crop Prot.* **1994,** *13* (2), 83–86.

Degen, T.; Stadler, E.; Ellis. P. R. Host-plant Susceptibility to the Carrot Fly, *Psila rosae.* 3. The Role of Oviposition Preferences and Larval Performance. *Ann. Appl. Biol.* **1999,** *134*, 27–34.

Delahaut, K. A.; Newenhouse, A. C. Growing Carrots, Beets, Radishes, and Other Root Crops in Wisconsin: A Guide for Fresh Worker Growers. University of Wisconsin Extension, 1998.

Deshmukh, P. S.; Chougale, A. K.; Shahasane, S. S.; Desai, S. S.; Gaikwad, S. G. Studies on Biology of Hadda Beetle, *Epilachna vigintioctopunctata* (Coleptera, Coccinillidae): A Serious Pest of Wild Bitter Gourd, *Momordica dioica. Trends Life Sci.* **2012,** *1* (3), 46–48.

Ferro, D. N.; Logan, J. A.; Voss, R. H.; Elkinton, J. S. Colorado Potato Beetle (Coleoptera: Chrysomelidae) Temperature-dependent Growth and Feeding Rates. *Environ. Entomol.* **1985,** *14*, 343–348.

Fletcher, T. B. Some South Indian Insects. Superintendent Government Press, Madras, 1914, p 565.

Flint, M. L. *Integrated Pest Management for Potatoes in the Western United States*; 2nd ed; Statewide-Integrated Pest Management Program University of California Division of Agriculture and Natural Resources Publication, 2006, p 3316.

Folcia, A. M.; Rodríguez, S. M.; Russo, S. Aspectos morfológicos, biológicos y de preferencia de *Epilachna vigintioctopunctata* Fabr. (Coleóptera: Coccinellidae). *Boletín de Sanidad Vegetal*, Plagas **1996**, *22*, 773–780.

Foottit, R. G.; Floate, K.; Maw, E. Molecular Evidence for Sympatric Taxa within *Pemphigus betae* (Hemiptera: Aphididae: Eriosomatinae). *Can. Entomol.* **2010**, *142*, 344–353.

Gianessi, L. The Benefits of Insecticide Use: Carrots. DOI: http://citeseerx.ist.psu.edu/viewdoc/download?doi=10.1.1.654.4078&rep=rep1&type=pdf.

Griffiths, G. C. D. Cyclorrhapha II (Schizophora: Calyptratae). Part 2, Anthomyiidae. In *Flies of the Nearctic Region*; Griffiths, G. C. D., Ed.; Vol. VIII. Schweizerbart'sche Verlagsbuchhandlung: Stuttgart, Germany, 1993; Vol. 8 (2/10); pp 1417–1632.

Groden, E.; Drummond, F. A.; Casagrande, R. A.; Haynes, D. L. *Coleomegilla maculata* (Coleoptera: Coccinellidae): Its Predation Upon the Colorado Potato Beetle (Coleoptera: Chrysomelidae) and its Incidence in Potatoes and Surrounding Crops. *J. Econ. Entomol.* **1990**, *83*, 1306–1315.

Hautier, L.; Jansen, J. P.; Mabon, N.; Schiffers, B. Building a Selectivity List of Plant Protection Products on Beneficial Arthropods in Open Field: A Clear Example with Potato Crop. *IOBC– WPRS Bull.* **2006**, *29*, 21–32.

Hazzard, R. V.; Ferro, D. N.; van Driesche, R. G.; Tuttle, A. F. Mortality of Eggs of Colorado Potato Beetle (Coleoptera: Chrysomelidae) from Predation by *Coleomegilla maculata* (Coleoptera: Coccinellidae). *Environ. Entomol.* **1991**, *20*, 841–848.

Hein, G. L.; Bradshaw, J. Ecology and Management of *Pemphigus betae* (Hemiptera: Aphididae) in Sugar Beet. *J. Integ. Pest Manag.* **2016**, *7* (1), 1–9.

Herman, T. Monitoring Shows Peaks of (potato tuber moth) PTM Generations. *Comm. Grow.* **1999**, *54* (1), 20.

Herman, T. J. B. Biological Control of Potato Tuber Moth, by *Apanteles subandinus* Blanchard in New Zealand. In *Integrated Pest Management for the Potato Tuber Moth, Phthorimaea operculella Zeller a Potato Pest of Global Importance. Tropical Agriculture 20, Advances in Crop Research*; Kroschel, J., Lacey, L., Eds.; Margraf Publishers: Weikersheim, Germany, 2008; Vol. 10, pp 73–80.

Horne, P. A. The Influence of Introduced Parasitoids on the Potato Moth, *Phthorimpa operculella* (Lepidoptera: Gelechiidae) in Victoria, Australia. *Bull. Entomol. Res*. **1990**, *80* (2), 159–163.

Horsfall, J. L. Life History Studies of *Myzus persicae* Sulzer. Pennsylvania Agric. Agricultural Experiment Station Bulletin 185, 1924, p 16.

Hossain, M. S.; Khan, A. B.; Haque, M. A.; Mannan, M. A.; Dash, C. K. Effect of Different Host Plants on Growth and Development of Epilachna Beetle. *Bangladesh J. Agr. Res.* **2009**, *34* (3), 403–410.

Hudson, A.; Richman, D. B.; Escobar, I.; Creamer, R. Comparison of the Feeding Behavior and Genetics of Beet Leafhopper, *Circulifer tenellus,* Populations from California and New Mexico. *Southwestern Entomol.* **2010**, *35* (3), 241–250.

Ivanova, T. S.; Borovaya, V. P.; Danilov, L. G. A Biological Method of Controlling the Potato Moth. Zashchita Rastenii (Moskva), 1994, Vol. 2, p 39.

Juliana J. S.; Lloyd M. D. Coping with Root Maggots in Prairie Canola Crops. *Prairie Soils Crops J.* **2011**, *4*, 24–31.

Kennedy, J. S.; Day, M. F.; Eastop, V. F. A Conspectus of Aphids as Vectors of Plant Viruses. *Commonwealth Institute of Entomology*, London, 1962, p 114.

Komi, K. Eradication of Sweet Potato Weevil, *Cylas formicarius* Fabricius from Muroto City, Kochi, Japan. *Ext. Bull. Food and Ferti. Tech. Cent.* **2000**, *493*, 15–22.

Korada, R. R.; Naskar, S. K.; Palaniswami, M. S.; Ray, R. C. Management of Sweet Potato Weevil [*Cylas formicarius* (Fab.)]: An Overview. *J. Root Crops* **2010**, *36* (1), 14–26

Kroschel, J.; Koch, W. Studies on the Population Dynamics of the Potato Tuber Moth [*Phthorimaea operculella* Zell. (Lep., Gelechiidae)] in the Republic of Yemen. *J. Appl. Entomol.* **1994**, *118* (4/5), 327–341.

Kumar, M. Tiwary, S. K. Variability of Varietal Resistance of Potato Against Cutworm *Agrotis ipsilon* Hufn. *Proc. Zool. Soc. Ind.* **2009**, *8* (1), 111–114.

Lakshman, L. Over Seasoning and Re-infestation Cycle of Potato Tuber Moth, *Phthorimaea operculella* (Zeller) (Lepidoptera: Gelechiidae) in North-Eastern hill region. *Ind. J. Hill Farm.* **1991**, *4* (2), 45–49.

Lal, L. Insect Pests of Potato and their Status in North-Eastern India. *Asian Potato J.* **1990**, *1* (1), 49–51.

Lal, L. Effect of Inter-cropping on the Incidence of Potato Tuber Moth, *Phthorimaea operculella* (Zeller). *Agr. Ecosyst. Environ.* **1991**, *36* (3–4), 185–190.

Langford, G. S.; Cory, E. N. Winter Survival of the Potato Tuber Moth, *Phthorimaea operculella* (Zell.). *J. Econ. Entomol.* **1932**, *27*, 210–213.

Logan, P. A.; Casagrande, R. A.; Faubert, H. H.; Drummond, F. A. Temperature-dependent Development and Feeding of immature Colorado Potato Beetles, *Leptinotarsa decemlineata* Say (Coleoptera: Chrysomelidae). *Environ. Entomol.* **1985**, *14*, 275–283.

Lugojja, F.; Ocenga Latigo, M. W.; Smit, N. E. J. M. Species Diversity and Activity of Parasitoids of the Sweet Potato Butterfly, *Acraea acerata*, in Uganda. *Afri. Crop Sci. J.* **2001**, *9* (1), 157–163.

MAFBNZ. Hadda Beetle Established in Auckland. Ministry of Agirculture and Food, Biosecurity, New Zealand (Wellington), 2010.

Markosyan, A. F. Effect of Temperature on the Development of the Potato Moth *Phthorimaea operculella* Zell. (Lepidoptera: Gelechiidae). *Edot over~ntomologicheskoe Obozrenie* **1992**, *71* (2), 334–338.

Metcalf, C. L.; Flint, W. P. *Destructive and Useful Insects Their Habits and Control*; 4th ed; Metcalf, R. L. Ed.; McGraw-Hill Book Company, Inc.: New York, San Francisco, Toronto, London, 1962; pp 646–647.

Misra, R. S.; Sriram, S.; Nedunchezhiyan, M.; Mohandas, C. Field and Storage Diseases of *Amorphophallus* and Their Management. *Aroideana* **2003**, *26*, 101–112.

Mrowczynski, M.; Wachowiak, H.; Boron, M. Cutworms a Dangerous Pest in the Autumn of 2003. *Ochr. Rosl.* **2003**, *47* (10), 24–26.

Mukherjee, A. K. Life-history and Bionomics of the Potato Tuber Moth, *Gnorimoschema operculella* Zell. at Allahabad (United Provinces), Together with Some Notes on the External Morphology of the Immature Stages. *J. Zoo. Soc. Ind.* **1948**, *1*, 57–67.

Munyaneza, J. E.; Upton, J. E. Beet Leafhopper (Hemiptera: Cicadellidae) Settling Behavior, Survival, and Reproduction on Selected Host Plants. *J. Econ. Entomol.* **2005,** *98* (6), 1824–1830.

Namba, R.; Sylvester, E. S. Transmission of Cauliflower Mosaic Virus by the Green Peach, Turnip, Cabbage, and Pea Aphids. *J. Econ. Entomol.* **1981,** *74,* 546–551.

Nedunchezhiyan, M.; Jata, S. K.; Ray, R. C.; Misra R. S. Management of Mealybug (*Rhizoecus amorphophalli*) in Elephant Foot Yam (*Amorphophallus paeoniifolius*). *Exp. Agr.* **2011,** *47* (4), 717–728.

OEPP/EPPO. EPPO Standard PP 2/22(1) Guideline on Good Plant Protection Practice: Umbelliferous crops, 2000, 01–22.

Okoroafor, E.; Amatobi, C. I.; Misari, S. M.; Onu, I. Field Assessment of Yam Beetle (*Heteroligus meles*) Damage on Yam Cultivars (Dioscorea spp.). *J. Res. Crops* **2009,** *10* (2), 398–401.

Okoroafor, E.; Onu, I.; Amatobi, C. I. Laboratory Evaluation of Botanical Effect on Yam Tuber Beetle (*Heteroligus Meles* Billberger) (Coleoptera: Dynastidae) Damage and Weight Loss in White Yam (Dioscorea Rotundata Poir). *IOSR J. Agr. Vet. Sci.* **2014,** *7* (5), Ver. II, 12–15.

Park, S. D.; Khan, Z. Occurrence of *Scutellonema unum* (Nematoda: Hoplolaimidae) on Yam (*Dioscorea batatas* Decne) in Korea. *Int. J. Nematol.* **2007,** *17* (1), 91–93.

Parker, B. L.; Hunt, G. L. T. *Phthorimpa operculella* (Zell.), the Potato Tuber Moth: New Locality Records for East Africa. *Am. Potato J.* **1989,** *66* (9), 583–586.

Patalappa, G.; Basavanna, P. C. Seasonal Incidence and Life History of *Pediobius foveolatus* (Hymenoptera: Eulophidae), a Parasite of *Henosepilachna vigintioctopunctata* (Fabricius) (Coleoptera: Coccinellidae). *Mysore J. Agric. Sci.* **1979,** *13,* 191–196.

Pepper, B. B. The Carrot Weevil, Listronotus latiusculus (Bohe), in New Jersey and Its Control, New Jersey Agricultural Experiment Station, Rutgets University, Bulletin 693, 1942.

Persoons, C. J.; Voerman, S.; Verwiel, P. E. J.; Ritter, F. J.; Nooyen, W. J.; Minks, A. K. Sex Pheromone of the Potato Tuberwonn Moth, *Phthorimaea operculella*: Isolation, Identification and Field Evaluation. *Entomol. Exp. Appl.* **1976,** *20,* 289–300.

Pillai, K. S, Palaniswami, M. S.; Rajamma, P.; Mohandas, C.; Jayaprakas, C. A. Pest Management in Tuber Crops. *Ind. Hort.* **1993,** *38* (3), 20–23.

Povolny, D. Morphology, Systematics and Phylogeny of the Tribe Gnorimoschemini (Lepidoptera, Gelechiidae). *Prirodovedne Prace Ustavu Ceskoslovenske Akademie Ved v Brne,* **1991,** *25* (9–10), 61.

Ragsdale, D. W.; Radcliffe, E. B.; Difonzo, C. D.; Connelly, M. S. Action Thresholds for an Aphid Vector of Potato Leaf Roll Virus. In *Advances in Potato Pest Biology and Management*; Zehnder, G. W., Jansson, R. K., Powelson, M. L., Raman, K. V., Eds.; American Phytopathological Society Press, St. Paul, MN, 1994; pp 99–110.

Richards, A. M. The *Epilachna vigintioctopunctata* Complex (Coleoptera: Coccinellidae. *Int. J. Entomol.* **1983,** *25* (1), 11–41.

Rondon, S. I. The Potato Tuber Worm: A Literature Review of its Biology, Ecology, and Control. *Am. J. Potato Res.* **2010,** *87,* 149–166.

Sankara, T.; Girling, D. J. The Current Status of Biological Control of the Potato Tuber Moth. *Biocont. News Inform.* **1980,** *1,* 207–211.

Sarfraz, M.; Andrew, B. K.; Lloyd, M. D. Biological Control of the Diamondback Moth, Plutella xylostella: A Review, Biocontrol Science and Technology **2005,** *15* (8), 763–789.

Schaefer, P. W. Natural Enemies and Host Plants of Species in the Epilachninae. (Coleoptera: Coccinellidae) – A World List. *Bull. Agric. Exp. Stn. Univ. Del.* **1983**, *445*, p 42.

Schroder, R. F. W.; Athanas, M. M.; Pavan, C. *Epilachna vigintioctopunctata* (Coleoptera: Coccinellidae), New Record for Western Hemisphere, with a Review of Host Plants. *Entomol. News* **1993,** *104*, 111–112.

Seaman, A. Production Guide for Organic Carrots for Processing. Cornell University Cooperative Extension, 2014, pp 30–34.

Shankar, U.; Kumar, D.; Gupta, S. Integrated Pest Management in Brinjal. Technical Bulletin No. 4. Sher-e-Kashmir University of Agricultural Sciences and Technology of Jammu, 2010, p 16.

Shelton, A. M. Management of the Diamondback Moth: de´ja` vu all over again? In *The Management of Diamondback Moth and Other Crucifer Pests*; Proceedings of the Fourth International Workshop, 26–29 November 2001. Melbourne. Melbourne, Australia: Department of Natural Resources and Environment, 2004, pp 3–8.

Shelton, A. M.; Nault, B. A. Dead-end trap Cropping: A Technique to Improve Management of the Diamondback Moth, *Plutella xylostella* (Lepidoptera: Plutellidae). *Crop Prot.* **2004,** *23* (6),497–503.

Shu-Sheng L.; Xin-Geng W.; Shi-Jian G.;B Jun-Hua H.; Zu-Hua S. Seasonal Abundance of the Parasitoid Complex Associated with the Diamondback Moth,*Plutella xylostella*(Lepidoptera: Plutellidae) in Hangzhou, China, 2002, Vol. 90 (3), pp 221–231.

Sikura, A. I.; Shendaraskaya, L. P. The Potential Area of Distribution of the Potato Moth in the Territory of the USSR. *Zashchita Rastenii* **1983,** *8*, 34.

Smit, N. E. J. M.; Lugojja, F.; Ogenga, L.; Morris, W. The Sweet Potato Butterfly (Acraea acerata): A Review. *Int. J. Pest. Manag.* **1997,** *43* (4), 275–278.

Sreerag, R. S.; Jayaprakas, C. A.; Nishanth Kumar S. Biology of the Mealy Bug, *Rhizoecus amorphophalli* Infesting Tubers of Majors Aroids. *J. Entomol. Nematol.* **2014,** *6* (6), 80–89.

Stathers, T.; Namanda, S.; Mwanga, R. O. M.; Khisa, G.; Kapinga, R. Manual for Sweet potato Integrated Production and Pest Management Farmer Field Schools in Sub-saharan Africa. International Potato Center, Kampala Uganda, 2005, xxxi, p 168.

Stoner, K. A. Effects of Straw and Leaf Mulches and Sprinkle Irrigation on the Abundance of Colorado Potato Beetle (Coleoptera: Chrysomelidae) on Potato in Connecticut. *J. Entomol. Sci.* **1993,** *28*, 393–403.

Stoner, K. A. Influence of Mulches on the Colonization by Adults and Survival of Larvae of the Colorado Potato Beetle (Coleoptera: Chrysomelidae) in Eggplant. *J. Entomol. Sci.* **1997,** *32*, 7–16.

Straub, C. S.; Snyder, W. E. Species Identity Dominates the Relationship Between Predator Biodiversity and Herbivore Suppression. *Ecology* **2006,** *87*, 277–282.

Szendrei, Z. The Impact of Plant Associations on Macrosteles Quadrilineatus Management in Carrots. *Entomologia Experimentalis et Applicata* **2012,** *143*, 191–198

Tayde, A. R.; Simon, S. Studies on Biology and Morphometris of Hadda Beetle, *Epilachna vigintioctopunctata* (Coleoptera: Coccinellidae) a Serious Pest of Bitter Gourd, *Momordica charantia*, in Eastern Uttar Pradesh, India. *Int. J. Agr. Sci.* Res. **2013,** *3* (4), 133–138.

Taylor, T. A. Studies on the Nigeria Yam Beetles II: Bionomics and Control. *J. West Afr. Sci. Assoc.* **1964,** *9*, 13–31.

Tooker, J. Black Cut Worm. Entomogical Notes. The Pennsylvania State University, 2009. DIO: http://ento.psu.edu/extension/factsheets/pdf/BlackCutworm.pdf.

Trivedi, T. P.; Rajagopal, D. Distribution, Biology, Ecology, and Management of Potato Tuber Moth, *Phthorimaea operculella* (Zeller) (Lepidoptera: Gelechiidae): A Review. *Trop. Pest Manag.* **1994,** *38* (3), 279–285.

Van Emden, H. F.; Eastop, V. F.; Hughes, R. D.; Way, M. J. The Ecology of *Myzus persicae*. *Ann. Rev. Entomol.* **1969,** *14*, 197–270.

Vishakantaiah, M.; Visweswara Gowda, B. L. Record of *Plutella xylostella* (Linnaeus) (*P. maculipennis* Curtis) as a New Pest of *Amaranthus viridis* in Karnataka. *Curr. Sci.* **1975,** *44*, 869.

Weber, D. C. Biological Control of Potato Insect Pests. In *Insect Pests of Potato Globed Perspectives on Biology and Management*; Giordanengo, P., Vincent, C., Andrei Alyokhin, A., Eds; Academic Press, Elsevier Publications: UK, 2013; pp 339–437.

Wilkerson, J. L.; Webb, S. E.; Capinera, J. L. Vegetable Pests I: Coleoptera-Hymenoptera. UF/IFAS CD-ROM. SW 180, 2005.

Yasuda, K. Occurrence of West Indian Sweet Potato Weevil, *Euscepes postfasciatus* (Fairmaire) (Coleoptera: Curculionidae) and Damage to Sweet Potato 14 (*Ipomoea batatas* (L.) Lam.) Fields. *Japan. J. Appl. Entomol. Zoo.* **1997a,** *41*, 83–88.

CHAPTER 2

INSECT PESTS OF GINGER AND TURMERIC AND THEIR MANAGEMENT

BUDHACHANDRA THANGJAM* and NAVENDU NAIR

Department of Entomology, College of Agriculture, Lembucherra, West Tripura 799210, India

Corresponding author. E-mail: budhathangjam@gmail.com

ABSTRACT

Ginger and turmeric are two powerful spices that have been used widely throughout history for both culinary and medical purposes. Using these spices in recipes provides a way to season a variety of foods without adding sodium or fat. Apart from its culinary uses, turmeric has been used widely in the traditional medicine in India, Pakistan, and Bangladesh because of its several beneficial properties. Although there is plenty of information about the use of turmeric powder as a spice in culinary and apart from its multiple medicinal uses, the plant is credited with interesting pesticidal properties against insects and fungi of agricultural significance, including repellent properties against some noxious mosquito species. Turmeric has anti-inflammatory analgesic, antibacterial, anti-tumor, anti-allergic, anti-oxidant antiseptic, antispasmodic, astringent, digestive, diuretic and stimulant properties.

Ginger (*Zingiber officinale* Rosc.) and turmeric (*Curcuma longa* L.) belonging to the family Zingiberaceae are one of the most important and most widely used spices worldwide. Due to its universal appeal, ginger has spread to most tropical and subtropical countries from the China–India region, where ginger cultivation was prevalent probably from the days of unrecorded history. Ginger is cultivated in most of the states in India.

However, states namely Karnataka, Orissa, Assam, Meghalaya, Arunachal Pradesh and Gujarat together contribute 65 per cent to the country's total production. Ginger and turmeric are prominent among the different spices grown in the northeastern India and their cultivation is undertaken as a cash crop mostly in *jhum* fields spread over the hills and tribal areas of the entire region. India is a leading producer of ginger in the world and during 2012–2013 the country produced 7.45 lakh tonnes of the spice from an area of 157,839 hectares. India is also the world's largest producer, consumer, and exporter of turmeric. During 2012–2013 the country produced 9.71 lakh tonnes from an area of 194,000 hectares.

2.1   INTRODUCTION

Ginger (*Zingiber officinale* Rosc.) and turmeric (*Curcuma longa* L.) belonging to the family Zingiberaceae are one of the most important and most widely used spices worldwide. Due to its universal appeal, ginger has spread to most tropical and subtropical countries from the China–India region, where ginger cultivation was prevalent probably from the days of unrecorded history. Ginger is cultivated in most of the states in India. However, states namely Karnataka, Orissa, Assam, Meghalaya, Arunachal Pradesh and Gujarat together contribute 65 per cent to the country's total production. Ginger and turmeric are prominent among the different spices grown in the northeastern India and their cultivation is undertaken as a cash crop mostly in *jhum* fields spread over the hills and tribal areas of the entire region. India is a leading producer of ginger in the world and during 2012–2013 the country produced 7.45 lakh tonnes of the spice from an area of 157839 hectares. India is also the world's largest producer, consumer, and exporter of turmeric. During 2012–2013 the country produced 9.71 lakh tonnes from an area of 194,000 hectares.

Ginger and turmeric are two powerful spices that have been used widely throughout history for both culinary and medical purposes. Using these spices in recipes provides a way to season a variety of foods without adding sodium or fat. Apart from its culinary uses, turmeric has been used widely in the traditional medicine in India, Pakistan, and Bangladesh because of its several beneficial properties (Chattopadhyay et al., 2004). Although there is plenty of information about the use of turmeric powder as a spice in culinary and apart from its multiple medicinal uses, the plant is credited with interesting pesticidal properties against insects and fungi of agricultural significance, including repellent properties against some noxious mosquito species. Turmeric has anti-inflammatory (Punithavathy et al., 2000), analgesic, antibacterial, anti-tumor, anti-allergic, anti-oxidant (Menon and Sudhir, 2007) antiseptic, antispasmodic, astringent, digestive, diuretic and stimulant properties.

Both the crops are exposed to various biotic and abiotic stresses resulting in low productivity. There are several reasons for low yield of the crops and among them infestation by insect-pest and infection by pathogen are considered as the major ones. A complete lists of insect pests associated with the crops are given by Devasahayam and Koya (2005, 2007). Information on insect pests of turmeric and ginger is available mainly from India and has been previously reviewed by Jacob (1980), Butani (1985), Koya et al. (1991), and Premkumar et al. (1994). The major insect pests infesting ginger and turmeric in field are described in details in this chapter.

2.2 MAJOR INSECT PESTS OF TURMERIC AND GINGER

2.2.1 SHOOT BORER, CONOGETHES PUNCTIFERALIS GUEN. (PYRAUSTIDAE: LEPIDOPTERA)

2.2.1.1 ORIGIN AND DISTRIBUTION

This species is found in southern and eastern Asia, Australia, Indonesia, and New Guinea (USDA, 1957). This species is also present in Hawaii (Nishida, 2002). In Asia, they are found in Bangladesh, Brunei, Cambodia, China, India, Indonesia, Iraq, Japan, Korea, Laos, Malaysia, Myanmar, Philippines, Sri Lanka, Taiwan, Thailand, and Vietnam. The pest is known to occur in most of the turmeric and ginger growing areas in the country. It is found to be recorded in the northern districts of Karnataka, namely at Raibag, Chikodi (Belgaum District), Jamakhandi, Indi (Bijapur District), Basavakalyan, and Humanabad (Bidar District) taluks (Kotikal and Kulkarni, 2000a). The shoot borer is known by many other common names general indicative of the crop and plant part infested. The shoot borer has been suggested to be a complex of more than one species, especially in Australia and South East Asia (Robinson et al. 1994; Boo, 1998). Species in this complex have very similar morphology, variable color morphs, and overlapping host ranges (Armstrong, 2010).

2.2.1.2   HOST RANGE

This species is highly polyphagous and has been recorded on 65 host plants from 30 different families. Many hosts are economically important (Devasahayam and Abdulla Koya, 2005). Shashank et al. (2015) state that the preferred host is cardamom followed by *Hedycium* spp., *Alpinia* spp., and *Ammomum* spp.

The host range include many important crops like castor, mango inflorescence, sorghum earheads, guava, litchi, peaches, cocoa, pear, avocado, cardamom, ginger, turmeric, mulberry, pomegranate, sunflower, cotton, tamarind, and hollyhock etc. It is a potential pest and occasionally becomes serious. It is active from September–February when crop is in flowering. Many of the hosts of the shoot borer are economically important plants, and the pest infests various parts of these plants, such as buds, flowers, shoots, and fruits

2.2.1.3   NATURAL ENEMIES

Various species of natural enemy are reported to be associated with shoot borer. *Angitia trochanterata* Morl. (Ichneumonidae), *Theromia inareolata* (Braconidae), *Bracon brevicornis* West., *Apanteles* sp. (Braconidae), *Brachymeria euploeae* West. (Chalcidae) (David et al., 1964), and *Microbracon hebetor* Say. (Braconidae) (Patel and Gangrade, 1971) were documented as natural enemies of the pest infesting castor. *Brachymeria nosatoi* Habu and *B. lasus* West. were recorded as parasitoids of the pest by Joseph et al. (1973). Various parasitoids have been recorded on the shoot borer infesting cardamom and they include *Palexorista parachrysops* (Tachinidae), *Agrypon* sp., *Apechthis copulifera*, *Eriborus trochanteratus* (Morl.), *Friona* sp., *Gotra* sp., *Nythobia* sp., *S. persimilis*, *Temeluca* sp., *Theronia inareolata*, *Xanthopimpla australis* Kr., *Xanthopimpla kandiensis* Cram. (Ichneumonidae), *B. brevicornis* West., *Microbracon hebator*, *Apanteles* sp., *Phanerotoma hendecasisella* Cram. (Braconidae), *Synopiensis* sp., *Brachymeria* sp. nr. *australis* Kr. and *B. obscurata* (Chalcidae) (Varadarasan, 1995). *Hexamermis* sp. parasitized larvae of shoot borer during July–November with a peak parasitisation of 72% during August.

The hymenopterous parasites (*Bracon* sp. and *Apanteles taragamme*) were observed during October–December with a peak parasitisation of 28% during November. Studies on host plant-insect pest-natural enemy inter- actions in ginger and turmeric indicated that the incidence of shoot borer was not significantly different on ginger and turmeric when these crops were grown individually and as mixed crops. The incidence of parasitism by hymenopterous parasitoids on shoot borer was also not significantly different in these crops.

2.2.1.4 DAMAGE AND YIELD LOSS

There are a limited number of reports describing the amount and type of damage caused by this species and most reports are limited to a specific host crop (Korycinska., 2012). In India and Sri Lanka, this species is a serious pest of castor, bean and fruit (USDA., 1957). Devasahayam and Abdulla Koya (2005) states that this species is the most serious pest of ginger, especially in India. Crop yield can be significantly affected when more than 45% of shoots in a clump are damaged (Devasahayam et al., 2010). Hill (1983) states this species is a major pest of *Curcuma domestica* (turmeric) and *Ricinus communis* (castor). The larvae of shoot borer bore into pseudostems and feed on the growing shoot of ginger plants, resulting in yellowing and drying of infested pseudostems. The presence of bore holes on the pseudostem, through which frass is extruded, and the withered central shoot are characteristic symptoms of pest infestation. The newly hatched larvae of shoot borer scrape and feed on the margins of unopened leaf or newly opened leaf of turmeric plants. Sometimes, the larvae also bore into the rhizome near the base of the pseudostem. This species has caused yield losses of 25% when 23–24% of ginger pseudostems were infested. Forty percent yield losses in ginger have been reported in certain parts of India (reviewed in Devasahayam and Abdulla Koya. 2005). According to Chong et al. (1991) "damage by *C. punctiferalis* is often isolated, and husks are frequently not bored through. In ginger, the larva feeding in the center of the stem causes death of the "heart", which is visible when the terminal shoot turns yellow. Usually, the larva is mature before it reaches the rhizome, and leaves the stem to pupate, occasion- ally it arrives at the rhizome and damages it. Boring by this species can predispose the fruits to secondary pathogens (Chong et al., 1991). Larvae begin feeding on the green contents of the leaves and later bore into the shoots, feeding on the inner core (Devasahayam et al., 2010). Usually, the larva is mature before it reaches the rhizome and leaves the stem to pupate: occasionally it arrives at the rhizome and damages it" (Chong et al., 1991). Studies on yield loss caused by the pest in Kerala indicated that when 50% of the pseudostems in a plant are affected, there was a significant reduction of 38 g of yield per plant (Koya et al., 1986). Yield losses of 25% have also been reported when 23–24% of a plant's pseudostems are infested and the pest was reported to cause 40% yield loss in Kottayam and Idukki districts in Kerala (Nybe, 2001).

2.2.1.5 MORPHOLOGY

Eggs

Eggs are round and light yellow in color, and 0.63 × 0.41 mm in size. After incubation of 6–7 days, the eggs turn dark brown with a dark head (reviewed in Shashank et al., 2015).

Larvae

The larva of the moth has a black head and a pale greenish body with a pinkish suffusion dorsally Coloration can vary by type of food (Chong et al., 1991). Fully grown larvae are 16–26 mm (Devasahayam and Abdulla Koya, 2005) and are rather stout, pale, or reddish-brown with numerous flattened horny warts that have short bristly hairs. The prothoracic shield is large and the head is reddish-brown (USDA, 1957).

Pupae

The pupa measures about 15 mm long (Chong et al., 1991) and is brown in color. The pupa is enclosed by a white silken cocoon (USDA, 1957).

Adults

Adults are pale straw yellow with numerous small black spots and a wing span of 18–24 mm (Chong et al., 1991).

Life Cycle

The life cycle of shoot borer on ginger have not been reported but its study on other Zingiberaeceous crops such as turmeric and cardamom have reported so far. After mating, females lay the small, oval eggs on or near fruit or seeds of hosts (USDA, 1957). Females lay from 20–30 eggs (reviewed in

MAF Biosecurity New Zealand, 2009). Once they hatch, larvae feed on or in seeds, seed capsules, and young shoots (USDA., 1957). The larvae go through five instars (Devasahayam and Abdulla Koya., 2005). Pupation usually occurs in the larval tunnels within a silk cocoon, surrounded by shelters of webbing and frass (Chong et al., 1991; MAF Biosecurity New Zealand, 2009).The adult emerges about 8 days later (Chong et al., 1991). In Australia, pupation can take 2–3 weeks (in summer) to as long as 8 weeks (in winter). Studies on turmeric conducted at Kasaragod (Kerala) under laboratory conditions (temperature range: 30–33°C; relative humidity range: 60–90%) indicated that the preoviposition and egg periods lasted for 4–7 and 3–4 days, respectively. The five larval instars lasted for 3–4, 5, 3–7, 3–8, and 7–14 days, respectively. The prepupal and pupal periods lasted for 3–4 and 9–10 days, respectively. Adult females laid 30–60 eggs during its lifespan, and 6–7 generations were completed during a crop season in the field. Variations were also observed in the life cycle (up to 30 days during August–October and up to 38 days during November–December) during various seasons (Jacob., 1981).

Ecology

No information is available on the seasonal population dynamics of the shoot borer on ginger. However, the damage was reported to be higher in the field during August, September, and October in Kottayam and Idukki districts in Kerala (Nybe, 2001). This species breeds throughout the year in India and parts of Australia (USDA., 1957). In Japan, this species has two to three generations per year. Overwintering occurs as full grown larvae within thick silken cocoons, spun inside loose scales of the trunk bark, or within the mummified fruit of the peach or chestnut. The larvae pupate within the cocoons in the spring, and moths emerge in mid-May (Púcat, 1995). In laboratory conditions, this species completes its lifecycle in about 28 days in castor, 31 days in cardamom, and 32 days in ginger. Damage caused by this species increased when relative humidity was increased (Stanley et al., 2009). A literature review of the bioecology can be found in (Shashank et al., 2015).

Management

Very few field trials have been conducted for the control of shoot borer on ginger and turmeric despite of its importance as one of the seriously damaging pest on the crop.

Cultural

Pruning and destroying freshly infested pseudo stems (at fort-nightly intervals) during July–August. Installation of light trap and Collection of the entire adult and destroy during mid May–June–July month for adult mass trapping. In the stem borer infested field collection of dead heart and destruction of the same help in reduction of the pest.

Biological

C. M. Senthil Kumar (National Research Centre on Spices, Calicut) described the role of hymenopteran insects and mermithid nematodes feeding on *Conogethes* larvae. He also stressed on the importance of biocontrol agents such as *Beauveria* sp., nematodes, bacteria and viruses for managing *Conogethes* spp. The isolates of entomopathogenic fungi (EPF) *Metarhizium* sp. and *Beauveria bassiana* and entomopathogenic nematode (EPN) *Heterorhabditis indica* successfully infected the larvae of *C. punctiferalis* in the laboratory. The biocontrol agents like *Bt* and EPF are promising for management of shoot borer. Two commercial products of *Bacillus thuringiensis,* namely Bioasp and Dipel, were evaluated, along with malathion for the management of the shoot borer in the field at Peruvannamuzhi. The trials indicated that all the treatments were effective in reducing the damage caused by the pest compared to control when sprayed at 21-day intervals during July–October. Spraying Dipel 0.3% was the most effective treatment, resulting in a significantly lower percentage of infested pseudostems on the crop (Devasahayam., 2000).

Pervez et al. (2012) tested the efficacies of eight native entomopathogenic nematodes (EPNs) against larvae and pupae of the shoot borer, *Conogethes punctiferalis* and reported that *Heterorhabditis* sp., *Steinernema* sp. and *Oscheius* sp caused 100% mortality of shoot borer larvae. *Oscheius* sp was the most virulent against the shoot borer pupae, causing 100% mortality, followed by *Steinernema* sp and *Oscheius* sp which killed 67% of the pupae. Choo et al. (1995) evaluated the pathogenecity of entomopathogenic nematodes against the shoot borer. *Steinernema* sp. and *Heterorhabditis* sp. caused 90 and 100% mortality, respectively, of test insects in the laboratory when 20 nematodes per larva were inoculated. Choo et al. (2001) later reported that the LC50 for *S. carpocapsae* Pocheon strain and *H. bacteriophora* Hamyang strain were 5.6 and 5.8, whereas their moralities were 96.9 and 96.5%, respectively, for these strains.

Botanicals

Evaluation of Neem oil (1%) and Nimbecidine (1%) indicated that these products were effective only on ginger when sprayed at fortnightly intervals during July–October (IISR 2003). An integrated approach consisting of mulching of neem leaves at 10 t ha-1 and spraying with NSKE (5%) during high pest incidence resulted in an increase in yield by 50%–72%; and this treatment was as effective as quinalphos (0.05%) in controlling the pest (Lalnuntluanga & Singh 2008).

Host Plant Resistance

Nybe and Nair (1979) studied the reaction of various types of ginger to shoot borer in the field and reported that among the 25 cultivars of ginger screened, the pest infestation was minimum in Rio de Janeiro and maximum in Valluvanad, although not significant. The reaction of various turmeric types to shoot borer in the field has been studied. Sheila et al. (1980) reported that among the 13 types of turmeric screened at Vellanikkara (Thrissur District, Kerala), Dindigam Ca-69 (an aromatica type) was the least susceptible and Amruthapani Kothapeta Cll-317 (a long type), the most susceptible. Philip and Nair (1981) reported that among the 19 turmeric types screened, Manuthy Local was the most tolerant. Velayudhan and Liji (2003) recorded the incidence of the shoot borer on 489 accessions belonging to 21 morphotypes and the lowest incidence was observed in morphotype II with a mean score of 2 on a 0–9 scale; 22 accessions were tolerant with a score of less than 3. Kotikal and Kulkarni (2001) screened eight genotypes of turmeric in the field at Belgaum (Karnataka) and reported that all of them were susceptible with more than 10% damage.

Bio-rational

Various literatures on the utility of pheromones on *C. punctiferalis* have been documented. Konno et al. (1982) identified the sex pheromone as (E)-10-hexadecenal. The addition of (Z)-10-hexadecenal le d to an increase trap catches (4 × more). Another minor component identified was hexadecanal. Liu et al. (1994) found that the most attractive blends were a mixture of 16: Ald, E10-16: Ald, and Z10-16: Ald at a ratio of 16:100:8 and a blend of E10-16: Ald and Z10 -16:Ald at a ratio of 100:8. Chakravarthy and Thyagaraj (1998) trapped for this species using a ratio of 9:1 for (E)-10-hexadecenal and (Z)-10-hexadecenal. This blend has been used for mass-trapping, monitoring, and mating disruption in Japan,

Korea, and China (Kimura and Honda, 1999). In Korea, an 80:20 ratio of (E)-10-hexadecenal and (Z)-10-hexadecenal had the highest attractiveness in several tests in orchard fields (Jung et al., 2000). Further work by Xiao et al. (2012) found that certain hydrocarbons had a synergistic effect on responses to pheromones.

Chemical

Field trials for the control of shoot borer pest on turmeric and ginger are very less documented in spite of the serious nature of damage caused by the shoot borer. The insecticides generally recommended for the management of shoot borer on ginger have also been recommended against the pest on turmeric (IISR, 2001) too. Koya et al. (1988) evaluated six insecticides at Peruvannamuzhi (Kerala) and found that all of them were effective in controlling the pest when sprayed at monthly intervals from July–October. Among the insecticides, malathion 0.1% resulted in minimum pest infestation on the pseudostems and was on par with monocrotophos 0.05%, quinalphos 0.05%, endosulfan 0.05%, and carbaryl molasses 0.05%. Koya et al. (1986) have evolved a sequential sampling strategy for monitoring the level of pest infestation in a field of ginger as guidance for undertaking control measures.

2.2.2 RHIZOME FLY, MIMEGRALLA COERULEIFRONS MACQUART (MICROPEZIDAE: DIPTERA)

2.2.2.1 ORIGIN AND DISTRIBUTION

The fly is found in India, Bali, Burma, China, Flores, Hainan Islands, Java, Malaya, Philippines and Sumatra. In India it is distributed throughout in different biogeographic zones of India viz., North-Eastern Region, Indo-Gangetic Plains and Ghats and Peninsular Region (Alfred et al., 2001) and state-wise in India it is widely found in Assam, West Bengal, Andhra Pradesh, Maharashtra, Tamil Nadu, Kerala, Karnataka Nadu, Kerala, and Karnataka.

2.2.2.2 HOST RANGE

No information is available regarding the host plants rather found feeding on other species of Zingiberaceae, *Colocasia* sp, wild turmeric and cardamom.

2.2.2.3   NATURAL ENEMIES

Ghorpade et al. (1988) recorded *Trichopria* (Hymenoptera: Diapriidae) from field collected pupae of rhizome fly with average parasitization rate of 30, 27, and 15% under field condition during 1978, 1979, and 1980 respectively.

2.2.2.4   DAMAGE AND YIELD LOSS

The fly has been reported as a pest of rhizomes of ginger and turmeric in association with soft rot of rhizome in the field (Karim, 1993; Nair, 1986). Sontakke (2000) reported that the maggots of rhizome fly feed inside the rhizomes by making tunnels. At early infestation no sign of injury was visible on shoots and leaves. In certain cases drying of central shoot (dead hearts) was observed initially when the maggots were entering through less thickened stems. After a week of rhizome infestation by the maggots, yellowing of lower leaves followed by drying of all the leaves including pseudo-stem was observed and such stems were easily detachable from the rhizomes. Under such conditions there was complete damage to rhizomes with bacterial and fungal infestation with foul smell. This fly breeds in decaying vegetation and excrement. The maggots live and feed in such decaying materials. Prem Kumar et al. (1980) observed the association of maggots with rhizome rot of ginger in Kerala, India. In the preliminary investigation they reported that maggots of this fly started feeding in the collar region and later migrated into both pseudo-stem and rhizomes, feeding voraciously on the soft tissues. Foliar yellowing symptom became severe when the maggots were attached to pseudo-stems as compared to rhizomes. Ghorpade et al. (1983) also reported on the incidence of rhizome fly, *M. coeruleifrons* on ginger and turmeric with crop losses averaged to 30.62% for ginger and 25.35% for turmeric.

2.2.2.5   MORPHOLOGY

Egg

The female laid eggs during the night time singly or in batches of 7–11 on rotted rhizomes. Freshly laid eggs are small, white, and oblong.

Larva

The newly hatched larva was apodous and colorless. The full-grown larva is creamy white, and measured on average 7.01 ± 0.06 mm in length and 1.65 ± 0.03 mm in breadth. The body is twelve segmented. The total larval period varied from 25.25 ± 0.85 to 38.25 ± 2.06 days.

Pupa

The full-grown maggots were pupated in the tunnels of infested rhizomes. The pupa was enclosed in a dark brown elongated coarctate puparium. It measured on an average 7.51 ± 0.03 mm in length and 1.76 ± 0.02 mm in breadth. The color changes from brown to black at the time of emergence. The pupal period lasted for 4.25 ± 0.95–7.25 ± 0.48 days.

Adult

The adult flies were fairly large with slender body and long legs. The flies hold the fore legs straight out in front like antennae. The tarsi of the fore legs were white in color. The body is blackish with transparent wings having three prominent ashy spots. The tarsal character of the fore legs and ashy spots on the wings could be the identifying characters of the fly. The last abdominal segments of the female were modified into a short blunder ovipositor. The average body length and breadth of male flies were 12.97 ± 0.11 mm and 1.55 ± 0.01 mm, respectively. The female flies were slightly larger than its male.

Life Cycle

The incidence of rhizome fly infesting ginger under field conditions happened during August–September in Orissa, India (Sontakke, 2000). The life history of the fly was studied under laboratory conditions. The peak period of *M. coeruleifrons* incidence was in the first week of September; the percentage of damaged ginger plants was 20–25% and 40–42% in protected and unprotected crops, respectively. Ghorpade et al. (1988) observed the peak period of infestation in endemic areas was from mid-August–mid-October. Ghorpade et al. (1988) studied life-history of *M. coeruleifrons* on ginger and turmeric in laboratory condition and reported that females laid 76–150 eggs in the soil. The egg stage lasted for 2–5 days, and the 3 larval instars lasted for 4–7, 4–8 and 3–10 days, respectively. Maggot completed its development in 13–25 days, while the pupal period and the adult lifespan were 5–15 days and

7–24 days, respectively. The total lifespan was 38–62 days. The sex ratio was about 1:1. The pupal stage appeared to be responsible for carrying over infestation in planting seed rhizomes from one season to another. Karim (1993) reported that in South India, the life cycle of the insect is completed in 38–62 days with 3.5, 16.8, and 13.5 days of egg, larval, and pupal periods, respectively. Sontakke (2000) reported that the fecundity of the female fly ranged from 84 to 146 eggs with an average of 79.2% egg hatching. The average egg incubation, larval, pupal, and adult periods were 60.4 h, 20.4 days, about 8 days, and about 16 days, respectively at Orissa, India.

Ecology

Maggots of this fly were reported in rotten rhizomes of ginger and turmeric in farmer's fields. The fly breeds in decaying vegetation and excrement. The maggots live and feed in such decaying materials. It is assumed by Karim (1993) that *M. coeruleifrons* actually breeds in the rotten rhizomes of ginger and turmeric as secondary pest. Ghorpade et al. (1983) also reported on the incidence of rhizome fly, *M. coeruleifrons* on ginger and turmeric. Their extensive survey in Maharashtra, India indicated that this fly species is an important pest of these plants. Infestation tended to be patchy, and less damage was caused in light, well-drained soils. It has been indicated that shady and high moisture conditions in the field are favorable for the soft rot disease and the attack by rhizome fly (Karim 1993).

Management

Many studies have so far been done on the management of rhizome fly. However, several author's priorities to control of the soil borne fungi and bacteria responsible for the soft rot of ginger rhizomes in the field (Karim, 1993), screening of germplasms (Rao et al, 1994) etc. but maximum of the management studies have been done on application of different pesticides.

Cultural

Preventive measures like destruction of stray plants in off season, selection of healthy rhizome for planting, removal and destruction of rotting rhizomes along with the maggots from the field after the harvest of the crop may help to check the breeding of the pest. Attacks of *M.*

coeruleifrons happened after the establishment of disease so to avoid attacks from these flies growing of healthy plant, free from wilt disease or other diseases should be used. Supriadi et al. (2000) suggested that strategies to control wilt disease can be done with disease free pathogens, using healthy seeds, "intercropping" and rotation, control nematodes and weeds, using cultivars tolerant / resistant, as well as soil improvement. Intercropping can reduce the risk of attack by the pests by inhibiting the pest to find its host. Ginger crop intercropping with Patchouli lower larvae and pupae population of *M. coeruleifrons*. In addition, ginger intercropping with coffee and soybean and corn can reduce populations of larvae and pupae in the ginger rhizome (Karmawati and Kristina, 1993). Overlapping shift (rotation) corn-corn-ginger onion and corn-+corn + ginger + peanut + upland rice can reduce the percentage of plants attacked by flies, but do not reduce larval populations in the rhizome (Baringbing and Gusmaini, 1999).

Natural Enemies and Biological Control

Until now publications on utilization of natural enemies to control *M. coeruleifrons* has not been studied for the satisfactory management of the pest and the studies regarding the natural enemies *of M. coeruleifrons* still limited. However, earwig (Dermaptera) often found in association with the ginger rhizome in field conditions. Most of these insects act as a predator (Kalshoven, 1981). Parasitoids of Hymenoptera also reported from the pupae of rhizome fly. In India, a pupal parasitoid *Trichopria* sp. has been reported (Jacob, 1980). In addition, the fungus *Beauveria bassiana* has been known to infect the larvae *M. coeruleifrons* naturally in the field. The adults were preyed upon by spiders (Araneae) and dragonflies (Odonata); eggs and larvae by earwigs (*Forficula auricularia* L.) and the pupae were parasitized by *Trichopria* sp. (Ichneumonidae) and *Spalangia* sp. (Pteromalidae) (Kotikal and Kulkarni, 2000c).

Host Plant Resistance

Screening of turmeric germplasm for resistance since 1986 in Andhra Pradesh, India has shown that the local high yielding varieties Armoor, Duggirala, and Mydukur are highly susceptible to the rhizome rot disease, while PCT8, PCT10, Suguna and Sudarshana performed well, even in disease prone conditions. In multi-locational trials PCT10, Suguna, and Sudarshana were free from rhizome rot in all the test locations and gave greater yields than Armoor and Duggirala. They were also free from *Colletotrichum* leaf spot, rhizome fly maggot and the root knot nematode. These high yielding cultivars mature in 195 days, with 20–22% dry recovery and contain 4–6% curcumin. PCT8 was not popular to farmers as it has small fingers (Rao et al., 1994). New fibreless cultivars, such as Suruchi, Suprabha, and Supriya, were more susceptible to the pest than traditional, local cultivars (Sontakke, 2000). The resistance of several varieties of turmeric to *M. coeruleifrons* was determined in field-plot tests in Maharashtra, India (Jadhav et al., 1982). Percentage infestation in a two years study showed that none of the varieties was completely resistant to the rhizome fly. However, Sugandham and Duggirala were the most resistant varieties, having less than 7.3% infestation.

Chemical

Dhoble et al., (1978) carried out a field-plot test in Maharashtra, India and evaluated the effectiveness of soil treatment with granules of 6 systemic insecticides at 0.75 or 1.00 kg a.i/ha for the control of *M. coeruleifrons* on turmeric. The granules were applied 3 times at 30 days interval starting when dead hearts were first noticed. Data on the percentage of infested rhizomes and yield following treatment showed that phorate was the most effective insecticide. In another field experiment, Dhoble et al., (1981) determined the effectiveness of some insecticides applied by foliar spray and soil drenching against the turmeric pest, *M. coeruleifrons*. At experimental sites, phorate, carbofuran, neem cake and pongamia cake caused the lowest *M. coeruleifrons* infestation (8–10%) compared to 26% in the untreated control. At Arabhavi, the weight of uninfested turmeric rhizomes was the greatest with phorate (16 kg plot-1) and pongamia cake (16 kg plot-1) and the fresh rhizome yield/hectare was the largest in the pongamia treatment at 25.1t/ha. At Ellihadalagi, the yield of fresh rhizomes was the greatest with the neem cake (25.8 t/ha), pongamia cake (27.3 t/ha) and phorate (26.2 t/ha) treatments. The lowest yields were achieved with vermicompost (Kotikal and Kulkarni, 1999).

Koya and Banerjee (1981) reported that aldicarb, carbofuran, and methyl parathion were effective in reducing the pest infestation in trials with various insecticides against *M. coeruleifrons* on ginger.

2.2.3   RHIZOME SCALE, ASPIDIELLA HARTII CKLL. (DIASPIDIDAE: HEMIPTERA)

2.2.3.1   ORIGIN AND DISTRIBUTION

Distribution of this scale includes the Caribbean Islands, Ecuador, Fiji, Ghana, Hawaii, Honduras, Hong Kong, India, Ivory Coast, Malaya, Nigeria, Panama, Papua New Guinea, Philippines, Sierra Leone, Solomon Is., Tonga, Trinidad, Vanuatu and Zambia. The first record of this scale in Hawaii was in 1986 from tumeric growing on the Big Island.

2.2.3.2   HOST RANGE

Aspidiella hartii has been recorded on hosts from the plant family Dioscoreaceae and Zingiberaceae. The species is associated with yams, especially tubers in storage, but also occurs on other root crops, especially Zingiberaceae when grown in proximity with yams. Hosts include species of: *Colocasia, Curcuma longa, Dioscorea, Ipomoea batatas* and *Zingiber*.

2.2.3.3   NATURAL ENEMIES

The aphelinid *Coccobius comperei* and the encyrtid *Adelencyrtus moderatus* were recorded from *Aspidiella hartii* on yams in India (Palaniswami, 1991). Another encyrtid parasitoid *Leptomastix dactylopii* was recorded from this scale insect (Japoshvili, 2010). Parasitization by *P. comperei* brought down the population of rhizome scale by about 80% in three months (Jacob, 1986). At Peruvannamuzhi (Kerala, India), apart from *Cocobius* sp., a predatory beetle and ant were observed to predate on the rhizome scale (Devasahayam, 1996).

2.2.3.4   DAMAGE AND YIELD LOSS

The injury of scale insect (*Aspidiella hartii* Sign.) to rhizomes is seen as encrustations on the rhizomes and severely infested rhizomes wither and dry. The scale insect feeds from the phloem of the host plant. Feeding damage due to an individual scale is small. However, when large populations are present, yellowing, defoliation, reduction in fruit set and loss in plant vigor result. Feeding sites are usually associated with discolorations, depressions and other host tissue distortions (Beardsley and

Gonzalez, 1975). The rhizome scale infests rhizomes of ginger both in the field and in storage. In the field, the pest is generally seen during the later stages of the crop. The pest infestation results in a weight loss of 14.0 and 22.5 %when stored for 128 days and 175 days, respectively (Hargreaves, 1930).

2.2.3.5   MORPHOLOGY

The tumeric root scale belongs to a family of scales called the armored scales. They are named this because they cover themselves with a shield or scale (puparium) composed of discarded skins (exuviae) and secreted matter. Scales belonging to the Aspidiella genus form a circular shaped shield that protects the body of the insect. This is contrary to other armored scales in which the shield is extended backwards from the scale.

Eggs: The elongate eggs, with ends equally rounded, are laid within the puparium.

Nymphs: The first stage larvae are active crawlers for a short time before selecting a feeding site. Once feeding begins the waxy protective shield starts to form. The second stage larvae occur after the first molt in which the discarded skin becomes incorporated into the puparium. This stage is similar in appearance to the adult except it does not have the grouped circumgenital glands.

The last stage is sometimes referred to as pupae. They have lost all traces of mouth organs and are thus a non-feeding stage. They possess rudimentary legs, antennae, wings and stylus (mouth).

Adults: *Adults* are moderately convex, approximately 1/20 inch in diameter, and brownish-gray with a slight purplish tint. Females are generally circular in shape where males are more oval. Females have rudimentary antennae and are unable to move about.

Female: scales are circular (about 1mm diameter) and light brown to gray and appear as encrustations on the rhizomes.

Male is orange colored with transparent wings, distinct head, thorax and abdomen.

Life Cycle and Ecology

Females are ovo-viviparous and also reproduce parthenogenetically. Little information is available on the life history of the pest on turmeric. A single female lays about 100 eggs and the life cycle from egg to adult is completed in about 30 d (Jacob, 1982, 1986). The pest completes its life cycle in 11–20 d on yams (*Dioscorea* spp.) (Palaniswami, 1994).

Crawlers, which are the first nymphal instar, are the primary dispersal stage and move to new areas of the plant, or are dispersed further by wind, or via contact with flying insects or birds. The crawlers can move up to a meter under their own locomotion (Watson 2011). At the end of the wandering period (dispersal phase), crawlers secure themselves to the host plant with their mouthparts. Once settled, the larvae draw their legs beneath the body and flatten themselves against the host (Koteja 1990). They then insert their piercing and sucking mouthparts into the plant tissue and start feeding on plant juices (Beardsley and Gonzalez 1975; Koteja 1990).

Management

Very few trials have been conducted for the control of rhizome scale on both ginger and turmeric. The management practices whichever documented are mostly of chemical control. Insecticides generally recommended for the management of the pest on ginger have also been recommended against the pest on turmeric.

Cultural

Discard and avoid of storing severely infested rhizomes, collection and destroy of damaged leaves and selection of healthy rhizomes free from scale infestation for seed materials helps in reduction of the scale population in the field. Application of well rotten sheep manure / poultry manure in two splits at 10 tons/ha, first before planting and the second at the time of earthing up helps in reducing the population of the scale insect.

Physical

Soaking of ginger rhizomes infested with *A. hartii* with hot water at a temperature of 50° C for 10 min results in the death of scales upto 50% (Vincent et al., 2003). To increase mortality of the scales the temperature of the hot water can be raised and extending the soaking time without affecting the viability of the seed.

Host Plant Resistance

Regupathy et al. (1976) studied the reaction of 191 turmeric types to rhizome scale at Coimbatore (Tamil Nadu, India) and found that 87 accessions were free from infestation. Velayudhan and Liji (2003) recorded the incidence of the rhizome scale on 489 accessions belonging to 21 morphotypes at Vellanikkara (Kerala). Eighty accessions were free of infestation and the lowest scale incidence was observed in morphotype six with a mean score of 0.6 on a 0–9 scale.

Chemical

Dipping the seed rhizomes of turmeric in quinalphos 0.1% for 5 min after harvest and before planting was found effective in controlling rhizome scale infestation (CPCRI, 1985). Discarding of severely infested and dipping the seed rhizomes in quinalphos 0.075% after harvest and before planting has also been recommended for the management of rhizome scale infestation (IISR, 2001). Drenching of soil with dimethoate 30 EC or phosalone 35 EC at 2 ml/l of water helps in reducing the population of rhizome scale.

2.2.4   THRIPS, PANCHAETOTHRIPS INDICUS BAGNALL (THRIPIDAE: THYSANOPTERA)

2.2.4.1   ORIGIN AND DISTRIBUTION

The insect is distributed throughout Bangladesh, Thailand, China and in India namely, Assam, Bihar, Goa, Haryana, Kerala, Manipur, Tamil Nadu, Uttar Pradesh, West Bengal (Tyagi and Kumar, 2016).

2.2.4.2   HOST RANGE

The thrips species is found infesting on turmeric, ginger, arrowroot leaves, banana and cotton (Ananthakrishnan and Sen, 1980).

2.2.4.3 NATURAL ENEMIES

Thripobius semiluteus has been recorded on thrips *Brachyurothrips anomalus* and *Panchaetothrips indicus* (Boucek, 1988).

2.2.4.4 DAMAGE AND YIELD LOSS

Different types of feeding damages were observed in thrips infested plants. Silvering, browning and discoloration of flowers and leaves were the most common types of damage observed. Nymphs are gregarious in nature and feed on the upper surface of young turmeric leaves. The adults are black, fast movers and suck the sap of the leaves which in turn become yellow and roll. The adults live inside the leaf roll and continue their feeding. Typical symptom of damage by the thrips is leaves become rolled up and turn pale and gradually dry-up of the whole plant. There is a considerable crop loss because following severe infestation, young plants die and the rhizome yield is proved to get decreased. The incidence of *P. indicus* in the field of northern districts of Karnataka namely at Gokak (Belgaum District), Indi (Bijapur District), Aland (Gulbarga District), and Basavakalyan (Bidar District) taluks was higher in the causing more than 10% damage to the crop (Kotikal and Kulkarni, 2000a).

2.2.4.5 MORPHOLOGY AND BIOLOGY

Field observation of *P.indicus* by Singh, et al. (2004) revealed that the thrips occurred on both abaxial and adaxial surfaces of the leaf their population varied from 10–60 thrips/plant from August–October. The male-female ratio varied from 1:2–1: 4. The females laid eggs within the leaf tissue with the help of saw-like ovipositor acid the fecundity ranged from 44–36 eggs/female, by calculating on the basis of emergence of young larvae from the leaf sheath. Presence of bean shaped eggs measuring 108 μm length and 25 μm width was detected within the leaf tissue. The eggs hatched in 4–6 days and the young pale white larvae could be easily noticed on the green leaf surface. There were two larval instars followed by a prepupal and pupal stage with a developmental duration of 7–9 days for larva and 3–4 days for pupa, of which the prepupal stage lasted for 15–20 h. The total period of development of immature stage till adult eclosion was 18 days. On mating the female began to lay eggs from 3rd day. Adult longevity was 10–12 days.

2.2.4.6   MANAGEMENT

Shanmugam et al. (2015) evaluated different pest management modules against major insect pests and diseases of turmeric at ten different locations of Dharmapuri district (Tamil Nadu), which reveals that the application of either imidacloprid 200 SL at 0.5 mL/L or thiomcthaxam 25 WG at 0.5 g /L (M4) found to be superior in managing the thrips incidence effectively. The two sprays of nimbecidine 1500 ppm at 3.0 mL/L on the incidence of thrips reduced the infestation up to 65.47–69.26 per cent in both the varieties. The spraying of dimethoate at 2 mL/L recorded 30 per cent incidence of thrips in both the varieties. Arutselvi et al. (2012) found that the combination of neem seed kernel and *Vitex negundo* leaf extracts significantly reduced incidence of leaf thrips in turmeric and recorded increased rhizome yield. Pachundkar et al. (2013) found that thiamethoxam 25 WG and imidacloprid 70 WG were effective in managing the sucking pest complex in cluster bean. Imidacloprid at 0.5 mL/L and thiamethoxam at 0.3 g/L were effective in controlling the thrips in cardamom (Prasath et al., 2013). Spraying of dimethoate 0.06%, fenpropathrin 0.02%, bendiocarb 0.08%, and methyl demeton 0.05% were more effective in controlling the pest (Balasubramanian, 1982).

2.3   MINOR INSECT PESTS OF TURMERIC AND GINGER

2.3.1   TURMERIC SKIPPER, UDASPES FOLUS CRAMER (HESPERIIDAE: LEPIDOPTERA)

The turmeric skipper, *Udaspes folus* is one of the most important leaf-feeding caterpillars on turmeric in India. The insect is distributed throughout Sri Lanka, South India to Saurashtra, Maharashtra, Gujarat, Madhya Pradesh, Uttar Pradesh, West Bengal, Himachal Pradesh to North East India and onto Myanmar, China, Thailand, Malaysian Peninsula, Singapore, Indonesia and Australia. The skipper found to feeds on ginger, turmeric, and their relatives *Curcuma aromatica, Curcuma decipiens,*

Hedychium sp. and other plants of Family Sctiaminae. It has also been recorded on *Zingiber* sp. and on Grasses.

The female normally sits on the underside of the leaf of the host plant and lays a single egg. The egg is reddish and appears smooth and dome shaped. The caterpillar is sluggish and comes out to feed only at night. The full grown larva has the habit of resting with the first three segments contracted, so as to give a humped appearance (Kalesh and Prakash 2007). The adult is a brownish-black butterfly with 8 white spots on forewings and one large patch on hind wing. The full-grown larva is dark-green and measures 36 mm in length. A female lays about 50 eggs on underside of the leaves which hatch in 3–4 days. The larva undergoes 5 instars during 12–21 days and pupates in leaf-fold for 6–7 days. The smooth green color larva with a black head pupates in December and emerges only in March. Longevity of males and females are 4 and 67 days respectively. The insect is present in abundance during August–October.

The larvae cut and fold leaves, remain within, and feed on them. Larvae of leaf roller (*Udaspes folus*) cut and fold leaves and feed from within, and are generally seen during the monsoon season. The adults are medium sized butterflies with brownish black wings with white spots; the larvae are dark green with black head. Entomopathogenic fungi, *Hirsutella citri-formis* followed by *Metrarhizium anisophilae* was found to significantly control *U. folus* (Arutselvi et al., 2012). The control measures undertaken against the shoot borer (spraying of malathion 0.1%) is adequate for the management of the pest. Studies on the biology of the pest conducted at Godavari Delta (Andhra Pradesh) indicated that egg, larva, prepupa, and pupal stages lasted for 5, 25–30, 2, and 7–8 d, respectively (Sujatha et al., 1992).

2.3.2 ROOT GRUBS, HOLOTRICHIA SPP (MELOLONTHIDAE: COLEOPTERA)

Root grubs (*Holotrichia* spp.) occasionally feed on tender rhizomes, roots and base of pseudostems causing yellowing and wilting of shoots. The initial symptoms are yellowing of leaves and wilting. The infected plants can be removed from soil easily. Seasonal activity shows that the adult whitegrubs become active with the onset of summer showers (May–June).

They come out of the soil during the night. The female lays the eggs in the moist soil. The pest can be controlled by drenching the soil around the rhizomes with chloropyriphos (0.075%).

Management practices such as provision of adequate irrigation, since under inadequate soil moisture conditions the pest appears in the root zone, collection and destroy of adult beetles when they are found feeding on the trees, installation of light traps to collect the adults (or) burn fire and adoption of crop rotation in the endemic areas can help in reduction of the population build up of white grubs. Application of Carbofuran 3G 10kg/ha near the root zone and another one more round at 30 days later also reduce the insect population. Seed treatment with chlorpyriphos 6–10 ml/kg seed and application of phorate 10 G 15 kg/ha at sowing time is effective against root grubs. Utilization of fungal pathogens like *Metarhizium anisopliae, Beauveria brongniartii* is now under consideration.

2.3.3 *SPILARCTIA (SPILOSOMA)* oblique, Walker *(ARCTIIDAE: LEPIDOPTERA)*

This pest damages the turmeric plants extensively in Bihar and West Bengal states. It is a highly polyphagous pest found to be feeding on sunflower, millets, cotton, jute, sunhemp, castor, cauliflower, cabbage etc. It has been reported to feed on 96 plant species in India. Damage is caused by the caterpillars. The larva defoliates the plants and move from one field to another. First two stages of larvae are the tiny caterpillars which feed gregariously and as they become older the larvae disperse widely in search of food. The adult is dull yellow with oblique line of black dots on hind wings. The dorsal side of the abdomen is red with dull yellow ventral side. The full grown larva is darkened with yellowish brown abdomen having numerous pale white brown and black hairs and measures about 43 mm. Pupation takes place inside the soil. Management practices includes collection and destruction of egg masses and gregarious larva, application of NSKE 5 % and foliar sprays particularly for grown up larva with phosalone 2 ml/l or chlorpyriphos 2 ml/l or dichlorvos 1 ml/l are effective. Moths can be trapped in artificial light and killed and young gregarious caterpillars can be collected and killed manually in kerosinized water.

2.3.4   FLEA BEETLE, LEMA PRAEUSTA FAB *(CHRYSOMELIDAE: COLEOPTERA)*

Both adults and grubs feed on leaf. The records in Orissa and Kerala. *L. praeusta* are observed to feed on leaves of cucurbits and sorghum in fields. Adult laid eggs singly on leaves. Incubation period is 8–10 days. Grub feeds on leaf tissue for 10–12 days and pupates in the soil. Adults emerge out form pupa, which lasts for 15–25 days. Adults are active during day time and feed on leaves. Longevity of the adults is 43–60 days.

2.3.5   MEALYBUG, FORMICOCOCCUS POLYSPERES *WILLIAMS (HEMIPTERA: PSEUDOCOCCIDAE)*

The mealybug was first reported to be infesting ginger in India from Northeast India (Meghalaya) (Annual Report 2014–15, ICAR). The mealybugs were found infesting ginger rhizomes in Meghalaya and caused up to 48.33% crop damage. Mealybug damage did not appear in the field until the end of July. Incidence of mealybug began after early August, and about 48.33% of the total rhizomes were found infested by this pest at the time of harvesting. Both nymph and adults of the mealybug suck the sap from the ginger rhizomes. Infested plants became yellow within 1–2 week and started drying from the tip downwards, and the infested rhizomes shriveled and eventually dried. Severe losses due to *F. polyspores* during storage of ginger have also been observed in adjoining areas.

KEYWORDS

- **pest management**
- **ginger**
- **turmeric**
- **integrated approaches**

REFERENCE

Alfred, J. R. B.; Das, A. K.; Sanyal, A. K. Ecosystems of India. ENVIS Centre, ZSI, Kolkata. 2001; 1–33, 93–122, 317–410.

Armstrong, K. DNA Barcoding: A New Module in New Zealand's Plant Biosecurity Diagnostic Toolbox. *Bull. OEPP/EPPO* **2010**, *40*, 91–100.

Arutselvi, R.; Balasaravanan, T.; Ponmurugan, P.; Adeena A. J. Effect of Various Biopesticides and Biocides on the Leaf Pest, *Udaspes folus* of Turmeric Plants. *J. Biopest.* **2012**, *5* (1) 51–56.

Balasubramanian, M. Chemical Control of Turmeric Thrips *Panchaetothrips indicus* Bagnall. *South Indian Hort.*, **1982**, *30*, 54–55.

Baringbing, B.; Gusmaini, S. Effect of Rotation Multi-Cropping Against Pests Ginger Rhizome. *In Proceeding ofNational Seminar on Role Entomology in the Pest Control Environmentally Friendly and Economical.Bogor.* 1999, 12–14.

Beardsley, J. W. Jr.; Gonzalez, R. H. The Biology and Ecology of Armored Scales. *Anl. Rew. Entomol.* **1975**, *20*, 47–73.

Boo, K. S. Variation in Sex Pheromone Composition of a Few Selected Lepidopteran Species. *J. Asia-Pacific Entomol.* **1998**, *1*, 17–23.

Boucek, Z. Australian Chalcidoidea (Hymenoptera). A Biosystematic Revision of Genera of Fourteen Families, with a Reclassification of Species, *C.A.B, Int.* **1988**, *832*.

Butani, D. K.; Spices and Pest Problems: Turmeric. *Pesticides.* **1985**, *19* (5), 22–25.

Chakravarthy, A. K.; Thyagaraj, N. E. Evaluation of Selected Synthetic Sex Pheromones of the Cardamom Shoot and Fruit Borer, *Conogethes punctiferalis* Guenee (Lepidoptera: Pyralidae) in Karnataka. *Pest Manage. Trop. Ecosyst.* **1998**, *4*, 78–82.

Chattopadhyay, I.; Biswas, K.; Bandyopadhyay, U.; Banerjee, R. K. Turmeric and Curcumin: Biological Actions and Medicinal Applications. *Curr Sci India*, **2004**, *87*, 44–53.

Chong, K. K.; Ooi, P. A. C.; Tuck, H. C. Crop Pests and Their Management in Malaysia. *Trop. Press Sdn. Bhd, Malaysia.* **1991**, 242.

Choo, H. Y.; Kim, H. H.; Lee, S. M.; Park, S. H.; Choo, Y. M.; Kim, J. K. Practical Utilization of Entomopathogenic Nematodes, *Steinernema carpocapsae* Pocheon Strain and *Heterorhabditis bacteriophora* Hamyang Strain for Control of Chestnut Insect Pests. *Korean J. Appl. Entomol.* **2001**, *40*, 69–76.

Choo, H. Y.; Lee, S. M.; Chung, B. K.; Park, Y. D.; Kim, H. H. Pathogenecity of Korean Entomopathogenic Nematodes (Steinernematidae and Heterorhabditidae) Against Local Agricultural and Forest Insect Pests. *Korean J. Appl. Entomol.***1995**, *34*, 314–320.

David, B. V.; Narayanaswami, P. S.; Murugesan, M. Bionomics and Control of the Castor Shoot and Capsule Borer *Dichocrocis punctiferalis* Guen. in Madras State. *Indian Oilseeds J.*; **1964**, *8*, 146–158.

Devasahayam, S.; Koya, K. M. A. Integrated Management of Insect Pests of Spices. *Indian J. Arecanut Spices Med. Plants,* **1999**, *1*, 19–23.

Devasahayam, S.; Koya, K. M. A. Insect Pests of Ginger. In *The Genus Zingiber.* Ravindran, P. N., Babu, K. N., Eds.; CRC Press: Boca Raton, Florida, USA. 2005, 367–389.

Devasahayam, S. Biological Control of Insect Pests of Spices. In *Biological Control in Spices*. Anandaraj, M., Peter, K. V., Eds; Indian Institute of Spices Research, Calicut, India. 1996, 33−45.

Devasahayam, S.; Koya, A. Insect pest of Turmeric, In *Turmeric the Genus Curcuma*. Ravindran, P. N., Babu, K. N., Sivaraman, K., Eds; *CRC Press. Boca Raton*, 2007, 169–191.

Devasahayam, S.; Jacob,T. K.; Abdulla, K. M.; Koya, A.; Sasikumar, B. Screening of Ginger (*Zingiber officinale*) Germplasm for Resistance to Shoot Borer (*Conogethes punctiferalis*). *J. Med. Arom. Pl. Sci.* **2010,** *32* (2), 137–138.

Dhoble, S. Y.; Kadam, M. V.; Dethe, M. D. Control of Turmeric Rhizome Fly by Granular Systemic Insecticides. *J. Maharashtra Agric. Univ.* **1978,** *3* (3), 209–210.

Dhoble, S. Y.; Kadam, M. V.; Dethe, M. D. Chemical Control of Turmeric Rhizome Fly, *Mimegralla coeruleifrons* Macquart. *Indian J. Entomol.* **1981,** *43* (2), 207–210.

Ghorpade, S. A.; Jadhav, S. S.; Ajri, D. S. Biology of Rhizome Fly, *Mimegralla coeruleifrons* Macquart (Micropezidae : Diptera) in India, a Pest of Turmeric and Ginger Crops. *Trop. Pest. Managt.* **1988,** *34* (1), 48–51.

Ghorpade, S. A.; Jadhav, S. S.; Ajri, D. S. Survey of Rhizome Fly on Turmeric and Ginger in Maharashtra. *J. Maharashtra Agric. Univ.* **1983,** *8* (3), 292–293.

Hargreaves, E. Annual Report on the Entomological Section. *Annual Report of Agriculture Department, Sierra Leone.* **1930,** 16–18.

Hill, D. S. Agricultural Insect Pests of the Tropics and Their Control (2nd Eds.). *Cambridge University Press, New York.* **1983,** *746*.

Jacob, S. A. Pests of Ginger and Turmeric and Their Control. *Pesticides.* **1980,** *14* (11), 36−40.

Jadhav, S. S.; Ghorpade, S. A.; Ajri, D. S. Field Screening of Some Turmeric Varieties Against Rhizome Fly. *J. Maharashtra Agric. Univ.* **1982,** *7* (3), 260.

Japoshvili, G.; Celik, H. Fauna of Encyrtidae, Parasitoids of Coccids in Golcuk Natural Park. *Entomologia Hellenica.* **2010,** *19*, 132–136.

Joseph, K. J.; Narendran, T. C.; Joy, P. J. Taxonomic Studies of the Oriental Species of *Brachymeria* (Hymenoptera: Chalcididae). *Report, PL 480 Research Project, University of Calicut, Calicut*. **1973,** 243.

Jung, J. K.; Han, K. S.; Choi, K. S.; Boo, K. S. Sex Pheromone Composition for Field Trapping of *Dichocrocis punctiferalis* (Lepidoptera: Pyralidae) males. *Korean J. Appl. Entomol.* **2000,** *39*, 105–110.

Kalesh, S.; Prakash, S. K.Additions of the Larval Host Plants of Butterflies of the Western Ghats, Kerala, Southern India (Rhopalocera, Lepidoptera): Part 1.*J. Bombay Nat. Hist. Soc.* **2007,** *104* (2), 235–238.

Kaomud, T.; Kumar, V. Thrips (Insecta: Thysanoptera) of India—An Updated Checklist. *Halteres*, **2016,** *7*, 64–98.

Karim, M. A. Vegetable and Spice Insect Pests and Their Control. In *Intensive Vegetable Growing and its Utilization. AVRDC-BARC/BARI-USAID.* **1993,** 135–168.

Karmawati, E.; Kristina, N. N. Effect of Intercropping on the Population Pest of Ginger Rhizome.*Communication Media Research and Development Plant Industry.* **1993,** *11*, 102–104.

Kimura, T.; Honda, H. Identification and Possible Functions of the Hair Pencil Scent of the Yellow Peach Moth *Conogethes punctiferalis* (Guenee) (Lepidoptera: Pyralidae). *Appl. Entomol. Zool.* **1999,** *34,* 147–153.

Konno, Y.; Arai, K.; Matsumato, Y. (E)-10-Hexadecenal, a Sex Pheromone Component of The Yellow Peach Moth, *Conogethes punctiferalis* (Guenee) (Lepidoptera: Pyralidae). *Appl. Entomol. Zool.* **1982,** *17,* 201–217.

Korycinska, A. Rapid Assessment of the Need for a Detailed Pest Risk Analysis for *Conogethes punctiferalis* (Guenée). *Food Environ. Res. Agency.* **2012,** 7–10.

Koteja, J. Developmental Biology and Physiology: Life History. In *Armoured Scale Insects–Their Biology, Natural Enemies and Control. World Crop Pests Volume 4A,* Rosen D.,Eds;. Elsevier: Amsterdam, Netherlands.1990; 243–254.

Kotikal, Y. K.; Kulkarni, K. A. Management of Rhizome Fly, *Mimegralla coeruleifrons* Macquart (Micropezidae : Diptera), a Serious Pest of Turmeric in Northern Karnataka. *Pest Managt. Hort. Ecosyst.* **1999,** *5* (1), 62–66.

Kotikal, Y. K.; Kulkarni, K. A. Incidence of Insect Pests of Turmeric (*Curcuma longa*) in Northern Karnataka, India. *J. Spices Arom. Crops.* **2000,** *9,* 51‒54.

Kotikal, Y. K.; Kulkarni, K. A. Studies on the Biology of Turmeric Rhizome Fly. *Karnataka J. Agric. Sci.* **2000,** *13,* 593‒596.

Kotikal, Y. K.; Kulkarni, K. A. Reaction of Selected Turmeric Genotypes to Rhizome Fly and Shoot Borer. *Karnataka J. Agric. Sci.* **2001,** *14,* 373‒377.

Koya, K. M. A.; Balakrishnan, R.; Devasahayam, S.; Banerjee, S. K. A Sequential Sampling Strategy for the Control of Shoot Borer (*Dichocrocis punctiferalis* Guen.) in Ginger (*Zingiber officinale* Rosc.) in India. *Trop. Pest Manage* **1986,** *32,* 343–346.

Koya, K. M. A.; Devasahayam, S.; Premkumar, T. Insect Pests of Ginger (*Zingiber officinale* Rosc.) and Turmeric *(Curcuma longa* Linn.) in India. *J. Plantation Crops.* **1991,** *19,* 1‒13.

Lalnuntluanga, J.; Singh, H. K. Performance of Certain Chemicals and Neem Formulations Against Ginger Shoot Borer (*Dichocrocis punctiferalis* Guen.). *Indian J. Entomol.* **2008,** *70* (2), 182–186.

Liu, M.Y.; Tian, Y.; Li, Y. X. Identification of Minor Components of the Sex Pheromone of Yellow Peach Moth, *Dichocrocis punctiferalis* Guenee, and Field Trials. *Entomol. Sinica* **1994,** *1,* 150–155.

M. A. F. Biosecurity New Zealand. Import Risk Analysis: Table grapes (Vitis vinifera) from China, **2009,** *322.*

Menon, V. P.; Sudheer, A. R. Antioxidant and Antinflammatory Properties of Curcumin. *Advances Exp. Med. Biol.,* **2007,** *595,* 105–125.

Nair, N. R. G. K. Insects and Mites of Crops in India. Indian Council of Agricultural Research Publication. New Delhi. 1986, 408.

Nishida, G. M. Hawaiian Terrestrial Arthropod Checklist. Fourth Edition. 2002, 313.

Nybe, E. V. Three Decades of Spices Research at KAU. *Kerala Agricultural University, Thrissur, India.* **2001,** 24–32.

Nybe, E. V.; Nair, P. C. S. Field Tolerance of Ginger Types to Important Pests and Diseases. *Indian Cocoa Arecanut Spices J.* **1979,** *2,* 109–111.

Kumar, T.; Singh, O.; Chochong, V. S.; Varatharajan, R. Biology of Turmeric Thrips, *Panchaetothrips indicus* Bagnall (Panchaetothripinae: Terebrantia: Thysanoptera. Short Scientific Report, *J. Plan. Crops* **2004,** *32* (1), 57–58.

Pachundkar, N. N.; Borad, P. K.; Patil, P. A. Evaluation of Various Synthetic Insecticides Against Sucking Insect Pests of Cluster Bean. *Int. J. Sci. Res.* **2013,** *3* (8), 1–5.

Palaniswami, M. S.Yam Scale Insects *Aspidiella hartii* Ckll. and Its Parasitoids. *J. Root Crops Res.* **1991,** *17*(1), 75–76.

Patel, R. K.; Gangrade, G. A. Note on the Biology of Castor Capsule Borer, *Dichocrocis punctiferalis. Indian J. Agric. Sci*, **1971,** *41*, 443–444.

Philip, J.; Nair, P. C. S. Field Reaction of Turmeric Types to Important Pests and Diseases. *Indian Cocoa Arecanut Spices J.* **1981,** *4*, 107−109.

Prasath, D.; Dinesh, R.; Srinivasan, V.; Kumar, C. M.; Anandaraj, M. Research Highlights 2013–14 of Indian Institute of Spices Research. 2014,p 20.

Prem Kumar, T.; Sharma, Y. R.; Gautam, S. S. Association of Dipteran Maggots in Rhizome Rot of Ginger. In *Proceedings of the National Seminar on Ginger and Turmeric.* Nair, M. K., Prem Kumar, T., Ravindran, P. N. Sharma, Y. R., Eds; 1982 CPCRI, Kerala, India. 1980, 128–129.

Premkumar, T.; Devasahayam, S.; Koya, K. M. A. Pests of Spice Crops. In *Advances in Horticulture, Vol. 10, Plantation and Spices Crops, Part 2.* Chadha, K.L., Rethinam, P., Eds; Malhotra Publishing House: New Delhi, 1994, 787−823.

Púcat,A. *Conogethes punctiferalis.* Yellow Peach Moth. *Canadian Food Inspection Agency Sci. Branch.* 1995, 27–32.

Punithavathi, D.; Venkatesan, N.; Babu, M. Curcumin Inhibition of Bleomycin-Induced Pulmonary Fibrosis in Rats. *Brazilian J. Pharmacol.* **2000,** *131*, 169–172.

Pervez, R.; Eapen, S. J.; Devasahayam, S.; Jacob, T. K. Efficacy of Some Entomopathogenic Nematodes Against Insect Pests of Ginger and Their Multiplication. *Nematol. Medit.* **2012,** *40*, 39–44.

Rao, P. S.; Krishna, M. R.; Srinivas, C.; Meenakumari, K.; Rao, A. M. Short Duration, Disease-Resistant Turmerics for Northern Telangana. *Indian Hort.* **1994,** *39* (3), 55–56.

Regupathy, A.; Santharam, G.; Balasubramanian, M.; Arumugam, R. Occurrence of the Scale *Aspidiotus hartii* C. (Diaspididae: Hemiptera) on Different Types of Turmeric *Curcuma longa* Lin. *J. Plantation. Crops*, **1976,** *4*, 80.

Robinson, G. S.; Tuck, K. R.; Shaffer, M. A Field Guide to the Smaller Moths of South-East Asia. *Malaysian Nature Society, Kuala Lumpur and Natural History Museum, London*, 1994, 308.

Shashank, P. R.; Doddabasappa, B.; Kammar, V.; Chakravarthy, A. K.; Honda, H. Molecular Characterization and Management of Shoot and Fruit Borer *Conogethes punctiferalis* Guenee (Crambidae: Lepidoptera) Populations Infesting Cardamom, Castor and Other Hosts. In *New Horizons in Insect Science: Towards Sustainable Pest Management.* A. K. Chakravarthy Ed; Springer: 2015; 207−227.

Sheila, M. K.; Abraham, C. C.; Nair, P. C. S. Incidence of Shoot Borer (*Dichocrosis punctiferalis* Guen. (Lepidoptera: Pyraustidae) on Different Types of Turmeric. *Indian Cocoa Arecanut Spices J.* **1980,** *3*, 59−60.

Sheo, G.; Chandra, R.; Karibasappa, G. S.; Sharma, C. K.; Singh, I. P. Research on Spices in NEH Region. ICAR Research Complex for NEH Region, Umiam. 1998, 9–22.

Sontakke, B. K. Occurrence, Damage and Biological Observations on Rhizome Fly, *Mimegralla coeruleifrons* Infesting Ginger. *Indian J. Entomol.* **2000,** *62* (2), 146–149.

Stanley, J. Chandrasekaran, S.; Preetha, G. *Conogethes punctiferalis* (Lepidoptera: Pyralidae) its Biology and Field Parasitization. **2009,** *79* (11), 906–909.

Sujatha, A.; Zaherudeen, S. M.; Reddy, R. V. S. K. Turmeric Leaf Roller, *Udaspes folus* Cram. and its Parasitoids in Godavari Delta. *Indian Cocoa Arecanut Spices J.* **1992,** *15,* 118−119.

Supriadi, K.; Sitepu, D. Strategy Controlling for Wilt Disease of Ginger Caused by *Pseudomonas solanacearum. J. Agric. Res. Dev.* **2000,** *19* (3), 106–111.

Ananthakrishnan, T. N.; Sen, S. Taxonomy of Indian Thysanoptera. Handbook Series No.1, *Zool. Survey of India,* 1980, 264.

U S D A. Insects Not Known to Occur in the United States. Yellow Peach Moth (*Dichocrocis punctiferalis* (Guenee) 1957, 37–38.

Varadarasan, S. Biological Control of Insect Pests of Cardamom. In *Biological Control of Social Forest and Plantation Crop Insects.* Ananthakrishnan, T. N. Ed; Oxford and IBH Publishing Company Private Limited: New Delhi. 1995, 109–111.

Velayudhan, K. C.; Liji, R. S. Preliminary Screening of Indigenous Collections of Turmeric Against Shoot Borer (*Conognethes punctiferalis* Guen.) and Scale Insect (*Aspidiella hartii* Sign.). *J. Spices Arom. Crops,* **2003,** *12,* 72−76.

Watson, G. W. Arthropods of Economic Importance–Diaspididae of the World. *Natural History Museum, London. World Diversity Database,* 2011.

Xiao, W.; Matsuyama, S.; Ando, T.; Millar, J. G.; Honda, H. Unsaturated Cuticular Hydrocarbons Synergize Responses to Sex Attractant Pheromone in the Yellow Peach Moth, *Conogethes punctiferalis. J. Chem. Ecol.* **2012,** 8–12.

CHAPTER 3

INSECT PESTS OF CUMIN AND THEIR MANAGEMENT

GEETANJLY*, RAHUL KUMAR CHANDEL, PREETI SHARMA, and VIJAY KUMAR MISHRA

Division of Entomology, Indian Agricultural Research Institute (IARI), New Delhi-110012, India

Corresponding author. E-mail: anjalygm@gmail.com

ABSTRACT

Cumin (Cuminum cyminum L.; Family: Umbelliferae) known as 'Jeera' or 'Zeera' in Hindi is an important spice of Indian kitchens for flavouring various food preparations. The flavour of cumin seeds is due to the presence of a volatile oil (2.5–3.5%), cuminaldehyde, cymene and terpenoids are the major volatile components of cumin. This spice is known under the various names in different countries Kreuzkumme Mutterkummel, weiser Kummel, Romischer Kummel, Welscher kummel, Kumin or Cumin in German; Cumino in Spanish; cumin (blanc), Cumin de Maroc, Faux Anis in French; Jeera in hindi; Cumino Romano in Italian; Romai kominyi in Hungarian; komijn in begium and the Netherlands; Spsskummen and Spisskarve in Norwegian Komin Rzymski on Polish; Spisskumin in Swedish; Kimyon in Turkish etc. Cumin seeds are extensively used in various ayurvedic medicines also especially for the conditions like obesity, stomach pain and dyspepsia. Nutritional value of cumin seeds is as follows: 17.7% protein, 23.8% fat, 35.5% carbohydrate and 7.7% minerals. Results of many studies conducted in India showed that cumin can be used as an antioxidant. The antioxidative potential is correlated with the phenol content of cumin. Cumin aldehyde has also antimicrobial and antifungal properties which could be shown e.g. with *Escherichia coli* and *Penicillium chrysogenum*.

3.1 INTRODUCTION

Cumin (*Cuminum cyminum* L.; Family: Umbelliferae) known as 'Jeera' or 'Zeera' in Hindi is an important spice of Indian kitchens for flavoring various food preparations. The flavor of cumin seeds is due to the presence of volatile oil (2.5–3.5%), cumin aldehyde, cymene, and terpenoids are the major volatile components of cumin. This spice is known under the various names in different countries Kreuzkumme Mutterkummel, Weiser Kummel, Romischer Kummel, Welscher kummel, Kumin or Cumin in German; Cumino in Spanish; cumin (blanc), Cumin de Maroc, Faux Anis in French; Jeera in Hindi; Cumino Romano in Italian; Romai kominyi in Hungarian; komijn in Begium and the Netherlands; Spsskummen and Spisskarve in Norwegian Komin Rzymski on Polish; Spisskumin in Swedish; Kimyon in Turkish, etc. Cumin seeds are extensively used in various ayurvedic medicines also especially for the conditions, like obesity, stomach pain, and dyspepsia. Nutritional value of cumin seeds is as follows: 17.7% protein, 23.8% fat, 35.5% carbohydrate, and 7.7% minerals. Results of many studies conducted in India showed that cumin can be used as an antioxidant. The antioxidative potential is correlated with the phenol content of cumin. Cuminaldehyde has also antimicrobial and antifungal properties that could be shown, for example, *Escherichia coli* and *Penicillium chrysogenum.* In southern India, popular drinks such as in *Kerala* and *Tamil Nadu* are called Jira water, which is made by boiling cumin seeds. It is believed that cumin is beneficial for heart disease, swellings, tastelessness, vomiting, poor digestion, and chronic fever.

3.2 ORIGIN AND DISTRIBUTION

Cumin is a native of the Levant and Upper Egypt and nowadays it is grown mainly in hot countries, especially India, North Africa, China, and America. India is one of the largest producers and consumers of cumin seed. Besides India, cumin seed is cultivated in Iran, Turkey, and Syria mainly for exports. It is widely used as a spice and for medicinal purpose all over the world. In the West, it is used mainly in veterinary medicine, as a carminative but it remains a traditional herbal remedy in the East. India is one of the major producers and consumers of cumin in the world. Almost 80% of the crop cultivated is consumed in India itself. The crop is exclusively cultivated in Rajasthan and Gujarat and both the states together contribute more than 95% of total country's cumin production with Gujarat contributing around 85% of total production (Table 3.1). Banaskantha and Mehsana in Gujarat and Barmer, Jalore, Jodhpur, and Nagaur in Rajasthan are the major Jeera producing areas. West Bengal, Uttar Pradesh, Andhra Pradesh, and Punjab, also make significant contributions to Indian output.

TABLE 3.1 India State-wise Production of Cumin Seed (in million kg).

State	2011–2012	2012–2013	2013–2014
Gujarat	221906	219215	283302
Rajasthan	80531	114925	177835
Total	303943	403744	462645

In terms of volume, cumin occupies the second position after chili during 2014–2015 with an export quantity of 1,55,500 tonnes that earned a foreign exchange worth 1838.20 crores. In 2013–2014, it was 1600.06 crores.

3.3 ECOLOGY OF CUMIN CROP

Seed spices come under the "high-value low volume crops," which are the important commodities of India's arid and semiarid regions. For the good growth of plants, temperature range from 10 to 24°C and below 5°C generally slows the growth of crop. Growing season of 3–4 months of cool, dry weather with full sunlight and low humidity, especially at flowering, produces the highest yield. Thus, cumin is suitable for dry areas. In Gujarat and Rajasthan, it is usually showing the middle 2 weeks of November when the day time temperature as fallen to within the optimum range. In Europe, it is showing mid-March to mid-April depending on the locality and cultivar.

Plants grow best on residual soil moisture rather than depending on rain falling during growth but cumin is generally not drought resistant. Cumin is usually shown toward the end of the rainy season in Iran, in Ethiopia after the main rains that occur at different times around the country, in Egypt and Sudan on a falling Nile flood. Cumin is basically a rabi (dry season) crop in India and planted to flower in February and March when humidity is low. Waterlogging is not tolerated, especially in the seedling stage. Cumin is considered a short-day plant but some local cultivars have extended its northern limits, to Scandinavia, for example, although long days are generally not suitable for the spice production. This crop does not stand with high humidity and heavy rainfalls. Well-drained, loamy soils that are rich in organic matter are best for cumin cultivation and after the crop is harvested, the cumin seeds are cleaned up through the winnowing process.

3.4 INSECT PEST INFESTING CUMIN CROP

In developing countries, the great loss to the cumin crop is due to insect pests infesting the crop in the field as well during the store. A list of the most common insect pest is shown below:

S. No.	Insect pest name		Order	Family	Damage (field/ stored seed)
	Common	Scientific			
1.	Tobacco caterpillar	*Spodoptera litura* Fabricius	Lepidoptera	Noctuidae	Field
2.	Cutworm	*Agrotis ipsilon* Hufnagel	Lepidoptera	Noctuidae	Field
3.	Cigarette beetle	*Lasioderma serricorne* Fabricius	Coleoptera	Anobiidae	Stored seed
4.	Drug store beetle	*Stegobium paniceum* Linnaeus	Coleoptera	Anobiidae	Stored seed
5.	Aphid	*Myzus persicae* Sulzer *Acyrthosiphon pisum* Harris and *A. craccivora* Koch	Hemiptera	Aphididae	Field
6.	Thrips	*Thrips tabaci* Lindeman	Thysanoptera	Thripidae	Field

3.4.1 TOBACCO CATERPILLAR

Distribution: Tobacco caterpillar is found all over tropical and subtropical parts of the world, widespread in India.

Host Range: It is polyphagous in nature feeds on tobacco and besides tobacco it feeds on tomato, cabbage, cumin, cotton, castor, groundnut, and other cruciferous crops.

Egg: Eggs are laid in clusters about 300 in number and covered by brown hairs. It takes around 3–5 days for hatching.

Larva: Caterpillar measures 35–40 mm in length when fully grown. It is velvety, black with yellowish-green dorsal stripes and lateral white bands with incomplete ring—like dark band on anterior and posterior end of the body. It passes through 6 instars. Larval stage lasts 15–30 days

Pupa: Pupation takes place inside the soil close to the plants. Pupal stage lasts 7–15 days.

Adult: Moth is medium-sized and stout-bodied with forewings pale grey to dark brown in color having wavy white crisscross markings. Hind wings are whitish with brown patches along the margin of wing. Pest breeds throughout the year. Moths are active at night. Adults live for 7–10 days. Total life cycle takes 32–60 days. There are eight generations in a year.

Biology: After adult emergence, peak oviposition occurs on the second night. Females mate three or four times during their whole lifetime, whereas males can mate up to ten times. In Andhra Pradesh, India, it completes 12 generations a year, each lasting more than a month in winter and less than a month in hot season.

Damage Symptoms:

- In early stages, the caterpillars are gregarious and scrape the chlorophyll content of leaf lamina giving it a papery white appearance.
- Irregular holes in leaves initially and later skeletonization leaving only veins and petioles.
- Heavy defoliation.

Management:

Cultural control:
- Deep plowing is required during summer months to expose the pupae.
- Avoid premonsoon sowing
- Sowing tolerant varieties viz., JS 80-21, PK 42, and PS 564.
- Use optimum seed rate 70–100 kg/ha.

Mechanical control:

- ✓ Collect and destroy infested plant parts, egg masses, and early-stage larvae found in clusters.
- ✓ Install one light trap (200 W mercury vapor lamp)/ha in the field to catch the *S. litura* adults.
- ✓ Install pheromone traps @ 10–12 Nos/ha at a distance of 50-m interval for early detection/mass trapping of *S. litura*.
- ✓ Erection of bird perches @ 10–12 Nos/ha

Biological control:

- ✓ Parasitoids: *Trichogramma sp, Tetrastichus sp, Telenomus sp, Bracon sp, Campoletis sp, Chelonus sp, Ichneumon sp, Carcelia sp,* etc.
- ✓ Predators**:** Lacewing, ladybird beetle, spider, red ant, dragonfly, robber fly, reduviid bug, praying mantis, king crow, etc.
- ✓ Release egg parasitoid *Telenomus remus* @ 50,000/ha.
- ✓ Spray SNPV@ 250 LE.
- ✓ Spray NSKE @ 5% to manage early-stage larvae.

Chemical control:

- ✓ Apply methomyl @ 2 L/ha, Ethofenprox 10 EC @ 1 L/ha, Triazophos 40 EC @ 625 mL/ha, Quinolphos 25 EC 1.5 L/ha and Lamdacyclothrin 50 EC @ 300 mL/ha for controlling tobacco caterpillars.

3.4.2   CUTWORM

Distribution: *A. ipsilon* is one of the most widely distributed species in the cutworm complex. It is generally considered to be worldwide in distribution.

Host range: It has a very wide host range, namely, carrot, sunflower, sweet potato, pea, mints, cotton, coffee, citrus; cumin, zinger, etc., but seedling crop plants are most seriously damaged.

Egg: Newly laid eggs are ribbed and whitish-yellow, become darker as hatching approaches.

Larva: The general body color of the larvae is light grey to black without distinct stripes or markings. The head is pale-brownish with black coronal stripes and reticulation. The skin bears convex, rounded, distinctly isolated, and coarse granules with smaller granules interspersed between the larger granules. The spiracles are black. There are six or seven generally.

Pupa: Pupae are brown to dark brown and approximately 17–25 mm in length and 5–6 mm in width. Pupae appear almost black in color just before the moth emerges.

Adult: The forewings are long and narrow, darker than the hind wings and marked with black dashes or "daggers:" The basal two-thirds of the forewing is dark, with the outer third pale grey to brown; orbicular is tear-shaped; reniform has a distinct black wedge- or dagger-shaped black marking on its outer margin; claviform is small, dark, oblong, and filled with dark scales. There is a zigzag line of pale scales on dark background in the subterminal area. The male antennae are plumose (feathered) and the female antennae are filiform. The wingspread is approximately 35–50 mm (Image 4).

Damage symptoms:

- Early instar *A. ipsilon* larvae can create "shot holes" while feeding on tender leaves of seedling plants.
- Cutting of young seedlings and sometimes causing death of the cut seedlings.
- Wilting is observed because of partial cutting (Image 5).

Management:

Prevention and cultural control:

- ✓ If possible, avoid planting crops in fields with a known history of cutworm problems.
- ✓ Plough in the autumn and use shallow tillage to keep down late autumn and early spring vegetation (where conservation practices allow).
- ✓ Monitor larvae with larval cutworm bait traps.
- ✓ Monitor adults to predict attacks.
- ✓ Monitor weather to predict attacks.
- ✓ Low mow grass to remove eggs, disposing of cuttings at a distance Topdressing with sand does not kill larvae but deters them from traveling.
- ✓ Encourage predators by encouraging their other prey species nearby, for example, by having conservation strips between fields or golf fairways.

Biological control:

- ✓ Kentucky bluegrass (*Poa pratensis*) inhibits the growth of *A. ipsilon* larvae.

✓ Endophyte enhanced perennial ryegrass (*Lolium perenne*) inhibits foraging behavior.
✓ Turfgrass composed of a mixture of these is resistant to *A. ipsilon* Endemic nematodes were investigated in India for effectiveness against *A. ipsilon*.
✓ Alginate formulations of entomopathogenic nematodes against *A. ipsilon* caused the maximum mortality.
✓ *Bacillus thuringiensis* was most effective against first- and second-instar larvae of *A. ipsilon*
✓ Volatile substances extracted by steam distillation from withered black poplar (*Populus nigra*) leaves showed strong attractive activity to *A. ipsilon* and other insects and could be used as a trap

Chemical control:

Natural Insecticides:

✓ Neem products were found effective.
✓ A methanol extract of *Melia azedarach* fruits was found to be toxic to *A. ipsilon*.
✓ Extract of *Bassia muricata* was found to be toxic to first-instar larvae.
✓ Leaf extracts of Lantana, Parthenium, Hyptis, and *Ipomoea carnea* were found to be toxic to *A. ipsilon* and other pests.

Synthetic Insecticides:

✓ Clothianidin, a new synthetic chloronicotinyl insecticide, has been found to be effective as a seed treatment against *A. ipsilon*.
✓ The effectiveness of diazinon 20 EC, quinalphos 25 EC, chlorpyrifos 20 EC, fenitrothion 50 EC, deltamethrin 2.8 EC, and malathion 5% dust against *A. ipsilon*.
✓ Common alum, aluminum potassium sulfate (solid) and aluminum oxide (liquid) were found to be toxic to *A. ipsilon* larvae and synergized the effectiveness of other insecticides.

3.4.3   CIGARETTE BEETLE

Distribution: The cigarette beetle is pan-tropical but can be found worldwide especially wherever dried tobacco in the form of leaves, cigars, cigarettes, or chewing tobacco is stored.

Host range: Drug store beetle feeds on stored spices and cereals.

Eggs: Females lay 10–100 eggs in the food and the larvae emerge in 6 to 10 days.

Larvae: Older larvae are white, scarab-like, and hairy. Hair is longer and the head is evenly rounded dorsally with a dark marking with a convex boundary that extends halfway up the frons. An arolium (pad-like structure between the tarsal claws) is also present and extends beyond the middle of the claw on each tarsus.

Adult: Cigarette beetles are quite small, measuring about 2–3 mm (about 1/8 of an inch), and are reddish-brown. They have a rounded, oval shape, and the head is often concealed by the pronotum. The elytra (wing covers) are covered with fine hairs. They prefer to reside in dark or dimly lit cracks, nooks, and crevices but become active and fly readily in bright, open areas, probably in an attempt to find refuge. They are most active at dusk and will continue activity through the night. Adults do not feed but will drink liquids.

Biology: The length of the cigarette beetle life cycle is highly dependent on temperature and the food source but usually takes 40–90 days.

Damage and symptoms:

- ✓ Larval feeding causes direct damage to foodstuffs and nonfood items.
- ✓ These products are contaminated by the presence of beetles, larvae, pupae, cocoons, frass (fecal material), and insect parts.
- ✓ Cocoons are often attached to a solid substrate and in severe infestations form large clusters (Image 6).

Management:

Preventions:

- ✓ Locating the source of infestation is the first and most important step.
- ✓ Heavily infested items should be wrapped in heavy plastic, taken outside and thrown away.
- ✓ All food containers and items should be checked for infestation and placed in the refrigerator or freezer (16 days at 36°F, 7 days at 25°F or 32°F for 4–7 days) to kill all stages.
- ✓ Uninfected items can be cold- or heat-treated to ensure that any undetected infestations are killed.
- ✓ To prevent reinfestation, clean up spilled flour, mixes, crumbs, etc., and thoroughly vacuum and clean areas where the contaminated items were stored.
- ✓ Store foods in airtight glass, metal, or plastic containers.

Mechanical control:

- ✓ Sticky traps baited with the female sex pheromone.
- ✓ Store grains in gunny bags with moisture-proof lining.
- ✓ Use commercially available cigarette beetle traps with synthetic serricornin.

Biological control:

- ✓ Cigarette beetle predators include *Tenebriodes* sp. (Tenebrionidae), *Thaneroclerus* sp. (Cleridae), and several carabids.
- ✓ Eggs may be eaten by predatory mites.
- ✓ Parasitoids include wasps in the families, for example, Pteromalidae, Eurytomidae, and Bethylidae.

Biological control for stored-product pests, however, has not been widely adopted. Part of the problem is that although it reduces the amount of pesticides used for control, the release of insects to control cigarette beetles increases contamination of foodstuffs with potentially more insects and insect parts.

Chemical control:

- ✓ The use of Methoprene on stored tobacco was one of the first uses of an IGR on a stored commodity.
- ✓ Sticky traps baited with the female cigarette beetle sex pheromone, serricornin, can be used to monitor for beetles.

3.4.4 DRUG STORE BEETLE

Distribution: Drugstore beetles have a worldwide distribution but are more abundant in warmer regions or in heated structures in more temperate climates. They are less abundant in the tropics than the cigarette beetle

Host range: Feed on food and nonfood material both. In food material, it has a vast host range, flours, dry mixes, bread, cookies, chocolates, and other sweets, and spices (cumin, caraway, and fennel).

Eggs: Females lay up to 75 eggs in the food or substrate.

Larvae: Small, white grubs; the later instars are scarab-like. They have shorter hairs and the marking on the head ends in a straight line across the frons just above the mouthparts.

Adults: The beetles are cylindrical, 2.25–3.5-mm (1/10–1/7 inch) long, and are a uniform brown to reddish-brown. They have longitudinal rows of fine hairs on the elytra (wing covers).

Biology: The entire life cycle is generally less than two months but can be as long as 7 months. The duration of the life cycle is highly dependent on the temperature and food source. Development occurs between 60°F and 93°F (~15–34°C) but is optimal at about 85°F (~30°C) and 60–90% relative humidity (Image 7).

Damage and symptom:

- ✓ The drugstore beetle attacks a wide variety of foods and material.
- ✓ It also feeds on flours, dry mixes, bread, cookies, chocolates, and other sweets, and spices. Nonfood material includes wool, hair, leather, horn, and museum specimens.
- ✓ It is found in pigeon nests and is known to bore into books, wooden objects.

Management:

Preventions: Preventing methods are same as cigarette beetle.

Mechanical control:

- ✓ Use commercially available traps and lures with the drugstore beetle sex pheromone, stegobinone (2, 3-dihydro-2, 3, 5-trimethy l-6-(1-methyl-2oxobutyl) -4H-pyran-4-one).
- ✓ Use sticky traps baited with the female sex pheromone, stegobi-none, for monitoring adult beetles.

Commercial method:

- ✓ Large-scale control for severe infestations can be achieved by fumigation.
- ✓ Heat treatment has been used with limited success. Effective control is obtained when a temperature of approximately 50°C (122°F) is maintained for 24–36 h.

3.4.5 APHID

Distribution: Aphids have a worldwide distribution. They were first reported on Oahu in 1910 and are now present on all islands within the State.

Host range: Aphids feed on wide range of host, for example, bitter gourd, cabbage, cauliflower, condol, chayote, eggplant, lemon, lettuce, loofah, melon, mustard, pechay, pomelo, potato, raddish, squash, tomato, tobacco, watermelon, spice crops, and on weeds like *Prunus persica, P. nigra, P. tanella,* and *P. Serotina*.

Eggs: In temperate regions, these aphids overwinter during the egg stage. The shiny black eggs are often laid on the bark of fruit trees.

Nymphs: Immature aphids are called nymphs. They are pale yellowish-green in color with three dark lines on the back of the abdomen that are not present on the adult. In Hawaii there are four nymphal stages. Nymphal development is completed in 6–11 days.

Adults: The wingless adult aphids vary in color from green to pale yellow. Winged adults are green with black or dark brown markings on their abdomens. Adults are small to medium-sized aphids from 1/25 to 1/12-inch long and their antennae are 2/3 as long as the body. Adult females give birth to approximately 50 nymphs.

Biology; There are many generations of this aphid throughout the year. The life span represents the period from birth of the nymphs to the death of the adult. Longevity may be affected by temperature, type of life cycle (egg-laying or live births), and plant host. Studies in cooler temperatures report the life cycle lasting up to 50 days. Population are larger during periods of adequate rainfall and smallest during hot, dry weather.

Damage symptoms:

- ✓ Aphids can attain very high densities on young plant tissue, causing water stress, wilting, and reduced growth rate of the plant.
- ✓ Prolonged aphid infestation can cause appreciable reduction in yield of root crops and foliage crops.
- ✓ Contamination of harvestable plant material with aphids, or with aphid honeydew, also causes loss.
- ✓ Blemishes to the plant tissue, usually in the form of yellow spots, may result from aphid feeding. Leaf distortions are not common except on the primary host.

Management:

Cultural control:

- ✓ This aphid develops on crop and noncrop hosts. Thus, it is important to remove crop residues and weed hosts prior to planting new crops.
- ✓ Spray the pressurized water on the crop.

Biological control:

- ✓ Natural enemies are the predatory syphid maggots, *Allograpta sp.,* lady beetles and parasitic wasps. Another effective parasite is *Diaretus chenopodiaphidis* Ashmead.
- ✓ Apply fish oil rosin soap or NSKE (3%), neem oil (2%), or tobacco decoction (0.05%).

Hostplant resistance:

- ✓ Host–plant resistance to insects is commonly based on secondary plant chemistry.
- ✓ *M. persicae* can attack plants in many unrelated botanical families, such resistance is hard to obtain, and the focus has been more on morphological plant characters.
- ✓ Glandular trichomes on potatoes are an important resistance factor, trichomes also release a sticky exudate, which immobilizes aphids, and contains toxic sucrose ester compounds, shown to inhibit settling and probing.
- ✓ Increased waxiness in brassicas decreased aphid colonization, mainly due to a nonpreference resistance mechanism.

Chemical control:

- ✓ Insecticides applicable for aphids include organophosphorus, pyrethroids, carbamates, and neonicotinoids. Apply insecticides only when necessary.
- ✓ Aphid has developed resistance to certain insecticides. It is important to test insecticides on local populations before making large purchases of pesticides.
- ✓ The use of chemicals to control the spread of virus diseases is usually not effective.

3.4.6   THRIPS

Distribution: Originated in the Mediterranean region, the onion thrips is cosmopolitan in distribution throughout most of the world.

Host range: Principal crop hosts include beans, broccoli, cabbage, cumin, carnation, carrot, cauliflower, Chinese broccoli, cotton, cucumber, garlic, head cabbage, leek, melon, onion, orchids, papaya, peas, pineapple, rose, squash, tobacco, tomato, and turnip, etc.

Eggs: Females have a saw-like structure that helps to make an incision in plant tissue for egg-laying. Eggs are placed singly just under the epidermis of succulent leaf, flower, stem, or bulb tissue. They are whitish at deposition and change to an orange tint as development continues. Hatching occurs in 4–5 days.

Larvae: Larvae are whitish to yellowish. There are two larval stages and besides the adults, they are the only damaging stages. Larval development is completed in about 9 days.

Pupae: There is two nonfeeding stages called the prepupa and pupa. They do not feed and occur primarily in the soil. Combined prepupal and pupal development is completed in 4–7 days.

Adults: *Adults* are 1/25-inch long. Their body color ranges from pale yellow to dark brown; wings are unbanded and dirty gray. In Hawaii, this species has a darker form during the rainy season. Males are wingless and exceedingly rare. Females live for about two to three weeks and each can lay about 80 eggs (Image 8).

Biology: The entire life cycle (egg to adult) requires about 19 days. There are many overlapping generations throughout the year.

Damage and symptom:

- ✓ Larvae and adults are found mainly in the narrow space between the tubular leaves, in flowers and on the underside of foliage of certain other plants.
- ✓ Thrips feed by piercing individual cells and sucking the contents.
- ✓ These cells lose their normal color, and when many adjacent cells are damaged, the tissue appears as whitish spots or silvery spots or streaks.
- ✓ In advanced injury, the leaves take on a blasted appearance. Substantial damage can be done to young plants especially to varieties grown in seedbeds (Image 9).

Management:

Cultural control:

- ✓ Sanitation techniques such as removing weeds in the field and outlying areas. These practices help eliminate alternative hosts on the thrips between crops.

✓ Crop rotations to prevent the successive plantings of several cumin crops and interplanting with nonhost crops can also be effective in deterring large populations.

Chemical control:

✓ While the onion thrips can be readily killed by many insecticides, they are often difficult to control because of their small size and cryptic habits.

✓ Insecticidal control of this pest depends on the choice of an effective chemical and adequate spray coverage on parts of the plant where the thrips inhabit. Many larvae and thrips are found in the leaf axils that often do not receive insecticide deposits.

✓ Thrips are shallow feeders that feed primarily on surface tissue. Contact-residual insecticides are more effective. Systemic insecticides that transport through the plant's vascular tissues are not as likely to be effective if they are not applied in a manner like other contact insecticides.

3.5   SUMMARY AND CONCLUSION

People usually have something common in their meals. One such is cumin which is used in different dishes. This crop is prone to attack by above-mentioned insect pests, the plant protection measures should be taken from the field to storage condition of cumin crop. Few of insects, for example, Aphid, leaf-eating caterpillar, thrips attack on the crop while standing in field; however, cigarette beetle, and drug store beetle damage the stored seeds of cumin crop. No insecticides should be applied directly on the dried seed material. It should be fumigated periodically by engaging authorized persons. Care should be taken in all stages of crop cultivation, harvesting, postharvesting handling, processing, packing, storage, and transportation. Needful practices should be followed to prevent contamination and deterioration of quality of produced cumin, which ensures customer satisfaction.

KEYWORDS

- **cumin insect pests**
- **integrated pest management**
- **cut worm**
- **tobacco caterpillar**
- **aphids**

REFERENCES

Andersch, W.; Schwarz, M. Clothianidin Seed Treatment (PonchoReg.) - The New Technology for Control of Corn Rootworms and Secondary Pests in US-corn Production. *Pflanzenschutz Nachrichten Bayer* **2003,** *56* (1), 147–172.

Annual report Model. AESA chart for Cumin from National Institute of Plant Health Management, Hyderabad, 2014.

Balikai, R. A.; Bagali, A. N.; Ryagi, Y. H. Incidence of Spodoptera litura Fab. on grapevine, Vitis vinifera L. *Insect Environ.* **1999,** *5* (1), 32.

Battu, S. Occurrence of *Parasarcophaga misera* (Walker) and Campoletis sp. as Parasites of *Spodoptera litura* (Fabricius) from India. *Curr. Sci.* **1977,** *46* (16), 568–569

Baur, F. J. Chemical Methods to Control Insect Pests of Processed Foods. In *Ecology and Management of Food-Industry Pests*, FDA Technical Bulletin 4, 1991; pp 427–440.

Baur, M. E.; Kaya, H. K.; Tabashnik, B. E. 1997. Efficacy of a Dehydrated Steinernematid Nematode Against Black Cutworm (Lepidoptera: Noctuidae) and Diamondback Moth (Lepidoptera: Plutellidae). *J. Eco. Entomol* **1997,** *90* (5), 1200–1206.

Bhagat, R. M.; Sharma, P. Status of Agrotis ipsilon Hufn. in Kangra Valley of Himachal Pradesh. *Insect Env.* **2000,** *5* (4), 166–167.

Bhatnagar, V. S. Cropping Entomology Annual Report 1979–80. Cropping Entomology Annual Report 1979-80. International Crops Research Institute for the Semi-arid Tropics. Patancheru, Andhra Pradesh India, [2+], 1981, p 22.

Blackman, R. L.; Eastop, V. F. *Myzus persicae* (Sulzer). In *Aphids on the Worlds Crops: An Identification and Information Guide*; John Wiley and Sons: Chichester, New York, Brisbane, Toronto, Singapore, 1984, p 466.

Burbutis, P. P.; Davis, C. P.; Kelsey, L. P.; Martin, C.E. Control of Green Peach Aphid on Sweet Peppers in Delaware. *J. Econ. Entomol.* **1972,** *65* (5), 1436–1438.

Chari, M. S.; Bharpoda, T. M.; Patel, S. N. Studies on Integrated Management of Spodoptera litura Fb. in Tobacco Nursery. Tobacco Research, **1985,** *11* (2), 93–98

Chelawat H. Jeera (Cumin Seed) Prices - On way to Scale Mount Everest. *Ind. J. Res.* **2015,** *4* (6), 2015, p 235–236.

Chiu, S. C.; Chou, L. Y. Hymenopterous Parasitoids of Spodoptera litura Fab. *J. Agr. Res. China* **1976,** *25* (3), 227–241

Clausen, C. P., Ed.; Onion Thrips, (Thrips tabaci Lindeman). pp 20–21. In *USDA Agriculture Handbook* #480: *Introduced Parasites and Predators of Arthropod Pests and Weeds: A World View*; 1078; p 545.

Flint, M. L. Green Peach Aphid, *Myzus persicae*. pp 36–42. Integrated Pest Management for Cole Crops and Lettuce. University of California Publication 3307, 1985, p 112.

Gatehouse, A. M. R.; Down, R. E.; Powell, K. S.; Sauvion, N.; Rahb, T. Y.; Newell, C. A.; Merryweather, A.; Hamilton, W. D. O.; Gatehouse, J. A. Transgenic Potato Plants with Enhanced Resistance to the Peach-potato Aphid Myzus persicae. *Entomologia Experimentalis et Applicata* **1997,** *79* (3), 295–307.

Granovsky, T. A. Stored Product Pests. In *Handbook of Pest Control*; Moreland, D., Ed.; (by Mallis A) 8th ed; Mallis Handbook and Technical Training Co., pp 635–728.

Guo, X. R.; Yuan, G. H.; Fan, C. L.; Zheng, Q.W.; Ma, J.S. 2001. Determination of Chemical Components of the Odor with Luring Activity from Withered Black Poplar Leaves by Gas Chromatography-mass Spectrometry. *J. Henan Agr. Univ.* **2001,** *35* (1), 24–25.

Hill, D. S. *Pests of Stored Products and Their Control*; CRC Press: Boca Raton, 1990; p 274.

Howe, R. W. A Laboratory Study of the Cigarette Beetle, *Lasioderma serricorne* (F.)(Col., Anobiidae) with a Critical Review of the Literature on its Biology. *Bull. Entomol. Res.* **1957,** *48,* 9–56.

HSchler, M.; Brunetti, R. Flight Prediction of the Back Cutworm, Agrotis ipsilon Hufn. (Lepidoptera, Noctuidae), a Pest of Seed Corns in the Tessin. *Revue Suisse d'Agriculture* **2002,** *34* (2), 45–53.

Hussaini, S. S.; Progress of Research Work on Entomopathogenic Nematodes in India. Current Status of Research on Entomopathogenic Nematodes in India: Workshop on the Entomopathogenic Nematodes in India held on 22 and 23rd January 2003, 2003, pp 27–68.

Hussaini, S. S.; Singh, S. P.; Shakeela, V. Variable Efficacy of Different Formulations of Entomopathogenic Nematodes Against Black Cutworm Agrotis ipsilon (Hufnagel) Larvae. Biological Control of Lepidopteran Pests Proceedings of the Symposium of Biological Control of Lepidopteran Pests July 17, 18 2002 Bangalore India, 2003, pp 193–197.

Shetty, R. S.; Singhal, R. S.; Kulkarni, P.R. Antimicrobial Properties of cumin. *World J. Microbiol. Biotechnol.* (Rapid Communications of Oxford Ltd) **1994,** *10,* 232–233.

Krischik, V.; Burkholder, W. Stored-product Insects and Biological Control Agents. In *Stored Product Management.*; Krischik, V., Cuperus, G., Galiart, D. Eds.; Oklahoma Cooperative Extension Service Circular Number E-912, 1995, pp 85–102.

Lockwood, S. Onion Thrips, Thrips tabaci. California Department of Agriculture, Bureau of Entomology. Loose-Leaf Manual of Insect Control, 1956.

Mayer, D. F.; Lunden, J. D.; Rathbone, L. Evaluation of Insecticides for Thrips tabaci (Thysanoptera: Thripidae) and Effects of Thrips on Bulb Onions. *J. Econ. Ent.* **1987,** *80* (4), 930–932.

Mishra, D. N. Chemical Control of Cut Worm, Agrotis ipsilon on Potato in Mid Hill Conditions. *Ann. Plant Prot. Sci.* **2002,** *10* (1), 151–153.

Munson, G.; Keaster, A. J.; Grundler, J. A. Corn Cutworm Control. Agricultural Guide 4150, University of Missouri-Columbia Extension Division, Columbia, MO, 1986.

Nassar, M. M. I. Assessment of Two Natural Marine Toxins (Microcystis aeruginosa and Parasicyonis actinostoloides) for the Control of Some Medical and Agriculture Insects with Reference to the Action on Mice. *J. Egypt. Soc. Parasitol.* **2000,** *30* (2), 631–642.

Richmond, D. S.; Shetlar, D. J. Black Cutworm (Lepidoptera: Noctuidae) Larval Emigration and Biomass in Mixtures of Endophytic Perennial Ryegrass and Kentucky Bluegrass. *J. Econ. Entomol.* **2001,** *94* (5), 1183–1186. Available online at http://www.entsoc.org/pubs/jee/jeetocs/.

Saini, R. K.; Dahiya, A. S.; Verma, A. N. Field Evaluation of Some Insecticides Against Onion Thrips, Thrips tabaci Lindeman (Thysanoptera: Thripidae). *Haryana Agr. Univ. J. Res.* **1989,** *19* (4), 336–342.

Salama, H. S.; Salem, S. A.; Zaki, F. N.; Abdel-Razek, A. The Use of Bacillus Thuringiensis to Control Agrotis ypsilon and Spodoptera exigua on Potato Cultivation in Egypt. *Arch. Phytopathol. Plant Prot.* **1999,** *32* (5), 429–435.

Shapiro, D. I.; Lewis, L. C.; Obrycki, J. J. Abbas, M. Effects of Fertilizers on Suppression of Black Cutworm (Agrotis ipsilon) damage with *Steinernema carpocapsae*. *J. Nematol.* **1999,** *31* (4, Supplement), 690–693.

Shelton, A. M.; North, R. C. Injury and Control of Onion Thrips (Thysanoptera: Thripidae) on Edible Podded Peas. *J. Econ. Ent.* **1987,** *80* (6), 1325–1350.

Shelton, A. M.; Hoy, C. W.; North, R. C.; Dickson, M. H.; Barnard, J. Analysis of Resistance in Cabbage Varieties to Damage by Lepidoptera and Thysanoptera. *J. Econ. Ent.* **1988,** *81* (2), 634–640.

Stoner, K. A. Density of Imported Cabbageworms (Lepidoptera: Pieridae), Cabbage Aphids (Homoptera: Aphididae), and Flea Beetles (Coleoptera: Chrysomelidae) on Glossy and Trichome-bearing Lines of Brassica oleracea. *J. Econ. Entomol.* **1992,** *85* (3), 1023–1030

Thippeswamy, N. B.; Naidu, K. A. Antioxidant Potency of Cumin Varieties—cumin, Black Cumin and Bitter Cumin—on Antioxidant Systems. *Eur. Food Res. Technol.* **2005,** *220,* 472–476.

Toba, H. H. Life-History Studies of *Myzus persicae* in Hawaii. *J. Econ. Entomol.* **1964,** *57* (2), 290–291.

Vallejo, R. L.; Collins, W. W.; Moll, R. H. Inheritance of A and B Glandular Trichome Density and Polyphenol Oxidase Activity in Diploid Potatoes. *J. Am. Soc. Hort. Sci.* **1994,** *119* (4), 829–832.

van Emden, H. F.; Eastop, V. F.; Hughes, R. D.; Way, M. J. The Ecology of *Myzus persicae*. *Ann. Rev. Ento.* **1969,** *14,* 197–270.

Viji, C. P.; Bhagat, R. M. Bioefficacy of Some Plant Products, Synthetic Insecticides and Entomopathogenic Fungi Against Black Cutworm, Agrotis ipsilon Larvae on Maize. *Ind. J. Entomol.* **2001,** *63* (1), 26–32.

Vikaspedia Home / Agriculture / Crop Production / Integrated Pest Managment / IPM for spice crops / IPM Strategies for Cumin / Cumin: Insect and Nematode Pests Management.

Waterhouse, D. F. Thrips tabaci Lindeman, Thysanoptera: Thripidae, onion thrips. In *Biological Control: Pacific Prospects*; Supp 1, Waterhouse, D. F., Norris, K.R., Ed.; Inkata Press: Melbourne, Australia, 1987.

Williamson, R. C. Potter, D. A. Nocturnal Activity and Movement of Black Cutworms (Lepidoptera: Noctuidae) and Response to Cultural Manipulations on Golf Course Putting Greens. *J. Econ. Entomol.* **1997,** *90* (5), 1283–1289.

Zhang, Y.; Ma, H.; Liu, W.; Yuan, T.; Seeram, N. P. New Antiglycative Compounds from Cumin (Cuminum cyminum) Spice. *J. Agr. Food Chem* **2015,** *63* (46), 10097–10102.

Zhou, Z. S. A Review on Control of Tobacco Caterpillar, Spodoptera litura. *Chinese Bull. Entomol* **2009,** *46* (3), 354–361.

Zimmerman, E. C. Thrips (Thrips) tabaci Lindeman. In *Insects of Hawaii*; A Manual of the Insects of the Hawaiian Islands, including Enumeration of the Species and Notes on the Origin, Distribution, Hosts, Parasites, etc. Apterygota to Thysanoptera. The University Press of Hawaii: Honolulu, 1948; Vol. 2, pp 422–425.

www.google.com (Images)

http://www.indianspices.com/spice-catalog/cardamom-large Indian Spice Board.

http://164.100.132.151/agriculture/crop-production/integrated-pest-managment/ipm-for-spice-crops/ipm-strategies-for-cumin/insect-and-nematode-pests-management.

CHAPTER 4

INSECT PESTS OF CUCURBITACEOUS VEGETABLES CROPS

HARI CHAND INGLE[1*] and DIPAK SHYAMRAO[2]

[1]*Department of Entomology, Dr. Rajendra Prasad Central Agricultural University, Pusa (Samastipur)-848125, India*

[2]*Banaras Hindu University, Varanasi (UP)-221005, India*

Corresponding author. E-mail: harichandento@gmail.com

ABSTRACT

India is the second largest producer of vegetables after China, about 75 million tons. The existing area under vegetable cultivation in India is around 4.5 million ha. Majority of Indians are vegetarian, with a per capita consumption 135 g per day as against the recommended 300 g per day. It is still very less than recommended diet level. In near future, there is a need of around 5–6 million tons of food to feed our 1.3 billion Indian population expected by the year 2020. Cucurbits are an excellent fruit in nature having composition of all the essential constituents required for good health of humans. Cucurbits belong to family Cucurbitaceous, includes about 118 genera and 825 species. In India, a number of major and minor cucurbits are cultivated, which share about 5.6 % of the total vegetable production. They are consumed in various forms, i.e., salad (cucumber, gherkins, long melon), sweet (ash gourd, pointed gourd), pickles (gherkins) and deserts (melons). It constitutes the largest group of summer vegetables grown all over the world. Cucumber and bitter gourd are the popular crops of the Cucurbitaceous family grown in low and mid-hills of Himachal Pradesh, Jammu & Kashmir and in plains throughout the India. Fruits of these crops are rich in iron, vitamins (A, B, C), proteins, minerals and have medicinal properties.

The major limiting factor, include the extensive crop devastations due to increased pest menace. In many cases, there is 100 per cent yield loss due to viral diseases vectored by insects. The insect pests inflict crop losses to the tune of 40 per cent in vegetable production. Cucurbits are infested by a number of insect pests right from the germination up to harvesting stage.

4.1 INTRODUCTION

India is the second-largest producer of vegetables after China, about 75 million tons. The existing area under vegetable cultivation in India is around 4.5 million ha. Majority of Indians are vegetarian, with a per capita consumption 135 g per day as against the recommended 300 g per day. It is still very less than the recommended diet level. In near future, there is a need of around 5–6 million tons of food to feed our 1.3 billion Indian population expected by the year 2020 (Paroda, 1999). Cucurbits are an excellent fruit in nature having composition of all the essential constituents required for good health of humans (Rahman, 2003; Duke, 1999). Cucurbits belong to family Cucurbitaceae, which includes about 118 genera and 825 species. In India, a number of major and minor cucurbits are cultivated, which share about 5.6% of the total vegetable production. They are consumed in various forms, that is, salad (cucumber, gherkins, long melon), sweet (ash gourd, pointed gourd), pickles (gherkins), and deserts (melons) (Rai et al., 2008). It constitutes the largest group of summer vegetables grown all over the world. Cucumber and bitter gourd are the popular crops of the Cucurbitaceae family grown in low and mid-hills of Himachal Pradesh, Jammu & Kashmir, and in plains throughout India. Fruits of these crops are rich in iron, vitamins (A, B, C), proteins, minerals, and have medicinal properties.

The major limiting factors include the extensive crop devastations due to increased pest menace. In many cases, there is a 100 percent yield loss due to viral diseases vectored by insects (Shivalingswami et al., 2002). The insect pests inflict crop losses to the tune of 40 percent in vegetable production (Srinivasan, 1993). Cucurbits are infested by a number of insect pests right from the germination up to harvesting stage.

Red pumpkin beetle *Aulacophora foveicollis* (Lucas) is polyphagous, feeds voraciously on leaves, flower buds, and flowers which may reach up to 35–75% at seedling stage (Rashid et al., 2014). Both larval and adult stages are injurious to the crop and cause severe damage to almost all cucurbits at seedlings, young, and tender leaves and flowers. (Rahamanand Prodhan, 2007; Rahman et al., 2008). It feeds underside the cotyledonous leaves by biting holes into them (Chandravadana and Pal, 1983).

Melon fruit fly, *Bactrocera cucurbitae* (Coq.) is one of the main tropical fruit flies causing considerable damage in cucurbits. It has been observed to infest a wide range of crops in the Cucurbitaceae family. Yield loss varies from 30–100% (Nath and Bhusan, 2006).

The serpentine leaf miner, *Liriomyza trifolii* (Burgess) (Diptera: Agromyzidae), an invasive pest was accidentally introduced into India from American subcontinent along with chrysanthemum cuttings (Anonymous, 1991). In India, it was initially recorded on 55 plant species (Viraktamath et al., 1993) and later on about 79 species (Srinivasan et al., 1995) that included pulses, oilseeds, vegetables, green manures and fodder and fiber crops.

There are a number of thrips species that attack cucurbits including melon thrips *Thrips palmi,* onion thrips *Thrips tabaci,* plague thrips *Thrips imaginis* and western flower thrips *Frankliniella occidentalis.* These are minor and frequent pests of cucurbits in temperate regions but major and regular pests in tropical and subtropical zones (Napier, 2009).

Mite species that attack cucurbits, including two-spotted (spider) mite *Tetranychus urticae,* bean spider mite *Tetranychus ludeni,* red-legged earth mite *Halotydeus destructor,* broad mite *Polyphagotarsonemus latus,* blue oat mite *Penthaleus major,* and clover mite *Bryobiacristata.* Two-spotted mites are the most important mite pest, with the other species being of less importance (Napier, 2009). According to an estimate, India will need to produce 215,000 tons of vegetables by 2015 to provide food and nutritional security at individual level and, being a large group of vegetables, cucurbits provide better scope to enhance overall productivity and production.

4.2   PESTS OF CUCURBITACEOUS VEGETABLE

Sr. no	Common name	Scientific name	Family	Order
Major pests				
1.	Red pumpkin beetle	*Aulacophora fove-icollis* and *A. lewisii*	Chrysomelidae	Coleoptera
2.	Serpentine leaf miner	*Liriomyza trifolii*	Agromyzidae	Diptera

Sr. no	Common name	Scientific name	Family	Order
3.	Fruit fly	*Bactrocera cucurbitae*	Tephritidae	Diptera
4.	Thrips	*Thrips palmi*	Thripidae	Thysanoptera
5.	Red Spider Mite	*Tetranychus neocaledonicus*	Tetranichidae	Acarina
Minor pests				
6.	Stem gall fly	*Neolasioptera falcate*	Cecidomyiidae	Diptera
7.	Stem borer/clear winged moth	*Melittia eurytion*	Aegeriidae	Lepidoptera
8.	Stem boring grey beetle	*Apomecyna saltator*	Cerambycidae	Coleopteran
9.	Spotted leaf beetle	*Epilachna vigintioctopunctata*	Coccinellidae	Coleoptera
10.	Flower feeder	*Mylabris pustulata*	Meloidae	Coleoptera

4.2.1 MAJOR PESTS

4.2.1.1 RED PUMPKIN BEETLE

Systematic position:

Phylum	:	Arthropoda
Class	:	Insecta
Order	:	Coleoptera
Family	:	Chrysomelidae
Genus	:	*Aulacophora*
Species	:	*Faveicollis, lewisii, intermedia and cincta*

Origin of distribution:

The pest is widely distributed in different parts of the world, especially in Asia, Africa, Australia, and South Europe. In India, it occurs throughout the country but is more common in Assam, Madhya Pradesh, Uttar Pradesh, and also north-western parts. Out of the several species of *foveicollis, intermedia,* and *cincta* is the commonest beetle found in India.

Host plants:

It is the most destructive pest of all cucurbitaceous vegetables. It infests a wide variety of vegetables like pumpkin, tinda, melon, ghiatori, cucumber, etc., but have a special liking for pumpkin.

Economic importance:

In India, the commonly known red pumpkin beetle is the most destructive, next in order of importance are black pumpkin beetle. It occurs in Uttar Pradesh, Bihar, Madhya Pradesh, Maharashtra, Gujarat, and Andhra Pradesh. Maximum damage in caused by overwintering beetles from March to May, reaching its peak in middle of April.

Nature of damage:

The young grubs, on hatching, bore through the root and gain entry into the base of the plant and feed, resulting in wilting and drying of plants. The grubs grow by boring into roots and also stems and leaves which lie on the ground. The adult beetles of these species feed extensively on leaves, flowers, and fruits. The feeding hole seen on the leaves is a characteristic symptom of damage by adult beetles.

Irregular holes on leaves.	Fruit with deep feeding damage by beetle.

Adult of red pumpkin beetle.

Life cycle:

Eggs: Eggs are elongated and brown in color and hatches into larva in 6–15 days (5–8 days in optimal conditions).

Grub: The whitish grub with brown head bores and feed upon plant roots, fallen leaves, and fruits lying on the surface of soil. Grub Molts four times during 13–25 days of their larval period. Molting occurs inside the soil. A fully-grown grub moves deep into the soil (1.3–25.4 cm deep) and pupates within a water-proof, thick-walled oval cocoon.

Pupation: Pupal period lasts for 7–17 days, after which it metamorphoses into adult beetle. The adult comes out of the soil to feed upon the host plants and to breed. After about 7 days of emergence, beetle starts laying eggs.

Adult: The adult weevils wake up after hibernation in early March. After mating, a female lay eggs singly or in batches of 8–9 in the moist soil at the base of the host plant. As many as 300 eggs are laid by a single female.

Five generations are completed from March to October. A complete life cycle takes about 25–37 days. The adults hibernate in November inside soil or among dry weeds and appear again in March.

Control measures:

- ➢ Summer plowing is done in the month of May–June to expose the eggs and pupa of this beetle.
- ➢ Furrow irrigation moistened the roots of cucurbit crops but not the soil immediately under the plants.
- ➢ Polythene bags protected cucumber seedlings effectively against infestation by *A. foveicollis* for up to 1 month after germination.
- ➢ Bacterium *Serratia marcescens* killed high percentages of adult beetle when smeared on their bodies.
- ➢ The fungus *Fusarium moniliforme subglutinans* killed all infected adult beetle in an average of 4 days. It was highly susceptible to infection by the nematode *Steinernema feltiae*, *Neoaplectana carpocapsae* when fed on foliage sprayed with suspensions of the nematode.
- ➢ If the pest incidence is very severe, spray Indoxacarb 14.5 SC @ 0.5 mL/L or Cabaryl 50 WP 4g/L or 25 EC @ 2mL/L or Chlorpyriphos 20 EC2.5mL/L (Anonymous, 2012).

4.2.1.2 SERPENTINE LEAF MINOR

Systematic position:

Phylum	:	Arthropoda
Class	:	Insecta
Order	:	Coleoptera
Family	:	Agromyzidae
Genus	:	*Liriomyza*
Species	:	*Trifolii, sativae, brassicae, pusilla*

Origin distribution:

L. trifolii has not yet been reported from many countries where it is actually present. It is generally recognized that all the countries bordering the Mediterranean have *L. trifolii* in varying degrees and that it occurs in all mainland states of the USA. *L. trifolii* has been recorded from the Juan Fernandez Islands.

Economic importance:

There are more than 330 *Liriomyza* species (Diptera: Agromyzidae) and many are economically important pests of field crops, ornamentals, and vegetables. Given the substantial economic losses associated with various aspects of *Liriomyza* feeding.

Host range:

The host range of *L. trifolii* includes over 400 species of plants in 28 families including both ornamental crops and vegetables. The main host families and species include Asteraceae, Brassicaceae, Caryophyllaceae, Chenopodiaceae, Cucurbitaceae, Fabaceae, Liliaceae, and Solanaceae.

Nature of damage:

Larvae mine between upper and lower leaf surfaces, creating winding, whitish tunnels that are initially narrow but then widen as the larvae grow. This may cause leaves to dry out, resulting in sun-burning of fruit and reduction and yield and quality. Excessive damage from larvae on young plants and seedlings can cause them to die due to lack of chlorophyll and reduced photosynthesis. Mines and feeding punctures also open up ways for pathogenic organisms to enter. Excessive leaf-mining may cause death of the plant.

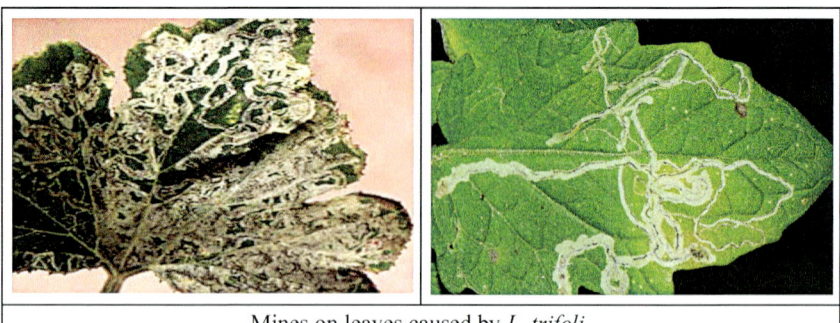

Mines on leaves caused by *L. trifoli.*

Life cycle and marks of identification:

Egg: *L. trifolii* eggs are 0.2–0.3 mm × 0.1–0.15 mm, off white and slightly translucent.

Larva: This is a legless maggot with no separate head capsule, transparent when newly hatched but coloring up to yellow–orange in later instars and is up to 3-mm long. *L. trifolii* larvae and puparia have a pair of posterior spiracles terminating in three cone-like appendages.

Puparium: This is oval and slightly flattened ventrally, 1.3–2.3 × 0.5–0.75 mm with variable color, pale yellow–orange, darkening to golden–brown. The puparium has posterior spiracles on a pronounced conical projection, each with three distinct bulbs two of which are elongate. Puparium occurs outside the leaf and in the soil beneath the plant.

Adult: *L. trifolii* is very small ranges 1–1.3 mm body length, up to 1.7 mm in female with wings 1.3–1.7 mm. The mesonotum is grey-black with a yellow blotch at the hind-corners. The scutellum is bright yellow; the face, frons, and third antennal segments are bright yellow. Male and female *L. trifolii* is generally similar in appearance.

Adults emerge from overwintering cocoons in early spring and lay their eggs in clusters just under the surface of the leaf epidermis. The larvae mine leaves for 1–3 weeks till maturity, then drops down to the soil and pupates for 2–4 weeks. The life cycle in complete warm environments is often 21–28 days; there are normally 2–3 generations per year and more in greenhouses.

| Eggs in the mine | Maggot | Pupa | Adult |

Control measures:

➢ Crop residue of fields infested with leaf miners should be destroyed as soon as possible after harvest.

➢ Shredding of crop residue before tillage is more effective in killing larvae in plant material than tillage alone.

➢ Deep plowing is also useful as it is difficult for adult flies to emerge from anything but a shallow layer of soil.

➢ Destruction of weeds in and around fields is recommended as these can serve as sources of adult infestations.

➢ Natural enemies, primarily parasitoid wasps can maintain leaf miner population below damaging levels.

➢ Use polythene bags fish meal trap with 5 g of wet fish meal + 1 mL Dichlorvos in cotton. Total 50 traps are required/ha, fish meal + Dichlorvos soaked cotton are to be renewed once in 20 and 7 days, respectively.

➢ The numerous parasites that attack leafminer, the most abundant is the parasitic eulophid wasp, *Solenotus intermedius* but *Diglyphus* spp. and *Chrysocharis* spp. are probably the most important for controlling leaf miners.

➢ The parasitoids *Diglyphus isaea* (Walker) (Hymenoptera: Eulophidae) and *Dacnusa sibirica* Telenga (Hymenoptera: Braconidae) are used for controlling leaf miners.

➢ Neem oil @ 3.0% as foliar spray as need-based soil application of neem cake @ 250 kg/ha, immediately after germination.

➢ If the incidence is high, first remove all severely infected leaves and destroy. Then mix neem soap 5 g and hostathion 1 mL/L and spray. Never spray the same insecticide repeatedly (Anonymous, 2012).

4.2.1.3   FRUIT FLY

Systematic position:

Phylum	:	Arthropoda
Class	:	Insecta
Order	:	Diptera
Family	:	Tephritidae
Genus	:	*Bactrocera*
Species	:	*Cucurbitae, tau* and *dorsalis*

Origin and distribution:

It is commonly called melon fruit fly. It is widely distributed and has been recorded from East Africa, some parts of the USA, northern Australia, Taiwan, Okinawa in Japan, South China, South-east Asia, and the Indian subcontinent. It is highly polyphagous having a wide range of host plants but its preferred hosts are musk melon, snap melon, bitter gourd, and snake gourd.

Host plant:

Melon flies use at least 125 host plants. They are major pests of beans, bittermelon, wintermlon, cucumbers, green beans, peppers, pumpkins, squashes, watermelon, etc.

Economic importance:

The genus *Bactrocera (*Macquart) comprises 651 described species. It is the most economically significant fruit fly genus with at least 50 species considered to be important pests, many of which are highly polyphagous (White, 1992; Dacine, 2015). The genus *Bactrocera* is widely distrib-uted throughout tropical Asia, the South Pacific, and Australia. Relatively few species exist in Africa, and only olive fly, *B. oleae* (Rossi), occurs in Southern Europe. Recently, *B. oleae* became established in California and two species in the *B. dorsalis* complex became established on two new continents: *B. carambolae,* Drew and Hancock, the carambola fruit fly, in South America (Suriname) and *B. dorsalis* (formerly *B. invadens,* Drew,

Tsuruta and White) in Africa (Kenya) (Drew, 2005; Rousse et al., 2005). The oriental fruit fly, *B. dorsalis* (Hendel), is native throughout tropical Asia, and has been recorded from over 270 host plant species (White, 1992; Dacine, 2015; Allwood, et al., 1999).

Nature of damage:

Maggots feed inside the fruits but at times also feed on flowers and stems. Generally, the females prefer to lay the eggs in soft tender fruit tissues by piercing them with the ovipositor. A watery fluid oozes from the puncture that becomes slightly concave with seepage of fluid and transforms into a brown resinous deposit. Sometimes pseudo punctures (punctures without eggs) have also been observed on the fruit skin. This reduces the market value of the produce. Young larvae leave the necrotic region and move to healthy tissue, where they often introduce various pathogens and hasten fruit decomposition.

Fruit damaged by maggot.

Life cycle and marks of identification:

Eggs: Eggs are generally laid in young fruit but are also laid in the succulent stems of host plants. The eggs are deposited in cavities created by the female using its sharp ovipositor. A female may lay more than 1000 eggs.

Maggot: These hatch within 1–2 days and the larvae feed for another 4–17 days (longest in thick-skinned fruits, such as pumpkin).

Pupa: Pupation is in the soil under the host plant for 7–13 days but may be delayed for several weeks under cool conditions.

Adults: Adults occur throughout the year and begin mating (at dusk) after about 10–12 days and may live 5–15 months depending on temperature (longer in cool conditions). Adult flight and the transport of infected fruit are the major means of movement and dispersal to previously uninfected areas. Many *Bactrocera* spp. can fly 50–100 km. This is one of the most common species attracted to cue lure.

| Eggs | Maggots | Pupa | Adult |

Control measures:

➢ Clean cultivation, that is, removal and destruction of fallen and infested fruits daily.

➢ Deep plowing to expose hibernating stages.

➢ The most effective cultural management technique is the destruction of all infested and unmarketable fruit and the disposal of crop residues immediately after harvest.

➢ Braconid parasitoids *Fopius arisanus* or *Pysttalia fletcheri* attacked both melon fly eggs and larvae at the same time suppression of development was as much as 56%.

➢ Fumigation, heat treatment (hot vapor or hot water), cold treatments, insecticidal dipping, or irradiation. Irradiation is not accepted in most countries and many have now banned methyl bromide fumigation. Heat treatment tends to reduce the shelf life of most fruits.

➢ Application of spray baits. Spraying with 0.05% malathion or 0.2% carbaryl at flowering.

4.2.1.4 THRIPS

Systematic position:

Phylum	:	Arthropoda
Class	:	Insecta
Order	:	Thysanoptera
Family	:	Thysanoptera
Genus	:	*Thrips*
Species	:	*palmi*

Origin and distribution: In recent years, it has spread from Southeast Asia to the rest of Asia and to many Pacific Ocean islands, North Africa, Australia, Central and South America, and the Caribbean. It has the potential to infest greenhouse crops widely but under field conditions, its distribution likely will be limited to tropical areas.

Host plants:

Melon thrips is a polyphagous species but is best known as a pest of Cucurbitaceae and Solanaceae. Among vegetables, injured are bean, cabbage, chili, Chinese cabbage, cowpea, cucumber, bean, eggplant, lettuce, melon, okra, onion, pea, pepper, potato, pumpkin, squash, and watermelon. Other crops infested include avocado, carnation, chrysanthemum, citrus, cotton, hibiscus, mango, peach, plum, soybean, tobacco, and others. Also, weeds can serve as important hosts, including species of nightshades (Solanaceae), legumes (Fabaceae or Leguminosae), and asters (Asteraceae or Compositae).

Nature of damage:

Melon thrips cause severe injury to infested plants. Leaves become yellow, white or brown and then crinkle and die. Heavily infested fields sometimes acquire a bronze color. Damaged terminal growth may be discolored, stunted and deformed. Feeding usually occurs on foliage, but on pepper, a less suitable host, flowers are preferred to foliage. Fruits may also be damaged; scars, deformities, and abortion are reported. Pepper and eggplant fruit are damaged when the thrips feed in blossoms.

| Thrips feeding damage on cucumber showing distortion and curling. | Damaged leaves by thrips. |

Life cycle:

A complete generation may be completed in about 20 days at 30°C but it is lengthened to 80 days when the insects are cultured at 15°C. Melon thrips are able to multiply during any season that crops are cultivated but are favored by warm weather. When crops mature, their suitability for thrips declines, so this thrips growth rate diminishes even in the presence of warm weather.

Eggs: Eggs are deposited in leaf tissue in a slit cut by the female. One end of the egg protrudes slightly. The egg is colorless to pale-white in color, and bean-shaped in form. Duration of the egg stage is about 16 days at 15°C, 7.5 days at 26°C, and 4.3 days at 32°C.

Larvae: The larvae resemble the adults in general body form though they lack wings and are smaller. There are two instars during the larval period. Larvae feed in groups, particularly along the leaf midrib and veins and usually on older leaves. Larval-development time is determined principally by the suitability of temperature but host plant quality also has an influence. Larvae require about 14, 5, and 4 days to complete their development at 15°C, 26°C, and 32°C, respectively.

Pupa: There are two instars during the "pupal" period. The prepupal instar is nearly inactive and the pupal instar is inactive. Both instars are nonfeeding stages. The prepupae and pupae resemble the adults and larvae in form except that they possess wing pads. The combined prepupal and pupal development time is about 12, 4, and 3 days at 15°C, 26°C, and 32°C, respectively.

Adult: Adults are pale-yellow or whitish in color but with numerous dark setae on the body. A black line resulting from the juncture of the wings runs along the back of the body. The slender fringed wings are pale. The hairs or fringe on the anterior edge of the wings are considerably shorter than those on the posterior edge. They measure 0.8–1.0 mm in body length. Unlike the larval stage, the adults tend to feed on young growth and so are found on new leaves. Adult longevity is 10–30 days for females and 7–20 days for males. Females produce up to about 200 eggs but on average about 50 per female.

| Eggs | 1st instar | 2nd instar | 3rd instar |

Control measures:

➢ Physical barriers, such as fine mesh and row cover material, can be used to restrict entry by thrips into greenhouses, and to reduce the rate of thrips settling on plants in the field.

➢ High levels of carbon dioxide, particularly in conjunction with high temperatures, can be a useful fumigation technique.

➢ Heavy rainfall is thought to decrease thrips numbers. However, there seems to be no evidence that overhead irrigation is an important factor in survival.

➢ Predatory mites (*Amblyseius) cucumeris, Amblyseius degenerans,* and *Hypoaspis* spp. and minute pirate bugs *(Orius insidiosus)* provide effective biological control of thrips.

➢ Need based application of imidacloprid 200 SL @ 100 mL/ha or methyl demeton 25 EC @ 500 mL/ha or NSKE 5% @ 2 kg/ha for controlling thrips in cucurbitaceous crops.

4.2.1.5   RED SPIDER MITE

Systematic position:

Phylum	:	Arthropoda
Class	:	Insecta
Order	:	Acarina
Family	:	Tetranychidae
Genus	:	*Tetranychus*
Species	:	*Urticae, neocaledonicus*

Origin of distribution:

T. urticae was originally native only to Eurasia. It has been spread throughout the temperate regions of the northern hemisphere by wind and throughout the world via the transport of plants by man. The species now has a cosmopolitan distribution. The strong colonizing potential of *T. urticae* may have led to connectivity between populations worldwide.

Host plant:

Mites are common pests in landscapes and gardens that feed on many fruit trees, vines, berries, vegetables, and ornamental plants. It is known to attack about 1200 species of plants, of which 150 are economically important. Web spinning spider mites include the Pacific spider mite, two-spotted spider mite, strawberry spider mite, and several other species. Most common ones are closely related species in the *Tetranychus* genus and cannot be reliably distinguished in the field.

Nature of damage:

Spider mite also called two spotter mite (TSSM) and feeds by puncturing cells with its stylets and draining the contents thereby producing a characteristic yellow specking of the leaf surface. This chlorotic damage, along with the webbings produced by protonymphs, deutonymphs, and adults, reduces the plants' ability to carry out photosynthesis resulting in reduction of the total yield of vegetable crops. The mites produce threads of silk, which they use to "balloon" into the wind, which sometimes carry them great distances *T. urticae* feeding causing aesthetic injuries, as well as morphological and biochemical alterations in leaf and fruit composition, has been reported earlier.

| Mites, *Tetranychus urticae* | Webbing of leaves by mites. |

Morphology:

To the naked eye, spider mites look like tiny, moving dots; however, you can see them easily with a 10× hand lens. Adult females, the largest forms are less than 1/20-inch long. Spider mites live in colonies, mostly on the undersurfaces of leaves; a single colony may contain hundreds of individuals. The presence of webbing is an easy way to distinguish them from all other types of mites and small insects, such as aphids and thrips, which can also infest leaf undersides. Adult mites have four pairs of legs and an oval body with two red eyespots near the head end. Females usually have a large, dark blotch on each side of the body and numerous bristles covering the legs and body. Immatures resemble adults (except they are much smaller) and the newly hatched larvae have only six legs. The other immature stages have eight legs. Eggs are spherical and translucent like tiny droplets becoming cream-colored before hatching.

Life cycle:

In colder areas and on deciduous trees that drop their leaves, web-spinning mites overwinter as red or orange mated females under rough bark scales and in ground litter and trash. They begin feeding and laying eggs when warm weather returns in spring.

Spider mites reproduce rapidly in hot weather and commonly become numerous in June through September. If the temperature and food supplies are favorable, a generation can be completed in less than a week. Spider mites prefer hot, dusty conditions and, usually, are first found on trees. Plants under water stress also are highly susceptible. As foliage quality declines on heavily infested plants, female mites catch wind currents and disperse to other plants. High mite populations may undergo a rapid decline in late summer when predators overtake them, host plant conditions become unfavorable and the weather turns cooler as well as following rain.

Control measures:

➢ Apply water to pathways and other dusty areas at regular intervals. Water-stressed trees and plants are less tolerant of spider mite damage. In cucurbitaceous crops, regular forceful spraying of plants with water often will reduce spider mite numbers adequately.

➢ Some of the most important are the predatory mites including the western predatory mite, *Galendromus occidentalis,* and *Phytoseiulus* mite species and are more effective under hot and dry conditions.

➢ Carbaryl, some organophosphates, and some pyrethroids apparently also favor spider mites by increasing the level of nitrogen in leaves. Insecticides applied during hot weather usually appear to have the greatest effect, causing dramatic spider mite outbreaks within a few days.

➢ There are also a number of plant extracts formulated as acaricides (a pesticide that kills mites) that exert an effect on spider mites. These include garlic extract, clove oil, mint oils, rosemary oil, and cinnamon oil.

4.2.2   CUCURBITS MINOR PESTS

4.2.2.1   STEM GALL FLY

Phylum	:	Arthropoda
Class	:	Insecta
Order	:	Diptera
Family	:	Cecidomyiidae
Genus	:	*Neolasioptera*
Species	:	*falcata*

Nature of damage:

Damage caused by maggots by boring into distal shoots. Thickening or gall are seen on shoot stem. Maggot feeds within distal stems of bitter-gourd, ribbed and smooth gourds causing formation of elongated galls in between nodes. Gall formation causes stunting of plants.

Galls on leaves.	Adult of stem gall fly.

Economic importance:

Family of small flies occurring throughout the world. The larvae of a great number are responsible for galls and other malformations on a large range of plants, whereas others are zoophagous, feeding on aphids, coccids or scale-insects, white-flies, and mites. Still, others feed on rusts, mildews, etc. and some are saprophytic.

Control measure:

Apply insecticides like Quinolphos or Carbaryl.

4.2.2.2   STEM BORER

Systematic position:

Phylum	:	Arthropoda
Class	:	Insecta
Order	:	Lepidoptera
Family	:	Aegeriidae
Genus	:	*Melittia*
Species	:	*eurytion*

Distribution:

It is a pest of snake gourd and occurs widely in India. Moth is clear winged with fan-like tufts of hairs on hind legs. White larvae bore into the stems producing galls. Frass comes out of the gall through a hole made on it. Plants stunted with poor foliage. Pupal period 20–24 days in earthen cocoon in soil.

Host: Snake gourd.

Nature of damage: Larva bores into the stem of snake gourd and produces galls.

Control measures: Collect and destroy the damaged plant parts with larvae.

Encourage activity of parasitoid: *Apanteles* spp.

Spray any insecticides like malathion 50 EC @ 500 mL or dimethoate 30 EC 500 mL or methyl demeton 25 EC @ 500 mL/ha.

4.2.2.3   STEM BORING GREY BEETLE

Phylum	:	Arthropoda
Class	:	Insecta
Order	:	Coleoptera
Family	:	Cerambycidae
Genus	:	*Apomecyna*
Species	:	*saltator*

In South India, it attacks *Coccinea* vines, which die as a result. Adult female is a white-spotted greyish-brown longicorn beetle and the male is smaller and black. Eggs are laid singly on internodes below the bark. Grubs bore into the long trailing stems at or near a node and tunnel inside. Adult beetles gnaw the leaf petioles and soft parts of the stem. A female lays 38–52 eggs, which hatch in 5–7 days. There are six larval instars covered in 31–35 days. Pupal period lasts for seven to nine days. Total life cycle takes 80–98 days. Adult has a longevity period of 37–43 days.

4.2.2.4   SPOTTED LEAF BEETLE

Phylum	:	Arthropoda
Class	:	Insecta
Order	:	Coleoptera
Family	:	Coccinellidae
Genus	:	*Epilachna*
Species	:	*vigintioctopunctata*

Biology:

Eggs are yellow, elongated-oval, and lay in clusters on the undersurface of leaves. Egg hatch ranged between 75.61–100 percent. There are four larval instars as reported earlier in the life cycle of *H. vigintioctopunctata* with a total larval period of 10–12 days. First instar larvae fed gregariously and they became nongregarious from second instar onward. The larvae scraps epidermal layer of leaves and sometimes skeletonize the entire leaf. Total larval period of 13–15 days and pupal duration is 4–6 days under natural condition. The adult beetle started feeding after hardening of elytra. *H. vigintioctopunctata* has been known to complete its life cycle in 22–30 days depending upon the host plant which it feeds and prevailing climatic conditions such as temperature and relative humidity.

Nature of damage:

The nature of damage by adults is the same as that of larvae, they also fed on the epidermal layer of leaves but sometimes make tattered holes on the leaves.

Eggs, grub, and adult of *Epilachna.*

Host plants:

Cucurbit and brinjal

Control measures:

Larvicidal bioassays with crude aqueous leaf extracts of three plants viz. *R. communis*, *C. procera,* and *D. metel* showed significant toxicity against the experimental *Epilachna* beetles.

Application of castor leaf extract is the most toxic against the beetles.

4.2.2.5   FLOWER FEEDER

Systematic position:

Phylum	:	Arthropoda
Class	:	Insecta
Order	:	Coleoptera
Family	:	meloidae
Genus	:	*Mylabris*
Species	:	*pustulata*

Host:

Adult beetles are phytophagous, feeding especially on plants in the families Amaranthaceae, Asteraceae, Fabaceae, and Solanaceae. Grub feed on eggs of grasshopper but adults eat only floral parts, but some particularly those of *Epicauta* spp. eat leaves as well.

Distribution:

Twelve species are more or less widely distributed in the central and/or eastern states. A third, weaker faunal link with the West Indies is represented by *Pseudozonitis longicornis*(Horn), which belongs to a group including one West Indian species and two relictual species in east Texas. No species is indigenous.

Life cycle:

Eggs are laid in masses in the ground or under stones (Meloinae). Larval development is hypermetamorphic with four distinct phases. In the first instar or triungulin (T) phase the larva reaches its feeding site on its own (most Meloinae) or attaches to an adult bee. After feeding to repletion, the larva, through ecdysis, becomes scarabaei form and enters a period of rapid growth that lasts until the end of instar five or six. In six or seven instars, the larva typically becomes heavily sclerotized and immobile. In this phase the musculature undergoes profound degeneration and respiration is reduced to an extremely low level, permitting survival for more than a year, if necessary.

| Young flower feed by adult beetle. | Adult, M*ylabris pustulata* |

Control measures:

Manual picking and destruction of adult blister beetles are often the only practical control measure.

KEYWORDS

- **insect pests**
- **cucumber**
- **bottle guard**
- **sponge guard**
- **bitter guard**
- **cucumber**
- **pest management**

REFERENCES

Allwood, A. J.; Chinajariyawong, A.; Drew, R. A. I.; Hamacek, E. L.; Hancock, D. L.; Hengsawad, C.; Jipanin, J. C.; Jirasurat, M.; Kong Krong, C.; Kritsaeneepaiboon, S. Host Plant Records for Fruit Flies (Diptera: Tephritidae) in Southeast Asia. *Raffles Bull. Zool.* **1999**, *4,* 1–92.

Anonymous. Castor: Annual Report. Directorate of Oilseeds Research, Hyderabad, 1991. p 137.

Anonymous. Proceedings of the IXth EUCARPIA Meeting on Genetics and Breeding of Cucurbitaceae (Pitrat M, ed), INRA, Avignon (France), 2008.

Anonymous. National Horticulture Mission, Department of Agriculture and Cooperation, Ministry of Agriculture, 2012, Vol. 6, p 17.

Dacine. Fruit Flies of the Asia-Pacific, 2015. DOI: *http://www.herbarium.hawaii.edu/ fruitfly.*

Doharey, K. L. Bionomics of Red Pumpkin Beetle, *Aulacophorafoveicollis*(Lucas) on some fruits. *Indian J. Entomol.* **1983**, *45,* 406–413.

Drew, R.; Tsuruta, K.; White, I. A New Species of Pest Fruit Fly (Diptera: Tephritidae) from Sri Lanka and Africa. *Afr. Entomol.* **2005**, *13,* 149–154.

Duke, J. A. *Handbook of Phytochemical and Constituents of Grass Herbs and other Economic Plants*; CRC Press: Boco Raton, FL, 1999; pp 98–119.

Napier, T. Insect Pests of Cucurbit Vegetables, 2009. www.DPI.NSW.Gov.Au

Nath, P.; Bhusan, S. Evaluation of Poison Bait Traps for Trapping Adult Fruit Fly. *Ann. Plant Prot. Sci.* **2006**, *14,* 297–299.

Paroda, R. S. For a Food Secure Future. The Hindu Survey of Indian Agriculture, 1999.

Rahaman, M. A.; Prodhan, M. D. H. Effects of Net Barrier and Synthetic Pesticides on Red Pumpkin Beetle and Yield of Cucumber. *Int. J. Sustain. Crop Prod.* **2007**, *2* (3), 30–34.

Rahaman, M. A.; Prodhan M. D. H.; Maula, A. K. M. Effect of Botanical and Synthetic Pesticides in Controlling *Epilachna* Beetle and the Yield of Bitter Gourd. *Int. J. Sustain. Crop Prod.* **2008**, *3* (5), 23–26.

Rahman, A. S. H. Bottle Gourd (*Lagenariasiceraria*) a Vegetable for Good Health. *Nat. Prod. Rad.* **2003**, *2,* 249–250.

Rashid, M. A.; Khan, M. A.; Arif, M. J.; Javed, N. Red Pumpkin Beetle, *Aulacophorafoveicollis* Lucas; A Review of Host Susceptibility and Management Practices. *Acad. J. Entomol.* **2014**, *7* (1), 38–54.

Rousse, P.; Harris, E. J.; Quilici, S. *Fopiusarisanus*, an Egg-pupalparasitoid of Tephritidae. Overview. *Biocontrol News Inform.* **2005**, *26,* 59–69.

Shivalingswami, T. M.; Satpathy, S.; Banergee, M. K. Estimationof Crops Losses due to Insect Pests in Vegetables. Sarath Babu, B., Varaprasad, K. S., Anitha, K., Prasada Rao, R. D. V. J., Chakrabarthy S. K., Chandurkar, P. S., Eds.; *Resource Management in Plant Protection*; 2002; Vol. 1, pp 24–31.

Srinivasan, K. Pests of Vegetable Crops and Their Control. In *Advances in Horticulture, 6: Vegetable Crops*; Chadha, K. L., Kalloo, G., Eds.; Malhotra Pub. House: New Delhi. 1993; pp. 859–886.

Srinivasan, K.; Viraktamath, C. A.; Gupta, M.; Tiwari, G. C. Geographical Distribution, Host Range and Parasitoids of Serpentine Leaf Miner, *Liriomyzatrifolii*(Burgess) in South India. *Pest Manag. Hort. Ecosys.* **1995**, *1*, 93–100.

Viraktamath, C. A.; Tiwari, G. C.; Srinivasan, K.; Gupta, M. American Serpentine Leaf Miner is a New Threat to Crops. *Indian Farm* **1993**, 10–12.

White, I. M.; Elson-Harris, M. M. Fruit Flies of Economic Significance: Their Identification and Bionomics. CABI International: Wallingford, UK, 1992.

CHAPTER 5

PHYTOPHAGOUS MITES OF VEGETABLE CROPS AND THEIR MANAGEMENT

PUSHPA SINGH[1*], R. N. SINGH[2], and C. P. SRIVASTVA[2]

[1]Department of Entomology, Dr. Rajendra Prasad Central Agricultural University, Pusa, Samastipur, Bihar, India

[2]Institute of Agricultural Sciences, Banaras Hindu University, Varanasi 221005, Uttar Pradesh, India

*Corresponding author. E-mail: pushpa8march68@gmail.com

ABSTRACT

A total of 660 mite species are known to occur on various plants of India, out of which 319 are plant feeding and predatory in nature (Banerjee 1988., Banerjee and Cranham, 1985., Bindra and Singh 1970., ChannaBasavanna 1966, 1981 and 1986., Gangwar and Lal 1988., Ghai 1964., Gupta 1974., 1985 and 1991., Gupta and Nahar 1981., Nassar and Ghai 1981., Prasad 1974., Singh et al 1987., and Singh and Mukherjee 1989). The plant feeding mites are represented by 169 species whereas 148 species belong to predatory group. The major groups of plant mites belong to spider mite (more than 69 species of Family Tetranychidae), false spider mite (more than 29 species of Family Tenuipalpidae), gall or bud mite (more than 64 species of Family Eriophidae) and yellow or broad mite (more than 4 species of Family Tarsonemidae). Seventeen species of mites are major pests on various plants. Besides this, there are some predatory mites feeding on injurious mites. Among predatory mites majority (102 species) belong to Phytoseiidae and remaining (46 species) belong to other families of predatory group. The knowledge of important detrimental species is necessary when there is continuous fall in stability and sustainability of useful species due to rapid changes in the cultural and pest control practices, including indiscriminate use of pesticides. This has given a new atmosphere to flare up the mite population in the absence of predatory fauna. Very limited effort has been made to document the severity of the mite problem with changes in agriculture scenario and compounded nature of atmosphere.

5.1   INTRODUCTION

In recent years, phytophagous mites have attracted the attention of acarologists and intensive farmers from all over the world (Ehara, 1975; Gupta, 1985, 1991; Helle and Sabelis, 1985; Jeppson et al., 1975; Keifer et al., 1982; Meyer, 1981; Pritchard and Baker, 1955; Singh and Mukherjee, 1989;Van de vrie et al., 1972; Wilson et al., 1998; Walsh et al., 1998). Predatory mites are more adversely affected than mite pests where broad-spectrum pesticides are used; consequently, an outbreak of mite pests occurs (Van de Vrie et al., 1972). Many species of mites which were rare and innocuous have assumed the status of major pests. The outbreaks of many phytophagous mites on several crops have also attracted the attention of agricultural scientists (Dhooria and Sagar, 1975; Meyer, 1981; Penman and Chapman, 1988; Sandhu et al., 1987). The species of spider mites are capable of developing resistance to certain pesticides (Goodwin et al., 1991; Thwaite, 1990; Thwaite et al., 1987).

A total of 660 mite species are known to occur on various plants of India, out of which 319 are plant feeding and predatory in nature (Banerjee, 1988; Banerjee and Cranham, 1985; Bindra and Singh, 1970; ChannaBasavanna, 1966, 1981, 1986; Gangwar and Lal, 1988; Ghai, 1964; Gupta, 1974, 1985, 1991; Gupta and Nahar, 1981; Nassar and Ghai, 1981; Prasad, 1974; Singh et al. 1987; Singh and Mukherjee, 1989). The plant feeding mites are represented by 169 species, whereas 148 species belong to predatory group. The major groups of plant mites belong to spider mite (more than 69 species of family Tetranychidae), false spider mite (more than 29 species of family Tenuipalpidae), gall or bud mite (more than 64 species of family Eriophidae), and yellow or broad mite (more than 4 species of family Tarsonemidae). Seventeen species of mites are major pests on various plants. Besides this, there are some predatory mites feeding on injurious mites. Among predatory mites, majority

TABLE 5.1 Important Mite Pests Causing Damage to Economic Plants.

S. No.	Family/Mite species and common name	Major host plants
Tetranychidae		
1.	*Tetranychus urticae* Koch [Two-spotted spider mite (TSM)]	okra, brinjal, beans, cucurbits, cotton, and marigold
2.	*Tetranychus, neoealidonicus* Andre' (Red vegetable mite)	vegetables and ornamental plants
3.	*Tetranychus ludeni* Zacher (Red-legged spider mite)	vegetables and beans
4.	*Tetranychus macfarlanei* Baker & Pritchard (Spider mite)	cucurbits
5.	*Eutetranychus orientalis* (Klein) (Oriental citrus mite)	Citrus
6.	*Eotetranychus hirsti* Pritchard & Baker (Fig spider mite)	Fig
7.	*Oligonychus indicus* (Hirst) (Sugarcane red spider mite)	sugarcane and sorghum
8.	*Oligonychus coffeae* (Nietner) (Red tea mite)	tea, coffee
9.	*Oligonychus mangiferus* (Rahman and Sapra) (Mango red spider mite)	mango
10.	*Schizotetranychus andropogoni* (Hirst) (White patchy/Sugarcane leaf spotted mite/Web mite)	Sugarcane and cereals
11.	*Panonychus citri* (McGregor) (Citrus red mite)	citrus
12.	*Petrobia latens* (Muller) (Brown wheat mite)	wheat, barley and coriander
Tenuipaldidae		
13	*Brevipalpus phoenicis* (Geijskes) (Scarlet mite / Reddish black flat mite/Leprosis mite)	fruit trees and tea
14	*Brevipalpus californicus* (Banks) (Citrus flat mite)	citrus and tea
15	*Larvacarus transitans* (Ewing) (Ber gall mite)	ber (*Zyziphus mauritiana*)
16	*Dolichotetranychus floridanus* (Banks) (Pineapple flat mite)	pineapple
17.	*Raoiella indica* Hirst (Coconut red mite)	coconut and areca nut
Eriophyidae		
18.	*Aceria mangiferae* Sayed (Mango bud mite)	mango
19.	*Aceria litchii* (Kiefer) (Litchi erineum mite)	litchi
20.	*Aceria cajani* ChannaBasavanna (Pigeonpea sterility mosaic mite)	pigeonpea
21.	*Aceria lycopersici* (Wolff) (Tomato/Brinjal erineum mite)	brinjal and tomato
22.	*Aceria guerrerronis* (Keifer) (Coconut eriophyid mite)	coconut
23.	*Acreia jasmine* ChannaBasavanna (Scarlet mite)	jasmine
24.	*Acaphylla theae* (Watt) (Tea pink mite)	tea
25.	*Eriophyes cernuus* Massee (Jujube (Ber) gall mite)	jujube (*ber*)
Tarsonemidae		
26	*Polyphagotarsonemus latus* (Banks) (silver mite/Yellow mite/Broad mite)	chilli, beans, potato, and jute brinjal
27.	*Steneotarsonemus spinki* Smiley (Rice sheath mite)	rice

(102 species) belong to Phytoseiidae and remaining (46 species) belong to other families of predatory group. The knowledge of important detrimental species is necessary when there is continuous fall in stability and sustainability of useful species due to rapid changes in the cultural and pest control practices, including indiscriminate use of pesticides. This has given a new atmosphere to flare up the mite population in the absence of predatory fauna. Very limited effort has been made to document the severity of the mite problem with changes in agriculture scenario and compounded nature of atmosphere. The view of the present chapter is to visualize the magnitude of the mite menace in vegetable crops and their management.

5.2   HOST RANGE OF PHYTOPHAOGUS MITES

In India, 17 species of mites are considered as major pests and 30 as minor pests on different crops (Gupta, 1991; Singh and Mukherjee, 1989). The major plant feeding mites belong to families of spider mite (Tetranychidae), false spider mite (Tenuipalpidae), gall mite (Eriophyidae), and yellow or broad mite (Tarsonemidae). The important injurious mites of these families are listed in Table 5.1.

5.2.1   TETRANYCHIDAE

The tetranychid mites are widely spread and polyphagous commonly known as spider mites. They are exclusively phytophagous in nature and many species are serious pests of agricultural crops, vegetables, ornamentals, and fruits. Many mites are potential pests and their pest status has been described (Gupta, 1991; Jeppson et al., 1975; Meyer, 1981; Mori et al., 1990; Singh and Mukherjee, 1989; Van de Vrie et al., 1972). Indian Tetranychidae consists of a total of 101 species with two subfamilies, six tribes, and 10 genera (Gupta and Gupta, 1994). Confusing taxonomic status of many tetranychid mites makes it difficult to differentiate them among themselves. Dosse and Boudreaux, 1963 mentioned that the economic importance of spider mites has made it necessary that good taxonomy be achieved. They further mentioned that the problems of mite taxonomy are still with us and workers must continue in making efforts at a better understanding of these mites.

During recent years, many species of tetranychid mites have also assumed the status of major pest. Many species are host specific, while others infest a wide variety of plants. Mostly tetranychid mites feed on both surfaces. Pillai and Palaniswamy (1985) reported that on cassava the red mites, *Tetranychus cinnabarinus* and *T. neocaledonicus*, usually feed on the lower surface, whereas *Eutetranychus orientalis* and *Oligonychus biharensis* are always found on the upper surface of the leaves. During feeding, mites damage the epidermis with their needle-like chelicerae to suck cell sap. Their feeding activity reduces chlorophyll content and leads to formation of numerous empty cells at a site resulting in formation of yellow or brown specks or stipples. Extensive feeding by large number of mites causes the leaves to appear yellow or brown. These leaves when severely injured, shrivel, die, and eventually drop off from the plant. Some species of tetranychid mites produce copious webs. Overcrowding of mites may also occur on leaf tip, fruit tip, and then they might migrate to other fields. The swarming phenomenon is characterized by hot spot of ballooning or foci.

The male and female sexes are common in most of the species of tetranychid mites. Males are very less in ratio. The unfertilized eggs produce only male offsprings, whereas fertilized eggs produce female only. Van de Vrie et al. (1972 reported that the mated females produce both females and males because every egg has not received a spermatozoon. Jeppson et al. (1975) mentioned that fertilized females produce both females and males. The development of tetranychid mites takes place through egg, larva, protonymph, deutonymph, and adult stages.

5.2.2   TENUIPALPIDAE

The false spider mites are known as tenuipalpid mites. Baker and Tuttle (1987 reported that false spider mites are widely distributed and are more numerous in warmer zones of the globe. Meyer (1979) reported 504 species in 21 genera for the world fauna. Many mites of this group are pests of economic plants, such as ornamentals and fruit trees (Jeppson et al., 1975). These mites look like spider mites but do not spin silken web. Majority of spider mites is of great economic importance and infests fruit trees, vegetable crops, etc. However, very little work has been done on these mites (Gupta 1985; Sadana, 1985; Sadana et al., 1990). The false spider mites are flat, pear shaped, mostly reddish bright colored, size 0.3–0.35 mm, slow moving and are normally found on the undersurface of the leaves near midrib or veins, twigs, and fruits. Most specialized members of this family form plant galls within which they feed (Jeppson et al., 1975). In India, genera *Brevipalpus* and *Tenuipalpus* are of great economic importance. At a time they cause serious injury to a wide variety of agricultural and horticultural crops. The first observable symptom on plant appears on the leaves as silvery areas that frequently become shrunken and brown. Seriously infested leaves become yellow and drop off from the plant.

5.2.3   ERIOPHYIDAE

Mites of family Eriophyidae are exclusively plant feeder and host specific are of economic importance. They are commonly known as gall, rust, bud, blister, and erineum or simply eriophyid mites. Members of this family have two body regions, the gnathosoma and idiosoma. They have elongated worm-like body shape with two pairs of legs. The body size varies from 0.1 to 0.5 mm. Eriophyid mite has simple life cycle but certain species has complicated cycle that includes alternation of generations. There are two forms, one is an overwintering female called deutogynes and the other consists of both sexes known as protogynes. The symptoms caused by eriophyid mites vary from simple russeting to coupled gall formation. The galls may be simple leaf galls, bud galls, and erineum. Witches broom, which is cluster of brush like eriophyid mites also cause growth of stunted twigs or branches on trees. Eriophyid mites also cause non-gall abnormalities such as leaf folding and twist blisters. Ghosh et al. (1989) described injury caused by gall mites. They recorded six types of plant injuries, that is, big buds, curling/crinkling and shrinkage of leaves, discoloration, erineum, russeting, and galls. Eriophyid mites also induce russeting and silvering of leaves. Kiefer et al. (1982) described various symptoms of injuries caused by eriophyid mites on buds, shoots, stems, twigs, flowers, and fruits. Large number of eriophyid mites is confined to one species of host plant. Some eriophyid mites act as vector of plant viral diseases. The important plant diseases transmitted by eriophyid mites are pigeon pea sterility mosaic disease, fig mosaic disease, and wheat streak mosaic disease and sugarcane streak mosaic virus. About 1859 eriophyids belonging to 156 genera have been reported from different parts of the world (Boczek et al, 1989; Davis et al., 1982; Keifer et al., 1982). Extensive work has been done on the Indian eriophyids (ChannaBasavanna, 1966, 1971; Das and Chakrabarti, 1988; Gupta, 1983; Mohanasundaram, 1981).

5.2.4 TARSONEMIDAE

Tarsonemid mites are very small, ranging in length from 100 to 300 μm. Mature forms have a relatively hard and shiny integument just like glistering white or shiny silver piece. The body and posterior legs are rather sparsely beset with setae. The anterior pairs of legs especially their terminal segments are more densely clothed with setae and are often equipped with specialized sensory setae of various configurations and sizes. The tiny glossy mites are found on the undersurface of leaves as well as in association of fungus. Males often carry female deutonymphs. Some of these mites are important crop pests (Gupta, 1985; Meyer, 1981; Singh et al., 1987).

5.3 IMPORTANT MITE PESTS OF VEGETABLE CROPS

5.3.1 TETRANYCHUS URTICAE KOCH OR TWO-SPOTTED SPIDER MITE

Two-spotted spider mite (TSSM), *T. urticae*, belongs to the group of acarines known as Acariformes, in the suborder Prostigmata and the family Tetranychidae. It is major pest of vegetables in India (Gulati, 2004; Geroh, 2007) and also worldwide (Gatarayiha et al., 2010). *T. cinnabarinus* is synonym of *T. urticae*. Gupta (1985) described 59 synonyms of *T. urticae* (Dosse and Boudreaux, 1963; Dupont, 1979; Hill et al., 1991; Meyer, 1981; Van de Vrie et al., 1972).

5.3.2 TETRANYCHUS NEOCALEDONICUS OR THE VEGETABLE MITE

Tetranychus cucurbitae is described as *Tetranychus neocaledonicus* (Jeppson et al., 1975). Gupta and Nahar (1981) recorded a serious attack by this mite in the brinjal crop. They also reported this mite from sponge gourd, bitter gourd, beans, tomato, and potato. Gupta (1985) reported this mite as one of the serious pests of vegetables in India. The vegetable mite produces white spots on leaves due to sucking of plant sap and as feeding continues the white spots coalesce, the leaves lose its green color, dry and drop. All this leads to decreased vitality, growth, flowering, and fruiting. The mites web profusely, covering the entire plant with thick sheath of webs. The eggs are spherical, translucent when laid, gradually turning brown. On hatching, the larva is light amber and slowly takes greeninsh tinge with dark lateral specks. The protonymph and deutonymph are green with dark specks on the dorsum and adult are carmine in color. The duration of different life stages in field conditions are as follows: incubation period 3–9 days, larva 3–5 days, protonymph 3–4 days, and deutonymph 2–5 days. The eggs of the vegetable mite are laid on the lower surface of the leaves, with female laying as many as 13 eggs daily and upto 60–90 eggs in lifetime (Jeppson et al., 1975).

The mite of all stages remains confined on the undersurface of the leaves covered with webs. The symptoms of damages are like other tetranychid mites. Sharma and Pande (1981) recorded lowest population during January and peak during May. Lal (1982) recorded very low population during December, which increased thereafter. The peak population was recorded from January–April. Singh and Singh (2015) reported that the mean population of *Tetranychus neocaledonicus* reached its maximum in first fortnight of September with 8.97 mite/leaf. It showed an increasing trend from 2nd fortnight of October with 8.33 mites/leaf to 11.17 mites/leaf and again with rising temperature from 2nd fortnight of February 10.6 mites/leaf to 22.37 mites in 1st fortnight of April. Mukherjee and Singh (1993) recorded this mite throughout the year on fig and mango plant.

5.3.3 TETRANYCHUS LUDENI *ZACHER OR RED-LEGGED SPIDER MITE*

Tetranychus ludeni is widespread in warmer parts of the globe. Singh and Mukherjee (1989) reported this species as most alarming mite pest of cowpea in Varanasi region. This species was earlier recorded as pest of vegetable crops from south India (Puttarudriah and ChannaBasavanna, 1959). ChannaBasavanna (1971) reported *Tetranychus ludeni* as an important pest of several vegetable crops in India. Meyer (1981) recorded the presence of this mite from about 70 plant species from South Africa. It was reported mostal from vegetable crops. In the early infestation, the yellowing of leaves can be observed and then necrotic patches. During heavy infestation of mites, bronzing effects on leaves are very clear. This occurs at later part of fruiting. The crowding of mites on tips of fruits is common at 40°C or above. Authors during field surveys have recorded the hot spots/balooning symptoms. This mite also occurs in mixed population with *T. urticae*. Identification of this mite is easy because it has dark red body with reddish legs. The females are oval in shape. The males are straw colored with pointed opisthosoma. The life cycle of mite is of short duration. It may be completed in 10 days during summer months. Coates (1974) recorded about 111 eggs by a fertilized female and that progeny consists on an average of 84 females and 27 males. This is the only tetranychid mite in India known to be vector of plant viral disease, Dolichos Enation Mosaic Virus—DEMV (Davis et al., 1982).

5.3.4 TETRANYCHUS MACFARLANEI *BAKER & PRITCHARD* *(SPIDER MITE)*

The tetranychid mite, *Tetranychus macfarlanei*, is a serious pest of vegetables and cotton in northern India. This mite builds up small colonies on the under surface of leaves. The sucking of plant cell sap by all the stages causes yellowing of leaves followed by drying. The infestation is more on mature crops than on the young ones. It disappears during the rainy season.

5.3.5 ACERIA LYCOPERSICI *(WOLFF)* OR TOMATO AND BRINJAL ERINEUM MITE

Aceria lycopersici (Wolff), tomato rust mite has a worldwide distribution and it attacks plants belonging to the family Solanaceae. It is an important pest of tomato and can easily kill the plants. Severe infestation may cause damage to chillies, potatoes, and brinjals. This orange yellow mite is fusiform and microscopically small. It thrives in warm weather and has no overwintering stage. In the winter it dies unless it has a suitable shelter. All stages of the rust mite are killed by temperature close to freezing. Favorable conditions for development are temperature of 26.5°C and a relative humidity of 30%. Light and photoperiod have little effect on the development of this mite (Jeppson et al., 1975; Meyer, 1981). The initial symptoms of damage to tomatoes are the curling of the lowest leaves and appearance of a silver gloss on the lower surfaces. Later these leaves become bronze-colored, withered, and eventually die. The mites also feed on the stem and before damage is noticeable on the top most leaves, the lower most portion of the stem loses its surface hair. The color of the stem changes from green to brown, and cracks appear on the surface. As the population builds up and the mites continue to spread, the browning of the leaves and stem increases and defoliation takes place (Gupta, 1985; Singh et al., 1987). The symptoms of tomato rust mite can easily be confused with those of late blight, but the leaves do not wilt in the same way as with blight (Meyer, 1981).

5.4   PEST STATUS AND DISTRIBUTION

Many species of *Tetranychus* are very serious mite pests in most parts of the world (Jeppson et al., 1975). It has a global distribution in the warmer parts and is an important pest of various crops like vegetables, ornamental plants, cereals, cotton, strawberry, deciduous fruits (Singh and Mukherjee, 1989). The population starts developing during late February to June and starts declining during July when monsoon comes (Gupta et al., 1976; Pande and Sharma, 1979; Singh, 1994, 1995; Singh and Singh, 1993). Mali et al. (1983), Singh and Singh (1992), and Mukherjee et al. (1992) recorded this mite on under surface of the leaves and found feeding by stippling the leaf tissue. The severely infested leaves become completely yellow to brown and fall off from the plants. Silver to yellowish appearance of leaves with webbing is a common symptom of this mite infestation. Van de Vrie et al. (1972) reported that neither *T. urticae* nor *T. cinnabarinus* produces silken threads or balooning threads for transportation. *T. urticae* attacked plants during hot and dry period, that is, April to June, resulting in significant yield loss (Gupta et al., 1971). On okra its peak activity was recorded in the months of April (Natarajan, 1989), January to April (Lal, 1982), May–June (Sharma and Pande, 1981; Singh and Singh, 1993), June (Putatunda and Tagore, 2003; Gulati, 2004), and August (Sunita, 1996). The peak population of red spider mite was reported on tomato from September to November and on brinjal, cucurbits, cowpea from May to

June (Dhooria, 2003); on brinjal in June and July (Khanna, 1991). Low to negligible mite population was encountered during winter.

5.5 DAMAGE AND YIELD LOSS

T. urticae is the most notorious pest responsible for significant yield losses in many economic crops, vegetables and fruit trees (Salman, 2007), ornamental and agronomic crops worldwide (James and Price, 2002). In India, *T. urticae* is reported as pest of vegetable crops from Haryana (Sunita, 1996;, Gulati, 2004; Kerala Lal, 1982), Punjab (Dhooria, 2003), Tamil Nadu (Nandagopal and Gedia, 1995), Uttar Pradesh (Singh and Singh, 1993), and West Bengal (Dhar et al., 2000). Apart from vegetable crops, the spider mites are also reported to cause economic losses in fruit crops like apple, citrus, pear (Chhillar et al., 2007), raspberry (Mariethoz et al., 1994), and strawberry (Congdon et al., 1993).

5.6 POPULATION DYNAMICS

Environmental factors play significant role in distribution and abundance of mites. It is well documented that warm and dry weather is favorable for the multiplication and spread of red spider mite (Jeppson et al., 1975). Temperature was found as regulatory factor for *T. urticae* build-up as most of the authors found positive correlation between mite population and temperature (Dhar et al., 2000; Putatunda and Tagore, 2003). Gulati (2004) revealed that *T. urticae* population showed positive correlation with maximum temperature and negative correlation with minimum temperature. But, Sunita (1996) reported positive correlation between mite population and minimum temperature. Haque et al. (2011) found that in vegetable plants temperature had direct positive impact on mite population on lady's finger, cucumber, brinjal, tomato, bottle gourd, bean, whereas negative on bitter melon, radish, and zinnia. As for relative humidity, positive correlation was reported by Pande and Yadav (1976) while nonsignificant negative correlation was observed by others (Singh and Singh, 1993; Dhar et al., 2000). Putatunda and Tagore (2003) also found no relation between *T. urticae* infesting okra and relative humidity. In brinjal, mite population increased significantly with increase of relative humidity but on radish mite number decreased significantly with increase of relative humidity

(Haque et al., 2011). Singh and Singh (1993) and Gulati (2004) reported that rainfall and sunshine hours to play no significant influence on mite population build up but Pande and Yadav (1976) recorded positive and negative correlation, respectively, between mite population and rainfall. Low wind velocity was also reported to favor the population build-up of tetranychid mites (Sadana and Kumari, 1987). These mites mostly remain in the field throughout the year on one or the other host but their number decreases during rainy and winter months (Natarajan, 1989; Gulati, 2004). In severe infestation, *T. urticae* density was recorded as 75–90 mites/leaf on cucumber (Souliotis, 1990). Afzal and Bashir (2007) recorded maximum mite population from brinjal (2.77) followed by tomato (2.55), pumpkin (1.1), and cucumber (0.91), respectively. Dutta et al. (2012) also recorded higher number of mites/leaf on cucumber (16.08), second to brinjal (32.27). On okra, it was recorded 11.9 mites/sq. cm leaf (Gulati 2004), 16.4 mites/sq. cm leaf (Sunita, 1996), 17.3 mites/sq. cm leaf (Khanna, 1991), 50.5 motile stages/sq. cm leaf (Dhar et al., 2000), 348 mites/leaf (Singh and Singh, 1993). Haque et al. (2011) studied seasonal abundance of spider mite *T. urticae* on vegetable and ornamental plants in Rajshahi and found that nine vegetable plants, viz., lady's finger, cucumber, brinjal, tomato, bottle gourd, bean, etc. recorded highest number of mites per leaf during the month of August, whereas Spinach recorded highest population in October, cowpea in November, pointed gourd and radish in December, and bitter melon in January, respectively. In cucumber, highest number of *T. urticae* eggs/female/day was recorded on the genotype Blackish Green while lowest on Winter Long Green (Ullah et al., 2006). Similarly, in okra, Pusa Sawani, Prabhani Kranti, and Varsha Uphar varieties were more prone to the attack than Arka Anamika (Gulati, 2004). In Varanasi, *T. neocaledonicus* showed positive correlation with maximum temperature ($r = +0.161$), negative correlation with minimum temperature ($r = -0.247$), significant negative correlation was found with morning relative humidity ($r = -0.581$) and evening relative humidity ($r = -0.717$) and the mite population was significantly negatively correlated with rainfall ($r = -0.576$) on brinjal crop (Singh and Singh, 2014).

5.7 HOST LEAF PREFERENCE

Maximum population of mites has been recorded from the middle strata followed by apical and bottom strata on okra (Gulati, 2004; Geroh, 2007) and brinjal (Gupta, 1991). Likewise, preference for mature leaves over young leaves by *T. urticae* was reported by Sunita (1996) and Sharmila et al. (1999). However, Onkarappa et al. (1999) observed more of mite population on bottom rather than on middle and top strata.

5.8   FEEDING HABITS AND NATURE OF DAMAGE

T. urticae is particularly dominant in intensive, high-yield cropping systems and affects crops by direct feeding, thereby reducing the area of photosynthetic activity and causing leaf abscission in severe infestations (Gorman et al., 2001). Puncturing of cells by mite stylets and injection of saliva causes mechanical damage, changes in cell cytology, physiological and biochemical processes of punctured as well as non-punctured adjacent cells. *T. urticae* feeding can damage protective leaf surfaces, stomata, and the palisade layer and the lowest parenchymal layer. It results in typical "stippling" damage, with white- or grayish-colored spots due to the punctures made by feeding (Park and Lee, 2002). Leaf bronzing, loss of chlorophyll, defoliation, and even plant death may occur due to direct feeding damage in severe infestation (Meyer and Craemer, 1999). The degree of leaf damage by *T. urticae* is a function of its stylet length and leaf thickness (Park and Lee, 2002). It mostly feeds on cell chloroplasts on the underside of the leaf, while the upper surface of the leaf develops whitish or yellowish stippling characteristic, which may join and become brownish as mite feeding continues. Indirect effects may include decreases in photosynthesis and transpiration (Park and Lee, 2002). This combination of direct and indirect effects often reduces the amount of harvestable material. Feeding causes significant changes in biochemical composition of leaves and fruits as reported in grapevine (Sivretepe et al., 2009) and beans (Farouk and Osman, 2012). Jayasinghe and Mallik (2010) recorded that tomato plants with early infestation and 12 weeks of mite feeding and late infestation with 9 weeks of mite feeding significantly lowered the total chlorophyll content (1.592, 1.597 mg/g) than other feeding durations. Plants on which 400 mites per plant were initially introduced had the lowest total chlorophyll, chlorophyll a, and chlorophyll b contents in 11th week (1.363, 0.986, and 0.377 mg/g) and 14th week (0.898, 0.693, and 0.205 mg/g), followed by 200, 100, and no mites (control plot).

5.9   BIOLOGY AND WEBBING BEHAVIOR

The two spotted spider mite, *T. urticae*, belongs to an assemblage of web spinning mites. The name "spider" highlights their ability to produce silk like webbing (Gerson, 1985). As mites move around, their webbing can be seen spun around leaves and stems. Heavy damage causes leaves to dry and drop, resulting in premature death of webbed plant (Park and Lee, 2005). Eggs are deposited beneath the webbing and larvae and nymphs develop within it. This webbing defines the colony boundaries, serves means of protection from rain, wind, and predators (Morimoto et al., 2006), and if dense enough to form a close knit, also protects from acaricide sprays. It is thought that the webbing and deposition of fecal pellets within the webbing is a mechanism to regulate humidity (Gerson, 1985). When the plant begins to decline, resulting in a reduced food supply, the mites enter a dispersal phase from sedentary phase and aggregate on the uppermost parts of the plants (Hussey and Parr, 1963), which includes both within plant and interplant movement. Aerial dispersal begins with the mites aggregating on the uppermost portions of the plants. The mites produce silken threads, which they use to "balloon" into the wind, carrying them long distances (Kennedy and Smitley, 1985). The egg-adult development of female *T. urticae* is completed in approximately 6.5 days at 30°C, while the males are reported to complete development earlier than females (Sabelis, 1981). Diapausing females or eggs are the most common overwintering stage for tetranychids in response to short day lengths and cooling temperatures. During diapause, *T. urticae* do not feed or oviposit, and they generally seek shelter in crevices in the bark of trees and shrubs, clods of dirt, and in leaf litter (Huffaker et al., 1969).

5.10   DAMAGE AND YIELD LOSS

Due to short life span, high reproductive potential, severe webbing, and resistance to insecticides, *T. urticae* is responsible for causing serious damage to various crops. According to Hussey and Parr (1963), cucumber crop may tolerate *T. urticae* damage up to 30% leaf surface without any yield loss. However, later it was reported that two-spotted spider mite infestation causes approximately 14% reduction of total leaf area resulting in significant yield loss in cucumber (Park and Lee, 2005). According to other reports, *T. urticae* is associated with 7.9–6.7 (Sugeetha and Sriniva-sana, 1999) and 26.5% to complete loss (Jeppson et al., 1975) in yield of okra crop. In brinjal, 28–31% reduction in yield was reported due to the attack of *T. cinnabarinus* (Palanisamy and Chelliah, 1987). In general, 5–11% yield loss in tea, 13–31% in brinjal, 23–25% in okra, 20–30% in cotton, and 36% in pointed gourd occur due to *T. urticae* (Gupta, 2003). Kropczynska and Tomczyk (1984) found that due to the feeding of *T. urticae*, growth of young chrysanthamum and carnation plants is distinctly inhibited. *T. urticae* infestation resulted in a reduction of 28.08, 20.53, and 14.37% in yield, fruit weight, and number of fruits per plant, respec-tively, in the greenhouse brinjal crop (Palanisamy and Chelliah, 1987). *T. urticae* caused significant losses in summer vegetables (Afzal and Bashir, 2007) and brinjal (Chhinniah et al., 2009). Kumar et al. (2010) reported that the two-spotted spider mite, *T. urticae*, assumes serious pest status, often causing 50–100% yield loss. A significant decrease in yield was also noticed when the mite level exceeded 0.2 mites/sq. cm leaf in brinjal (Patil, 2005).

5.11   HOST PREFERENCE AND RESPONSE

The growth of a spider mite population is modified within a genetically defined spectrum by various environmental factors. Among the environ-mental influences are climate, food availability, predation, and inter and intraspecific competition (Wermelinger et al., 1991). Plant species vary on the basis of their suitability as hosts for pests (insects and mites) in terms of survival and longevity. Host plants of spider mites differ in the degree of food quality, which either depend on the level of primary plant metabolites or on the quantity and nature of secondary metabo-lites. Many secondary metabolites found in plants have a responsibility in defense against herbivores, pests, and pathogens. These compounds can perform as toxins, deterrents, digestibility reducers, or act as precur-sors to physical defense systems (Balkema-Boomstra et al., 2003). Van den Boom et al. (2003) found that the plant species vary in their degree of acceptance by *T. urticae* population. Their results indicated that the plants including soybean, hop, golden chain, and tobacco are highly acceptable to the spider mite, because almost 100% of the spider mites stayed on the plant while eggplant and cowpea had a lower incidence of spider mite. The findings of Greco et al. (2006) also showed a high preference and a better performance of *T. urticae* on strawberry leaves than on onion, leek, and parsley leaves. This was shown not only in fecundity but also in the maximum number of offsprings settled, as well. Several potential mechanisms could be responsible for this phenomenon including plant nutritional quality of the host plant and morphological or allelochemical features (Balkema-Boomstra et al., 2003). Razmjou et al. (2009) evaluated the population growth characteristics of *T. urticae* with common bean, cowpea, and soybean as host plants. They found soybean as the most favorable host for two-spotted spider mites with rm= 0.296 (offsprings/female/day), followed by cowpea (0.242) and bean (0.230). The slowest population growth was observed on the bean species with rm = 0.214, thus indicating the choice of host plant species as a major factor in affecting how fast spider mite populations reach damaging levels in a culture. The effects of plant mineral nutrition was reviewed by Van de Vrie et al. (1972) as concerning spider mites, and found that N and K content of cucurbit leaves positively influenced populations of *T. neocaledonicus* while P content could not be related to the mite population size (Sharma and Pande, 1986). Of the macro-nutrients (nitrogen, phosphorus, and potassium), nitrogen shortage had the most distinct influence on mite population growth (Wermelinger et al., 1991). Fecundity was positively correlated with N and carbohydrate content of the leaves. The life table analysis showed a gradual decline in the intrinsic rate of natural increase (rm) with N and P deficiency. However, K contents yielded only minor differences in spider mite biology. Phenolic compounds as secondary plant metabolites are known to adversely affect pathogens and insect/mite populations. Total pheno-lics were strongly negatively related to spider mite development and oviposition as indicated by rm (Larson and Berry, 1984). Wermelinger et al. (1991) also reported negative correlation between spider mite fecun-dity and the phenolic content of leaves of apple trees. Upon infestation with *T. urticae*, lima bean plants respond with the emission of a mixture of volatiles attracting the predatory mite *Phytoseiulus persimilis*, which effectively eliminates local populations of spider mites. The induced production of carnivore-attracting volatiles has been recorded for over 20 plant species in 13 families (Dicke, 1999).

5.12   INTEGRATED MITE MANAGEMENT

The integrated pest management (IPM) programme minimizes pesticide use and manages pest population by other means like utilization of biological agents, resistant varieties cultural practices, and also by identifying ecofriendly pesticides. Integrated mite control (IMC)/integrated mite management (IMM) is an effective programme to keep mite population below economic injury level. The organic farming is another concept being augmented by acarologists for vegetable production in India due to high demands of farm produce by consumers in developed world. Breeding for resistance, multiple resistances, gene protection, and genetic engineering are important components of integrated production and should be included in IPM. The main hurdles for adoption of organic farming are lack of scientific information and fear for production security. IPM is one of the tools to substantiate the requirements of organic farming.

Synthetic acaricides have been widely used to manage *T. urticae* (Van Leeuwen et al., 2006). Indiscriminate use of pesticides has led to problems of resistance and environmental pollution and there is an increasing demand for sustainable, environmentally friendly control methods. These control methods based on the use of synthetic acaricides sometimes fail to manage mites population below economic threshold levels (Tirello et al., 2012).

Selective acaricides compatible with natural enemies can minimize negative effects on the environment (Steiner et al., 2011). Biological control of mites has been successfully achieved in glasshouses as an alternative method to chemical control (Osborne et al., 1985; Kropczynska et al.; 1999). Plant extracts-based acaricides, allelochemicals, which are chemical mediators used for interspecies communication between plants and herbivores (Regnault-Roger, 1997), that is, plant–insect interactions, are used as an alternative or complementary approach to synthetic insecticide treatments (Heuskin, 2011). These have evolved as broad spectrum chemical and physical defenses against diverse groups of insects (Ryan and Byrne, 1988) as when used as pesticides, these plant extracts affects pest behavior by repelling the pest or prohibiting feeding activity, and pest physiology, including molting and respiratory inhibition, growth and fecundity reduction, and also cuticle disruption (Saxena, 1989; Isman, 2000; Gokce et al., 2011) as well as varying mixtures of biosynthetically different compounds found in plant extracts can delay the evolution of resistance (Isman, 2000). These extracts are also biodegradable, showing minimal side effects on non-target organisms as well as on the environment (Isman, 2000, 2001; Tewary et al., 2005; Pontes et al., 2010; Cavalcanti et al., 2010) Many plants extract have been tested for the control of spider mites showing acaricidal properties.

5.13   CULTURAL CONTROL

High humidity reduces the reproductive potential of tetranychids whose optimal environment is provided by hot and dry air (Sabelis, 1986; Duso et al., 2004). Host plant nutrition may have a positive effect on the reduction of the mite population by varying the fertilizer regime applied to the crops (Markkula and Tiittanen, 1969), large amount of nitrogen or a deficiency of potassium can increase the amount of soluble nitrogen available in the plant, resulting sharp increases in the populations of mites (Sabelis, 1986). Since the plant response to such extreme regimes is economically unviable, so variations in host nutrition have not been used for pest control (Helle and Sabelis, 1985).

5.14   USE OF PREDATORY MITES FOR MANAGEMENT OF PHTYOPHAGOUS MITES

Gupta (1987) listed 139 phytoseiid mites from India with their taxonomic categories. Hoy (1981) and Hoy et al. (1982) reviewed the latest Phytoseiidae and their biological control by mites. Biological control provides an environmentally safe, cost-effective, and energy-efficient means of pest control either alone or as a component of IPM (Hoy et al., 1983). Phytoseiid mites are well accepted bioagents for phytophagous mite control. However, the families of Stigmaeidae, Erythraeidae, Tydeid, Anystidae, Cheyletidae, and Hemisarcoptidae also contain species that are predacious (Laing and Knop, 1983). McMurtry (1983) reported that most of the efficient predators of phytophagous mites in orchards belong to family Phytoseiidae. Rijn and Houten (1991) reported about successful biocontrol of thrips by the predatory mite, *Amblyseius cucumeris,* whereas another mite Amblyseius barkeri is less successful in controlling thrips in sweet pepper crops. Hluchy and Pespisil (1991) recorded tetranychid mites, *Tetranychus urticae*, and *Panonychus ulmi* as a serious pest in Moravian/ Czechoslovakia Vineyards and predatory mite *Typhlodromus pyri* was used successfully for biological protection of grone vines. Baier (1991) successfully used *Amblysieus barkeri* and *A. cucumeris* for the control of thrips and spider mite in Germany. McMurtry (1991) emphasized the use and augmentative release of *Phytoseiulus persimilis* for mite control on high value crops. This mite is most widely used species of predator for augmentative release and effectivity for a wide range of situations. Laing and Knop (1983) described the predatory mites from families Stegmaeidae, Erythraeidae, Tydeidae, Anystidae, Chetyletidae, and Hemisarcoptidae. Lo et al. (1990) reported more than 50 species of spider mites in Taiwan and recorded 30 species of native predators as natural enemies of spider mite. Mori et al. (1990) described the use of native predator *Amblyseius longispinosus* and introduced predator *Phytoseiulus persimilis* against spider mite and concluded that combined action of both native and introduced predators gave better suppression of tetranychid population under experimental conditions. McMurtry (1989) reported the use of Coccinellid, *Stethorus picipes*, as a main predator regulating abundance of *Oligonychus punicae*. Singh and Ray (1977) identified *Stethorus* beetle as a predator of *Tetranychus neocaledonicus*. Rasmy (1989) recommended biocontrol using predacious mites in mite pest management programme and suggested to follow IPM approach: (1) by utilizing naturally occurring predacious mites, (2) by releasing predatory mites, and (3) by utilizing pesticide-resistant strains of predacious mites.

Spooner Hart (1991) reported about the alarming situation of the two-spotted mite, *Tetranychus urticea,* a major pest of ornamental, fruit, and vegetable crops. He mentioned that withdrawal of miticides in 1987 due to high level of resistance paved way for the use of IMC. This utilizes predatory mites as a major tool of mite control and reduced usage of pesticides. Use of the Chilean predatory mite, *Phytoseiulus persimilis,* a monophagous predator of tetranychid has revolutionized IPM in Australia and Europe. He further mentioned that within a period of 12 years, *P. persimilis* has become one of the most widely used biological control agents in Australia and the backbone of many IMC programmes in horticulture. Osmelak and MacFarlane (1991) described the importance of *P. persimilis* for IPM of two-spotted mite in roses and carnations. Parker (1991) also emphasized on the widespread use of IMC in flower crops.

Borthakur and Das (1987) reported the occurrence of three species of predacious mites, viz., *Agistemus* sp., *Exothorhis caudata*, and *Cunaxa* sp. of tea mites. They further reported *Agistemus* sp. as an abundant predator on tea leaves in northeast India and expected it to be a very promising biocontrol agent for tea mite complex. Kumari and Sadana (1991) and Sadana (1985) identified *Amblyseius alstoniae* as a potential biological control agent for *Brevipalpus phoenicis* on grapevine, citrus, and guava. Sadana and Sharma (1988) recorded *Amblyseius finlandicus* predating on *Oligonychus mangiferus* and *Eutetranychus orientalis,* which infest grapevine and citrus. Mallik et al. (1989) recorded interaction between *Tetranychus ludeni* and its two phytoseiid predators, *Amblyseius longispinosus* and *Amblyseius tetranychivorus*. They further recorded that *A. longispinosus* prefers egg followed by the younger immature stages, whereas *T. tetranychivorus* prefers adult and older nymphal stages. Jose et al. (1989) found predatory mite, *Amblyseius alstoniae,* predatory thrips, *Scolothrips indicus*, and beetle, *Stethorus pauperculus*, feeding on all stages of spider mites. Singh et al. (1989) listed natural enemy complex of *A ceria litchii* and found *Amblyseius* as most prevalent followed by species of *Typhlodromus* and *Agistemus.*

Singh (1994) reported that *Phytoseiulus persimilis* Athias Henroit is very important and commercialized predatory mite and widely used for control of common injurious tetranychid mites. This mite was imported in India from the Glasshouse Crops Research Institute, Littlehampton, U.K. during 1984. Since then culture of this mite is being maintained at Project Directorate of Biological Control, Bangalore on *Tetranychus urticae,* which is most ideal. The bean plants are used for multiplication of *Tetranychus urticae.* The most preferred host of this predatory mite is *Tetranychus urticae,* followed by *Tetranychus cinnabarinus, Tetranychus ludeni*, and *Tetranychus neocaledonicus.* Krishnamoorthy (1989) found that introduced predator, *Phytoseiulus persimilis* feeds readily on *Tetranychus cinnabarinus, T. fifiensis, T. ludeni*, and *T. neocalidoniclls.* Borah and Rai (1989) recorded *Amblyseius ovalis* for the first time as a predator of *Bemisia tabaci.* Bhagat and Patnaik (1989) recorded a neuropteran, *Micromus timidlls* as a predator of spider mite. Sufficient information is available on predatory fauna of mites but unfortunately none of the predatory mites/insects are commercialized so far in India for the use as bioagents at farmer's field. Singh (1988) reported the methods for mass multiplication of predatory mites for large-scale production of *Amblyseius tetranychivorus.* Recently, Department of Entomology & Agricultural Zoology, Banaras Hindu University, has organized training programme on IPM in collaboration with Hawkesbury IPM service, University of Western Sydney, Australia (Anonymous, 1994).

5.15   ECOCHEMICAL CONTROL

As an alternative strategy, chemical mediators, especially allelochemicals, are used in interspecies communication. These non-nutritional molecules, produced by one organism, can modify the behavior or the biology of an organism from another species. Consequently, plant allelochemicals exert a wide range of influence on insects: they can be repellent (Ojimelukwe and Adler, 1999; Huang et al., 2002), deterrent, or antifeedant (Tapondjou et al., 2005), and they also seem to be important agents of interspecific communication, as they favor pollination by attracting insects. In addition, they may increase oviposition or, conversely, decrease reproduction by ovicidal and larvicidal effects (Kim, 2003; Obeng-Ofori and Amiteye, 2005), classified as secondary plant metabolites and have chemical structures that identify them as alkaloids, polyphenolics, terpenes, isoprenoids, cyanogenic glycosides (Strebler, 1989), or tannins (Chiasson and Beloin, 2007). The use of these allelochemcials as an alternative strategy based on the identification of plant insecticidal molecules was traditionally used in order to protect crops (Golob and Webly, 1989) as several active molecules have been extracted from plants, nicotine from tobacco, rotenone from *Derris elliptica* (Fabaceae), and pyrethrum from Chrysanthemum (Compositeae)

5.16   THE ESSENTIAL OILS

Among the molecules extracted from plants, essential oils are obtained by hydrodistillation, steam distillation, dry distillation, or mechanical cold pressing of aromatic plants (Bakkali et al., 2008). Plant essential oils are produced commercially from several botanical sources, particularly those belonging to the mint family (Lamiaceae). They can be synthesized by all plant organs, such as buds, flowers, leaves, stems, twigs, seeds, fruits, roots, wood, or bark, and are, for some species, stored in secretory cells, cavities, canals, epidermic cells, or glandular trichomes (Isman, 2001). Repellence is property of essential oils, as some contain numerous secondary metabolites that can deter attacks from insects and general herbivores (Isman, 2000).

Contact acaricidal activity of plant extracts have been well documented against several species of mites (Isman, 2000, 2001; Gokce et al., 2011), including *T. urticae* (Roh et al., 2011).

5.17 ECOFRIENDLY APPROACHES IN MITE MANAGEMENT

Jeppson et al. (1975) described that sulphur has been extensively used since 1920 to control mites. Eriophyidae and Tenuipalpidae are susceptible to sulphur. Most of the tetranychid mites are not readily controlled by sulphur. Still sulphur has been regarded as one of the very safe acaricides. The dinitrophenyl acaricides were used during 40s. DNOCHP that is popularly known by growers as DN-dry mix No. 1 had been used in Florida. Another commercial formulation DN 111 had been used for red spider mites.

The DN acaricides had many disadvantages like sulphur because they produced phytotoxic effects when applied during hot weather. During the late 40s, many chlorinated hydrocarbon acaricides appeared. DCPM (neotran) was found more effective than DN 111. The ovex and aramite were extensively used throughout the world for control of tetranychid mite infesting vegetables. The dicofol was introduced in 1952. This acaricide has been extensively used in tea gardens of India. Commercially, this formulation is available as kelthane, hilfol, etc. The acaricides like mores tan, cyhexatin (plictran), and chlorodimeform (galecron) which emerged after 60s are not being frequently used in India. Phosalone (zolone), dinocap (karathane), ethion (phosmite), tetradiform (tedion) are also not very popular acaricides or used moderately.

Insectoacaricides from the group of organophosphorus compounds are often used in India for mite control. Synthetic pyrethroid (danitol) is also as effective as insectoacaricides. Ray and Rai (1981) found phosphamidon as better insecticide for control of *Tetranychus neocalidonicus.* Raja et al. (1995) tested efficacy of certain chemicals to control sugarcane sheath mite, *Aceria sacchari,* a pest of ratoon sugarcane crop in Tamil Nadu. They found ethion and monocrotophos as effective chemicals to control sheath mite.

The mite control through acaricides is not very common in India. Still many farmers are using pesticides for wide coverage of pest control including mites. Singh and Singh (1992) evaluated some pesticides against *T. cinnabarinus*. Mukherjee et al. (1992) evaluated many insecticides/acaricides against *T. cinnabarinus*. Singh and Mukherjee (1989) reported that due to continuous use of dimethoate for pest coverage on cowpea resulted in outbreak of *Tetranychus ludeni* in Varanasi region. One of the components of integrated mite control is identification of ecofriendly acaricides, which can give safe coverage to predatory fauna also. This can be achieved by monitoring carefully the fauna of crop agroecosystem for several weeks after using these chemicals to make sure that sufficient predators survive. Therefore, use of safe miticide is always recommended at time of operation flux of outbreaks. Mansour and Ascher (1983) reported the effect of neem seed kernal extract on *T. cinnabarinus* particularly on fecundity and mortality. Mansour et al. (1987) reported the effect of neem seed kernal extract on *Phytoseiulus persimilis* and *T. cinnabarinus*. All extracts of neem were much more toxic to phytophagous mite in comparison to predatory mite. Patel et al. (1993) compared the efficacy of different pesticides including botanical pesticides and found that plant products gave better results against *T. cinnabarinus* and was comparable with the most effective pesticides/acaricides tested. Frequent application of acaricides from same chemical group hastens the development of resistance.

Most of the chlorine-based acaricides are hazardous to the predators. Waiting period between last spray and predator release should not be less than 14 days and should not be used until predators have established in the plants. Kelthane and synthetic pyrethroids are very hazardous to predators and should not be used at all. The use of these chemicals should be restricted to spot spraying only.

5.18 ACARICIDAL MANAGEMENT

Fortunately, the acaricides resistance has not developed so far due to their limited use in India. However, many reports are available on pesticides resistance in spider mites 17 (Dittrich, 1975; Helle, 1965; Jeppson, et al., 1975; Van de Vrie, 1973). Herne et al. (1979) mentioned that spider mites are potential danger for many crops because of their ability to develop resistance to chemicals, which initially give effective control.

Cranham and Helle (1985) discussed the pesticides resistance in Tetrany-chidae. Goodwin et al. (1991) reported about the miticide resistance in two-spotted mite which had caused serious difficulties in maintaining product quality of flowers in Australia. They further pointed out that reliance of growers on miticides to provide economic control of two-spotted mite was a failure due to development of resistance. They suggested that chemical companies must undertake testing of their products against predatory mites and other biological control agents keeping in view the resistance problem and deleterious effects on non-target organisms. Hoy et al. (1980) and Hoy and Knop (1981) reported increased tolerance in phytoseiid mites against pyrethroids. Carolyn and Spooner Hart (1991) reported tolerance in *Phytoseilus persimilis* against improved organophosphate.

During recent years, the development of plastic culture particularly for growing vegetables and flowers under plastic cover or glass house conditions possibly have accelerated the problem of resistance in spider mites. Therefore, continuous watch is required to avoid the failure of a chemical product. Many times cross-resistance may also occur. Cardwell et al. (1987) suggested a rapid bioassay method that can detect build-up of resistance and help growers to avoid the use of acaricides. In India, tea research units may follow up the bioassay method to detect acaricides resistance against *Oligonychus coffeae,* which is one of the serious pests in tea gardens (Banerjee and Cranham, 1985).

5.19   MITE OUTBURST

Mite tumults are most often associated with blossoming and fruiting stages of the host plants rather than with the vegetative growth stage (Huffaker et al., 1969). Cadapan (1976) reported mite population increase most rapidly during pod and seed development period, attaining population level of over 1000 individuals per leaf. The commotion phenomenon has often been related directly to plant nutrition (Cannon and Connell, 1965). The mite injury and outbursts are associated with hot and dry weather on numerous field crops. Singh and Mukherjee (1989) recorded outbreak of *Tetranychus ludeni* on cowpea during May and June months in Varanasi region. Coppious webbing is characterized with hot spot symptoms in *T. ludeni*. Poe (1980) reported mite outbreak in soybean.

Mite outbursts may occur due to indiscriminate use of insecticides which destroy natural predatory fauna (Duncombe, 1972; Ghobril and Dittrich, 1980; Meyer, 1981). The use of pyrethroid insecticides in recent years has stimulated spider mite outbreaks on cotton. Banerjee and Cranham (1985) reported outbreak of *Oligonychus coffeae* on tea in northern India more commonly than in southern India and Sri Lanka. They further mentioned that carbaryl increased spider mite. Singh and Mukherjee (1989) reported *T. ludeni* as an alarming mite on cowpea in Varanasi region due to continuous use of dimethoate, a wrong insecticide for mite control. Dhooria and Sagar (1975) reported an outbreak of *Tetranychus cinnabarinus* on squah melon. Carbaryl and endrin application on squash-melon crop increased the fecundity and killed the natural enemies, respectively, of the mites. Husseine (1958) reported induction of outbreaks of *T. ludeni* on cotton in Egypt as a result of over use of carbryl and endrin. Phosalone, endosulphon, and neem oil showed no resurgence of mite pest, whereas cypermethrin, fenvalerate, and carbaryl did so on brinjal. Singh et al. (1987) recorded spider mite, *T cinnabarinus* as induced pest of cotton in Punjab. On cotton, Sandhu et al. (1987) reported the occurrence of secondary pests, due to insecticidal sprays. Penman and Chapman (1988) reported pyrethroid-induced spider mite outbursts. The problem of pyrethroid-induced mite outbursts has been recognized since early trial works with these chemicals (Hoyt et al., 1978; Plaut and Cohen, 1978; Zwich and Fields, 1978). Huffaker et al. (1970) reported that there are other factors also responsible for mite outbursts in varied situations. Kropczynska et al. (1988) described that one of the factors responsible for outbursts of *Eotetranychus tiiiarium* on *Telia* trees in towns was due to change in predaceous mite population in contrast to its natural forest habitats. They identified *Paraseiuius soieiger* as most effective predaceous mite. The use of nitrogenous fertilizers and other management practices (irrigation) may also enhance the population of spider mites on cotton (Leigh et al., 1969).

KEYWORDS

- integrated mite management
- phytophagous mite
- cucurbitaceous vegetable
- solanaceous vegetable
- cruciferous vegetable,

REFERENCES

Afzal, M.; Bashir, M. H. Influence of Certain Leaf Characters of Some Summer Vegetables with Incidence of Predatory Mites of the Family Cunaxidae. *Pak. J. Bot.* **2007,** *39* (1), 205–209.

Agarwal, L.; Singh, J. Present Status of *Eriophyes mangiferae* in Eastern Uttar Pradesh, India. In *Progress in Acarology*; ChannaBasavanna, G. P.; Viraktamath, C. A., Eds, Vol. II, Oxford & IBH, New Delhi, pp. 109–114.

Agarwal, R. A. The Surgarcane Mite *(Schizotetranychus andropogoni* H) and Its Control. *Ind. Sugar* **1957,** *7,* 394–399.

Ahuja, D. B. Seasonal Incidence and Chemical Control of Oriental Mite, *Eutetranychus orientalis* (Klein) on Castor. *Ind. J. Ent.* **1994,** *56,* 1–5.

Anonymous. Mites of Agricultural Importance in India and Their Management. *Technol. Bull. No.1, AICRP on Agricultural Acarology, Bangalore.*

Anonymous. A Guide to the Agriculturally Important Mites of India with Illustrated Keys and Field Keys for Their Easy Identifications. *Technol. Bull. No.2, AlCRP on Agricultural Acarology, Bangalore.*

Anonymous. Training Programme on IPM of mites. *Pestology* **1994,** *18,* 53.

Baier, B. Relative Humidity—A Decisive Factor for the Use of Oligophagous Predatory Mites in Pest Control. In: *Modent Acarology*; Dusbabek, F.; Bubva, V., Eds; Academia: Prague, pp. 661–665.

Baker, E. W.; Tuttle, D. M. The False Spider Mites of Mexico (Tenuipalpidae: Acari). *U.S. Department of Agriculture, Technical Bulletin No. 1706,* 1987.

Bakkali F.; Averbeck, S.; Averbeck, D.; Idaomar, M. Biological Effects of Essential Oils—A Review. *Food Chem. Toxicol.* **2008,** *46,* 446–475.

Balkema-Boomstra, A. G.; Zijlstra, S.; Verstappen, F. W. A.; Inggamer, H.; Mercke, P. E.; Jongsma, M. A.; Bouwmeester, H. J. Role of Cucurbitacin-C in Resistance to Spider Mite (*Tetranychus urticae*) in Cucumber (*Cucumis sativus* L.) *J. Chem. Ecol.* **2003,** *29* (1), 225–235.

Banerjee, B. *An Introduction to Agricultural Acarology.* Associated Publishing Co., New Delhi, 1988.

Banerjee, B.; Cranham, J. E. Control of Tetranychidae in Crops-Tea. In: *World Crop Pests, Spider Mites, Their Biology, Natural Enemies and Control*; Helle, W.; Sabelis, M. W., Eds; Elsevier, Amsterdam, 1985, pp. 371–374.

Bhagat, K. C.; Patnaik, M. C. Biological Studies and Host Range of *Micromus finidus* (Neuroptera: Hemerobiidae). A Predator of the Common Spider Mite, *Tetranychus telarius* (Acari: Tetranychidae). In: *Progress in Acarology*; ChannaBasavanna, G. P.; Viraktamath, C. A., Eds; Oxford & IBH: New Delhi, pp. 381–384.

Bindra, O. S.; Singh, G. Mite Pests of Ornamental Plants in India. *Pesticides* **1970**, *4*, 17–24.

Bindra, O. S.; Verma, G. C. A Study in the Control of the Fig Mite, *Eotetranychus hirsti* Pritchard & Baker. *J. Res. Punjab Agric. Univ.* **1966**, *3*, 417–420.

Boczek, J. H.; Shevtchenko, V. G.; Davis, R. *Genetic Key to Worldfauna of Eriophyid Mites* (Acarina: Eriophyoidea). Warsaw Agricultural University Press, Warsaw, Poland, 1989.

Borah, D. C.; Rai, P. S. Potentiality of *Amblyseius ovalis* (Acari: Phytoseiidae) as a Biological Control Agents on *Bemisia tabaci* (Homoptera: Aleyrodidae). In: *Progress in Acarology*; ChannaBasavanna, G. P.; Viraktamath, C. A., Eds; Oxford & IBH, New Delhi, 1989, pp. 375–379.

Borthakur, M.; Das, S. C. Studies on Acarine Predators of Phytophagous Mites on Tea in North-East India. *Two and a Bud* **1987**, *34*, 21–24.

Brandenburg, R. L.; Kennedy, G. G. Ecological and Agricultural Considerations in the Management of Two Spotted Spider Mite *(Tetranychus urticae* Koch). In: *Agricultural Zoology Reviews* Russell, G. E., Ed.; Intercept, Wimbome, 1987, pp. 185–236.

Bull, D. *A Growing Problem: Pesticides and the Third World Poor.* Oxford, U.K., OXFAM, 1982.

Cadapan, E. P. The Effect of the Two Spotted Spider Mite and Several Insects on the Yield of Soybean. Ph.D. Thesis, University of California, Berkeley, 1976.

Cannon, W. N.; Connell, W. A. Populations of *Tetranychus atlanticus* McG. (Acarina: Tetranychidae) on Soybean Supplied with Various Levels of Nitrogen, Phosphorous and Potassium. *Entomol. Exp. Appl.* **1965**, *8*, 153–161.

Capoor, S. P. Observations on the Sterility Disease of Pigeonpea Pea. *Ind. J. Agric. Sci.* **1952**, *22*, 271–274.

Cavalcanti, S. C.; Niculau Edos, S.; Blank, A. F.; Caˆmara, C. A.; Arau´jo, I. N.; Alves, P. B. Composition and Acaricidal Activity of *Lippia sidoides* Essential Oil Against Two-Spotted Spider Mite (*Tetranychus urticae* Koch). *Biores. Technol.* **2010**, *101*, 829–832.

ChannaBasavanna, G. P., Ed. *Contribution to Acarology in India.* Acarological Society of India, Bangalore, India, 1981.

ChannaBasavanna, G. P. *A Contribution of the Knowledge 'of Indian Eriophyid Mite (Eriophyoidea: Trombidiformes, Acarina).* Univ. Agric. Sci. Bull. Bangalore, India, 1966.

ChannaBasavanna, G. P. In The Present Status of Our Knowledge of Indian Plant Feeding Mites. Proceedings of the 3rd International Congress of Acarology, Prague, pp. 201–204.

ChannaBasavanna, G. P. Plant Feeding Mites in India. In: *Souvenir; VlllInternational Congress of Acarology,* Bangalore, India, 1986, pp. 21–30.

Chen, C. N.; Cheng, C. C.; Heiao, K. C. Bionomics of *Steneotarsonemus spinki* Smiley Attacking Rice Plants in Taiwan. *Recent Adv. Acarol.* **1979,** *1,* 111–117.

Chinniah, C.; Kumar, S. V.; Muthiah, C.; Rajavel, D. S. Population Dynamics of Two Spotted Spider Mite, *Tetranychus urticae* Koch in Brinjal Ecosystem. *Karnataka J. Agric. Sci.* **2009,** *22,* 734–735.

Coates, T. J. D. The Influence of Some Natural Enemies and Pesticides on Various Populations of *Tetranychus cinnabarinus* (Boisduval.), *T. lombardinii* Baker & Pritchard and *T. ludeni* Zacher (Acari: Tetranychidae) with Aspects of Their Biologies. *Entomol. Mem. Dep. Agric. Tech. Servo Repub.* S. *Afr.* **1974,** *42,* 1–40.

Cranham, J. E.; Helle, W. Pesticide Resistance in Tetranychidae. In: *World Crop Pests, Spider Mites, Their Biology, Natural Enemies and Control,* Helle, W.; Sabelis, M. W., Eds; Elsevier, Amsterdam, 1985, pp. 405–432.

Das, B.; Chakrabarti, S. Eriophyid Mites (Acari: Eriophyoidea) of North-East India—Some Aspects of Their Evolution and Host Associations. In *Progress in Acarology;* ChannaBasavanna, G. P.; Viraktamath, C. A., Eds; Oxford & IBH: New Delhi, 1988, pp. 391–394.

Das, G. M. Pests of Tea in North-East India and Their Control. *Tocklai Expt. Stn. Mem.* **1965,** 27.

Das, G. M.; Sengupta, M. Observations on the Pink Mite *Acaphylla theae* (Watt) Keifer, of Tea in North-East India. *J. Zool. Soc. India* **1958,** *10,* 39–48.

Davis, R.; Flechtmann, C. H. W.; Boczek, J. H.; Barke, H. E. *Catalogue of Eriophyid Mites (Acari: Eriophyoidea),* Warsaw Agricultural University Press, Warsaw, Poland, 1982.

Desai, M. K.; Chavda, D. E. "Mite" *(Oligonychus),* as a Causal Agent for *"Ratada"* Disease of Sorghum. *Poona Agri. Coll. Mag.* **1955,** *45,* 138–141.

Dhar, T.; Dey, P. K.; Sarkar, P. K. Influence of Abiotic Factors on Population Build-Up of Red Spider Mite, *Tetranychus urticae* on Okra *vis a vis* Evaluation of Some New Pesticides for their Control. *Pestology* **2000,** *24* (9), 34–37.

Dhooria, M. S. Development of Citrus Mite, *Euteranychus orientalis* (Acari: Tetranychidae) as Influenced by Age and Surface of Leaves of Different Hosts. *Ind. J. Acarol.* **1984,** *9,* 82–88.

Dhooria, M. S. Effect of Releasing Different Levels of Citrus Mite, *Eutetranychus orientalis* (Klein) (Acarina: Tetranychidae) on Growth of French Bean. *J. Insect Sci.* **1994,** *7,* 210.

Dhooria, M. S.; Bhutani, D. K. Seasonal Incidence of Citrus Mite, *Eutetranychus orientahus* and its Predators. *Ind. J. Acar.* **1983,** *7,* 59–62.

Dhooria, M. S.; Sagar, P. An Outbreak of *Tetranychus cinnabarinus* (Acarina: Tetranychidae) on Squash-melon and Its Control. *Int. J. Acarol.* **1975,** *1,* 6–9.

Dicke, M. Evolution of Induced Indirect Defence of Plants. *In The Ecology and Evolution of Inducible Defenses;* Tollrian, R.; Harvell, C. D., Eds; Princeton University Press, Princeton, 1999, pp. 62–88.

Ditchfield, C.; Spooner-Hart, R. Improved Organophosphate Tolerance in *Phytoseiulus persimilis* Athian-Henriot for Integrated Mite Control in Horticultural Crops. Proceedings of the 1st National Conference, Australian Society of Horticultural Science, Macquarie University, Sydney, 1991, pp. 487–496.

Dittrich, V. Acaricide Resistance in Mites. *Z. Angew. Entomol.* **1975,** *78,* 28–45.

Dosse, G.; Boudreaux, H. B. Some Problems of Spider Mite Taxonomy Involving Genetics and Morphology. In *Advances in Acarology,* Vol. I, Cornell University Press, 1963, pp. 343–349.

Duncombe, W. G. Red Spider Mite on Cotton and Its Control. *Rhodesia Agric. J.* **1972,** *69,* 7–10.

Dupont, L. H. On Gene Flow Between *Tetranychus urticae* Koch 1836 and *Tetranychus cinnabarinus* (Boisdmal) Boudreaux 1956 (Acari: Tetranychidae): Synonym Between the Two Species. *Entomol. Exp. Appl.* **1979,** *25,* 297–303.

Duso, C.; Malagnini, V.; Pozzebon, A.; Castagnoli, M.; Liguori, M.; Simoni, S. Comparative Toxicity of Botanical and Reduced-Risk Insecticides to Mediterranean Populations of *Tetranychus urticae* and *Phytoseiulus persimilis* (Acari Tetranychidae, Phytoseiidae). *Biol. Control* **2008,** *47,* 16–21.

Dutta, N. K.; Alam, S. N.; Uddin, M. K.; Mahmudunnabi, M.; Khatun, M. F. Population Abundance of Red Spider Mite in Different Vegetables Along With Its Spatial Distribution and Chemical Control in Brinjal, *Solanum melongena* L. *Bangladesh J. Agril. Res.* **2012,** *37* (3), 399–404.

Ehara, S. *A Guide to the Tetranychid Mites of Agricultural Importance in Japan.* Represented JEBP Synthesis, Tokyo, 1975.

Farouk, S.; Osman, M. A. Alleviation of Oxidative Stress Induced by Spider Mite Invasion Through Application of Elicitors in Bean Plants. *Egyptian J. Biol.* **2012,** *14,* 1–13.

Gangwar, S. K.; Lal, L. Phytophagous Mites in the North Eastern Hill Region of India. *Trop. Pest Mang.* **1988,** *34,* 438–440.

Geroh, M. Ecology and Management of *Tetranychus urticae* Koch on Okra, *Abelmoschus esculentus* L. M.Sc. Thesis, CCS HAU, Hisar, 2007.

Gerson, U. Webbing. In: *Spider Mites, Their Biology, Natural Enemies and Control*; Helle, W.; Sabelis, M. W., Eds; Elsevier: Amsterdam, The Netherlands, 1985, pp. 223–232.

Ghai, S. Mites. In: *Entomology in India.* Entomological Society, India, New Delhi, 1964, pp. 385–396.

Ghai, S.; Baker, E. W. Notes on the Gall Forming Mite, *Larvacarus transitans* (Tenuipalpidae) on *Ziziphus mauritiana* in India. In: *Progress in Acarology*; ChannaBasavanna, G. P.; Viraktamath, C. A.; Oxford & IBH: New Delhi, 1989, pp. 81–89.

Ghai, S.; Shenhmar, M. A Review of the World Fauna of *Tenuipalpidae* (Acarina: Tetranychidae). *Oriental Insects* **1984,** *8,* 99–172.

Ghobrial, A.; Dittrich, V. Early and Late Pest Complexes on Cotton, Their Control by Aerial and Ground Application of Insecticides and Side Effects on the Predator Fauna. *Z. Angaw. Entomol.* **1980,** *90,* 306–313.

Ghosh, N. K.; Das, B.; Chakrabarti, S. Injury by Gall Mites (Acari: Eriophoidea) to Plants in North East India. In *Progress in Acarology*; ChannaBasavanna, G. P.; Viraktamath, C. A.; Oxford & IBH: New Delhi, 1989, pp. 135–140.

Gokce, A.; Isaacs, R.; Whalon, M. E. Ovicidal, Larvicidal and Antiovipositional Activities of *Bifora Radians* and Other Plant Extracts on the Grape Berry Moth *Paralobesia viteana* (Clemens). *J. Pest. Sci.* **2011,** *84,* 487–493.

Golob, P.; Webly, D. J. The Use of Plants and Minerals as Traditional Protectants of Stored Products. *Trop. Prod. Inst.* **1989,** *138,* 1–32.

Goodwin, S.; Gough, N.; Herron, G. Miticide Resistance and Field Control in TwoSpotted Mite, *Tetranychus urticae* Koch, Infesting Roses. Proceedings of the 1st National

Conference, Australian Society of Horticultural Science, Macquarie University, Sydney, 1991, pp. 325–328.

Gorman, K.; Hewitt, F.; Denhoim, I.; Devine, G. New Developments in Insecticide Resistance in the Greenhouse Whitefly (*Trrialeurodes vaporariorum*) and the Two-Spotted Spider Mite (*Tetranychus urticae*). *UK Pest Manag. Sci.* **2001,** *58,* 123–130.

Grafton-Cardwell, E. E.; Granett, J.; Dennehy, T. J. Quick Tests for Pesticide Resistance in Spider Mites. *Calif. Agric.* **1987,** 8–10.

Greathead, D. J.; Waage, T. K. Opportunities for Biological Control of Agricultural Pests In Developing Countries, Washington DC, USA. World Bank Technical paper No. 11, 1983.

Gupta, S. K. Fruit Mites of India. *Pesticides* **1974,** *8,* 46–52.

Gulati, R. Incidence of *Tetranychus cinnabarinus* Infestation on Different Varieties of *Abelmoschus esculentus. Ann. Pl. Protec. Sci.* **2004,** *12,* 45–47.

Gupta, S. K. The Mites of Agricultural Importance in India with Remarks on Their Economic Status. *Modern Acarol.* **1991,** *1,* 509–522.

Gupta, S. K. Mite Pests of Agricultural Crops in India, Their Management and Identification. In *Mites, Their Identification and Management*; Yadav, P. R.; Chauhan, R.; Putatunda, B.N.; Chhillar, B. S., Eds; CCSHAU, Hisar, 2003, pp. 48–61.

Gupta, S. K. Mites of the Genus *Amblyseius* (Acarina: Phytoseiidae) from India with Descriptions of Eight New Species. *Int. J. Acarol.* **1975,** *1,* 26–45.

Gupta, S. K. *Hand Book-Plant Mite of India.* Zoological Survey of India, Calcutta, 1985.

Gupta, S.K. *A Taxonomic Review of Oriental Phytoseiidae with Key to Genera and Species.* Records of the Z.S.I. Miscellaneous Pub. Occasional paper No. 95. Zoological Survey of India, Calcutta, 1987.

Gupta, S. K. 1991. The Mites of Agricultural Importance in India with Remark on Their Economic Status. In: *Modern Acarology*; Dusbabek, F.; Bukva, V.; Academia: Prague, 509–522.

Gupta, S. K.; Gupta, Y. N. A Taxonomic Review of Indian Tetranychidae (Acari: Prostigmata) with Description of Known Species and Keys to Genera and Species. *Memoirs of the Zoological Survey India* **1994,** *18,* 196.

Gupta, S. K.; Nahar, S. C. Plant Mites (Acari) of Agricultural Importance in Bihar. In *Contribution to Acarology in India*; ChannaBasavanna, G. P., Ed.; Acarological Society of India: Bangalore, 1981, pp. 6–11.

Gupta, S. K.; Dhooria, M. S.; Sidhu, A. S. Seasonal Abundance of *Tetranychus telarius* (Linn.) on Castor in the Punjab. *Oilseed J.* **1976,** *6,* 16–18.

Gupta, S. K.; Sidhu, A. S.; Singh, G. Occurrence of a Tenuipalpidae Mite on Citrus in the Punjab and its control. *Indian J. Ent.* **1971,** *33,* 30–33.

Gutierrez, J. Monitoring Techniques. In: *World Crop Pests, Spider Mites, Their Biology, Natural Enemies and Control*; Helle, W.; Sabelis, M. W., Eds; Elsevier: Amsterdam, 1985, pp. 351–353.

Haque, M.; Islam, T.; Naher, N.; Haque, M. M. Seasonal Abundance of Spider Mite *Tetranychus urticae* Koch on Vegetable and Ornamental Plants in Rajshahi. *Univ. J. Zool. Rajshahi Univ.* **2011,** *30,* 37–40.

Helle, W. Resistance in the Acarina: Mites. In *Advances in Acarology*; Naegle, J. A., Ed.; Comstock Publishers, Ithaca: New York, pp. 71–93.

Helle, W.; Sabelis, M. W., Eds. *World Crop Pests, Spider Mites, Their Biology, Natural Enemies and Control.* Elsevier, Amsterdam, 1985.

Helle, W.; Overmeer, W. P. J. Rearing Techniques. In: *World Crop Pests, Spider Mites, Their Biology, Natural Enemies and Control*; Helle, W.; Sabelis, M. W.; Elsevier: Amsterdam, pp. 331–335.

Herne, D. H. C.; Cranham, J. E.; Easterbrook, M. A. New Acaricides to Control Resistant Mites. In: *Recent Advances in Acarology*; Rodriguez, J. G., Ed.; Academic Press, New York, USA, 1979.

Heuskin, S. Contribution to the Study of Semiochemical Slow Release Formulations as Biological Control Devices. P.hD Thesis, Universite´ de Lie`ge, Gembloux, Agrobiotech, Belgium, 2011.

Hill, R. L.; Donnell, D. J.O. Reproductive Isolation Between *Tetranychus lintearius* and Two Related Mites, *T. urticae* and *T. turkestani* (Acarina: Tetranychidae). *Exp. Appl. Acar.* **1991**, *11*, 241–251.

Hluchy, M.; Pospisil, Z. Use of the Predatory Mite *Typhlodromus pyri* Schenter (Acari: Phytoseiidae) for Biological Protection of Grape Vine from Phytophagous Mites. In *Modem Acarology* Dusbabek, F.; Bukva, V., Eds; Academia: Prague, 1991, pp. 655–660.

Hoy, M. A., Ed. *Recent Advances in Knowledge of the Phytoseiidae.* Division of Agricultural Science, University of California, 1981.

Hoy, M. A.; Knop, N. F. Selection for and Genetic Analysis of Permethrin Resistance in *Metaseiulus occidentalis:* Genetic Improvement of a Biological Control Agent. *Entomol. Exp. Appl.* **1981**, *30*, 10–18.

Hoy, M. A.; Castro, D.; Cahn, D. Two Methods for Large Scale Production of Pesticide Resistance Strains of the Spider Mite *Metaseiulus occidentalis* (Nesbitt.) (Acarine: Phytoseiidae). *Z. Angew. Entomol.* **1982**, *94*, 1–9.

Hoy, M. A.; Knop, N. F.; Joos, J. L. Pyrethroid Resistance Persists in Spider Mite Predator. *Calif. Agric.* **1980**, *34*, 11–12.

Hoy, M. A.; Westigard, P. H.; Hoyt, S. C. Release and Evaluation of a Laboratory Selected, Pyrethroid-Resistant Strain of the Predacious) 1lite *Metaseiulus occidentalis* (Acari: Phytoseiidae) in Southern Oregon Pear Orchards and a Washington Apple Orchard. *J. Econ. Entomol.* **1983**, *76*, 383–388.

Hoyt, S. C.; Westigard, P. H.; Burts, E. C. Effects of Two Synthetic Pyrethroid Insecticides on the Codling Moth, Pear Psylla, and Various Mite Species in North West Apple and Pear Orchards. *J. Econ. Entomol.* **1978**, *71*, 431–434.

Huang, Y.; Ho, S. H.; Lee, H. C.; Yap, Y. L. Insecticidal Properties of Eugenol, Isoeugenol and methyleugenol and Their Effects on Nutrition of *Sitophilus zeamais* Motsch. (Coleoptera: Curculionidae) and *Tribolium castaneum* (Herbst) (Coleoptera: Tenebrionidae). *J. Stored Prod. Res.* **2002**, *38*, 403–412.

Huffaker, C. B.; Van de vrie, M.; McMurtry, J. A. Ecology of Tetranychid Mites and Their Natural Enemies: A Review. II. Tetranychid Populations and Their Possible Control by Predators: An Evaluation. *Hilgardia* **1970**, *40*, 391–458.

Huffaker, C. B.; Van de vrie, M.; McMurtry, J. A. 1969. The Ecology of Tetranychid Mites and Their Natural Control. *Ann. Rev. Entomol.* **1969**, *14*, 125–174.

Husseine, K. K. The Effect of Insecticides on Outbreaks of Spider Mites on Cotton. *FAO Pl. Proto Bull.* **1958**, *6*, 155–157.

Hussey, N. W.; Parr, W. J. The Effect of Glasshouse Two-Spotted Spider Mite on the Yield of Cucumber. *J. Hort. Sci.* **1963,** *38,* 255–263.

Isman, M. B. Plant Essential Oils for Pest and Disease Management. *Crop Prot.* **2000,** *19,* 603–608.

Isman, M. B. Pesticides Based on Plant Essential Oils for Management of Plant Pests and Diseases. *In Symposium on Development of Natural Pesticides from Forest Resources.* Republic of Korea, Seoul, 2001, pp. 1–9.

Jain, P. C.; Yadav, C. P. S. New Record of Brown Wheat Mite, *Petrobia latens* (Muller) on Coriander. *Ind. J. Ent.* **2001,** *50,* 396.

Jayasinghe, G. G.; Mallik, B. Growth Stage Based Economic Injury Levels for Two Spotted Spider Mite, *Tetranychus urticae* Koch (Acari, Tetranychidae) on Tomato, *Lycopersicon esculentum* Mill. *Tropical Agric. Res.* **2010,** *22* (1), 54–65.

Jeppson, L. R.; Keifer, H. H.; Baker, E. W. *Mites Injurious to Economic Plants.* University of Calfornia Press, 1975, pp. 614.

Jose, V. T.; Shah, A. H.; Patel, C. B. Feeding Potentiality of Some Important Predators of the Spider Mite, *Tetranychus macferlanei,* A Pest of Cotton. In *Progress in Acarology* ChannaBasavanna, G. P.; Viraktamath, C. A.; Oxford & IBH: New Delhi, 1989, pp. 357–360.

Keifer, H. H. Eriophyid Studies. VIII. *Bull. Calif. Dept. Agric.* **1940,** *29,* 21–46.

Keifer, H. H.; Baker, E. W.; Kono, T.; Delfinado, M.; Styer, W. E. *An Illustrated Guide to Plant Abnormalities Caused By Eriuphyid Mites in North America.* U.S.D.A Hand book No. 573, 1982.

Khan, M. Q.;Murthy, D. V. *Dicanthium annulatum* Stapf. An Important Alternate Host Plant of Jowar and Sugarcane Mites. *Ind. J. Ent.* **1956,** *18,* 190–199.

Khan, R. M.; Doval, S. L.; Joshi, H. C. Biology of Brown Wheat Mite *Petrobia latens* (Muller). *Ind. J. Ent.* **1969,** *31,* 258–264.

Khanna, A. *Bionomics of Some Important Mite Pests of Vegetable Crops at Hisar.* M.Sc. Thesis, CCS HAU, Hisar, 1991.

Kim, S. Contact and Fumigant Activities of Aromatic Plant Extracts and Essential Oils Against *Lasioderma serricorne* (Coleoptera: Anobiidae). *J. Stored Prod. Res.* **2003,** *39,* 11–19.

Krishnamoorthy, A. Development of *Phytoseiulus persimilis* (Acari: Phytoseiidae) on the Carmine Spider Mite, *Tetranychus cinnabarinus* (Acari: Tetranychidae) at Two Temperature Regimes. In *Progress in Acarology*; ChannaBasavanna, G. P.; Viraktamath, C. A.; Oxford & IBH: New Delhi, pp. 369–374.

Kropczynska, A.; Pilko, A.; Witul, A. Asshleb, A. M. Control of Two-Spotted Spider Mite with *Amblyseius californicus* on Cotton. *IOBC/WPRS Bull.* **1999,** *22,* 133–136.

Kropezynska, D.; Tomczyk, A. Some Feeding Effects of *Tetranychus urticae* Koch on Productivity of Selected Plants. *J. Acarol.* **1984,** *2,* 747.

Kropezynska, D.; Van de vrie, M.; Tomezyk, A. Bionomics of *Eotetranychus tiliarium* and Its Phytoseiidae Predators. *Exp. Appl. Acarol.* **1988,** *5,* 65–81.

Kulkarni, G. S. The "Murda" Disease of Chilli *(Capsicum). Trop. Agr. Peredenya* **1922,** *58,* 237.

Kumari; Meena; Sadana, G. L.Influence of Temperature and Relative Humidity on the Development of *Amblyseius alstoniae* (Acari: Phytoseiidae). *Exp. Appl. Acarol.* **1991,** *11,* 199–203.

Lal, L.; Mukherjee, S. P. Observations of the Injury Symptoms Caused by the Phytophagous mites. *Zoologische Beitrage* **1979**, *25*, 13–17.

Lal, S. S. Influence of Weather Factors on the Population of Spider Mites (Acari: Tetranychidae) on Cassava in Kerala. *Ind. J. Acar.* **1982**, *7*, 5–10.

Lang, J. E.; Knop. N. E 1983. Potential Uses of Predaceous Mite Other Than Phytoseiidae for Biological Control of Orchard Pests. In *Biological Control of Pests by Mites*; Hoy, M. A., Cunningham, G. L., Knutson, L., University of California, Berkely: 1983, pp. 28–35.

Larson, K. C.; Berry, R. E. Influence of Peppermint Phenolics and Monoterpenes on Two Spotted Mite (Acari: Tetranychidae). *Environ. Entomol.* **1984**, *13*, 282–285.

Latif, A.; Wali, M. Distribution, Bionomics and Description of *Larvacarus transitans* (Ewing) (Acarina: Phytoptipalpidae). *Pakistan J. Sci. Res.* **1961**, *13*, 77–87.

Latif, A.; Wali, M. Bionomics of "Ber" Mite, *Larvacarus transitans* (Ewing). In Proceedings of the 8th Pak Science Conference, Lahore, Pakistan, 1956.

Leigh, T. E; Grimes, D. W.; Yamada, H.; Stockton, J. R.; Basset, D. Arthropod Abundance in Cotton in Relation to Some Cultural, Management Variables. In Proceedings of Tall Timbers Conference on Ecological Animo Control by Habitat Management; Komarek, R., Ed.; Tall Timbers Res. Sta., Tallahassee, 1969, pp. 76–78.

Lo, Kang-Chen; Lee, Wen-Tai.; Wu, Tze-Kann; Ho, Chyi-Chen. Use of Predators to Control Spider Mites (Acarina: Tetranychidae) in the Republic of China on Taiwan. In *The Use of Natural Enemies to Control Agricultural Pests*; FFfC Book Sr. No. 40, 1990, pp. 166–178.

Mali, A. R.; Gandhali, D. N.; Patil, A. S. Heavy Incidence of Two Spotted Spider Mite *(Tetranychus urticae* (Acarina: Tetranychidae) on Grapevine and Roses. *Acar. Newslett.* **1983**, *12*, 6–7.

Mallik, B.; Krishnaswamy, H. S.; ChannaBasavanna, G. P. Mathematical Models for the Interaction Between *Tetranychus ludeni* and its Phytoseiidae Predators. In *Progress in Acarology*; ChannaBasavanna, G. P.; Viraktamath, C. A.; Oxford & IBH: New Delhi, 1989, pp. 343–355.

Manjunatha, M. Mites Reduces Jasmine Yield. *The Hindu* (March 18), 1999, P. 24.

Mann, H. H.; Nagpurkar, B. D.; Kulkarni, G. S. The "Tambera" Disease of Potato. *Agric. J. India* **1920**, *15*, 282.

Mansour, E.; Ascher, K. P. S.; Omari, N. Effect of Neem *(Azadirachta indica)* Seed Kernel Extracts from Different Solvents on the Predacious Mite *Phytoseiulus persimilis* and the Phytophagous Mite *Tetranychus cinnabarinus. Phytoparasitica.* **1987**, *15*, 125–130.

Mansour, E. A.; Ascher, K. R. S. Effects of Neem *(Azadirachta indica)* Seed Kernel Extracts from Different Solvents on the Carmine Spider Mite, *Tetranychus cinnabarinus.* In Proceedings of the 2nd International Neem Conference Rauischholzhausen, 1983, pp. 461–470.

Markkula, M.; Tiittanen, K. Effect of Fertilisers on the Reproduction of *Tetranychus telarius* (L.), *Myzus persicae* (lz) and *Acyrthosiphon pisum* Harris. *Ann. Agric. Fenn.* **1969**, *8*, 9–14.

McMurtry, J. A. The Use of Phytoseiid for Biological Control Progress and Future Prospects. In *Recent Advances in Knowledge of the Phytoseiidae*; Hoy, M. A., Ed.; Univ. of California Publication: USA, 1983, pp. 23–48.

McMurtry, J. A. Utilizing Natural Enemies to Control Pest Mites on Citrus and Avocado in California, U.S.A. In *Progress in Acarology*; ChannaBasavanna, G. P.; Viraktamath, C. A.; Oxford & IBH: New Delhi, 1989, pp. 325–336.

McMurtry, J. A. Augmentative Releases to Control Mites in Agriculture. In *Modem Acarology*; Dusbabek, F.; Bukva, V.; Academia: Prague, 1991, pp. 151–157.

Meyer, M. K. P. Mite Pests of Crops in Southern Africa. *Sci. Bull. Dept. Agric. Fish Repub. S. Afr. No.* 397, **1981**, 92.

Meyer, M. K. P.; Smith. The Tenuipalpidae (Acari) of Africa, with Keys to the World Fauna. *Entomology Memp. Dep. Agric. tech. Servo Repub. Afr.* **1979**, *50*, 135.

Mitra, M. Report of the Imperial Mycologist. *Sci. Rep. Agr. Res. Inst., Pusa* **1931**, 58–71.

Mohanasundaram, M. New Gall Mites of the Subfamily Nothopodinae (Acarina: Eiophyidae) from India. *Oriental Insects* **1981**, *15*, 145–166.

Moraes, B. M.; Birkett, M. A.; Gordon-Weeks, R.; Smart, L. E.; Martin, J. L.; Pye, B. J.; Bromilow, R.; Pickett, J. A. Cis-Jasmone Induces Accumulation of Defence Compounds in Wheat, *Triticum aestivum*. *Phytochemistry* **2008**, *69*, 9–17.

Mori, H.; Saito, Y.; Nakao, H. Use of Predatory Mites to Control Spider Mites (Acarina: Tetranychidae) in Japan. In *The Use of Natural Enemies to Control Agricultural Pests*, FFfC Book, 1990, Sr. No. 40, pp. 142–156.

Morimoto, K.; Furuichi, H.; Yano, S.; Osakaba, M. H. Web-Mediated Interspecific Competition Among Spider Mites. *J. Econ. Entomol.* **2006**, *99*, 678–684.

Mukherjee, I. N.; Singh, J. Records of Phytophagous and Predatory Mites Associated with Fruit Plants in Uttar Pradesh. *J. Insect Sci.* **1993**, *6*, 134–136.

Mukherjee, I. N.; Singh, R. K.; Singh, J. Biology and Chemical Control of Carmine Spider Mite, *Tetranychus cinnabarinus* (Boisd.) (Acarina: Tetranychidae) on Greengram *(Vigna radiata)* in Varanasi. *Pestology* **1992**, *16*, 18–24.

Mukherjee, I. N.; Singh, R. K.; Singh, J. Incidence and Control of Jujube Gall Mite *(Eriophyis cernuus)* at Varanasi. *Ind. J. Agric. Sci.* **1994**, *64*, 343–345.

Mukherjee, I. N.; Singh, R. N.; Singh, R. K.; Singh, J. Incidence of Fig Mite, *Eotetranychus hirsti* (Acari: Tetranychidae) in Relation to Weather Factors in Varanasi. *J. Acarol.* **1995**, *13*, 63–68.

Nangia; Neelu; Muniyappa, V. Screening for Resistance to Pigeonpea Sterility Mosaic and Incidence of *Aceria cajani* (Acari: Eriophyidae). *J. Acarol.* **1995**, *13*, 75–79.

Narayanan, E. S.; Ghai, S. Malformation of Mango Trees in India. *Proc. Indian Sci. Congr.* **1961**, *48*, 502.

Nassar, O. A.; Ghai, S. Taxonomic Studies on Tetranychid Mites Infesting Vegetable and Fruit Crops in Delhi and Surrounding Areas. *Oriental Ins.* **1981**, *15*, 333–396.

Natarajan, K. Studies on Seasonal Incidence of Tetranychid Mites on Bhendi and Brinjal. *AICRP Rep. Agric. Acarol.* **1989**, *6*, 181–184.

Nene, Y. L. A Survey of the Viral Diseases of Pulse Crops in Uttar Pradesh. *G. B. Pant Univ. Agric & Tech., Pantnagar Exp. Stn. Bull. No.4*, 1972.

Obeng-Ofori, D.; Amiteye, S. Efficacy of Mixing Vegetable Oils with Pirimiphos-Methyl Against the Maize Weevil, *Sitophilus zeamais* Motschulsky in Stored Maize. *J. Stored Prod. Res.* **2005**, *41*, 57–66.

Onkarappa, S.; Mallik, B.; Kumar, H. M. Spatial Distribution of *Tetranychus urticae* on Open Cultivated Rose. *J. Acarol.* **1999**, *15* (1–2), 44–46.

Osborne, L. S.; Ehler, L. E.; Nechols, J. R. Biological Control of the Two-Spotted Spider Mite in Greenhouses. *Fl. Agric. Exp. Stn. Bull.* **1985**, *853*,40

Osmelak, J. A.; MacFarlane, J. Integrated Pest Management of Two Spotted Mite in Roses and Carnations. In Proceedings of the 1st National Conference, Australian Society of Horticultural Science, Macquarie University, Sydney, 1991, pp. 337–344.

Page, F. D.; Bower, C. C.; Thwaite, W. E.; Mimmo, P. R.; Heaton, J. B. Progress in Implementing Mite Control in Apples in New South Wales and Quensland. In Proceedings of the 1st National Conference, Australian Society of Horticultural Science, Maquarie University, Sydney, 1991, pp. 445–450.

Pande, Y. D.; Sharma, A. K. Evaluation of Some Pesticides for the Control of *Tetranychus neocaledonicus.* In *Contribution to Acarology in India*; ChannaBasavanna, G. P., Ed.; Acarological Society of India, Bangalore, 1979, pp. 207–208.

Pande, Y. D.; Yadav, S. R. S. A New Host Record of *Tetranychus macfarlanei* Baker and Pritchard (Acari: Tetranychidae). *Labdev J. Sci. Tech.* **1976**, *13*(B), 75.

Park, Y. L.; Lee, J. H. Leaf Cell and Tissue Damage of Cucumber Caused by Two-spotted Spider Mite (Acari:Tetranychidae). *J. Econ. Entomol.* **2002**, *95*, 952–957.

Park, Y. L.; Lee, J. H. Impact of Two Spotted Spider Mite (Acari:Tetranychidae) on Growth and Productivity of Glasshouse Cucumber. *J. Econ. Entomol.* **2005**, *98* (2), 457–463.

Parker, R. The Widespread Use of Integrated Mite Control in Flower Crops in S.E. Queensland. In Proceedings of 1st National Conference, Australian Society of Horticultural Science, Macquarie University, Sydney, 1991, pp. 345–350.

Patel, C. B.; Rai, A. B.; Patel, M. B.; Patel, A. J.; Shah, A. A. Acaricidal Tests of Botanical Pesticides in Comparison to Conventional Acaricides/Pesticides Against Red Spider Mites (Acari: Tetranychidae) on Okra, Brinjal and Indian Bean. *Indian J. Ent.* **1993**, *55*, 184–190.

Patil, R. S. Investigation on Mite Pests of Solanaceous Vegetable with Special Reference to Brinjal. Ph.D. Thesis, Univ. Agril. Sci., Dharwad, 2005.

Penman, D. R.; Chapman, R. B. Pesticide-Induced Mite Outbreaks: Pyrethroids and Spider Mites. *Experimental mid Appl. Acarol.* **1988**, *4*, 265–276.

Pillai, K. S.; Palaniswamy, M. S. Spider Mites of Cassava. *Tech. Bull.* Sr.1. Central Tuber Crops Research Institute, Trivandrum, 1985.

Plaut, H. N. ; Cohen, M. Trials on the Control of *Lithocolletis blancardella* F. (Lep: Gracillaridae) on Apple Trees with Diflubenzuron and Permethrin. *Alon Hanotea* **1978**, *33*, 5–11.

Poe, S. L. Sampling Mites On Soybean. In *Sampling Methods in Soybean Entomology*; Kogan, M.; Herzog, D. C., Eds; Springer Verlag: New York, pp. 312–323.

Pontes, W.; Silva, J.; Da Camara, C.; Gondim-Junior, M.; Olivera, J.; Schwartz, M. Chemical Composition and Acaricidal activity of the Essential Oils from Fruits and leaves of *Protium bahianum* Daly. *J. Essent Oil Res.* **2010**, *22*, 279–282.

Prasad, A. H.; Singh, H.; Shukla, T. N. Present Status of Mango Malformation Disease. *Ind. J. Hort.* **1965**, *22*, 254.

Prasad, V. G.; Singh, R. K. Prevalence and Control of Litchi Mite, *Aceria litchii* Keifer in Bihar. *Indian J. Ent.* **1981**, *43*, 67–75.

Prasad, Y. *A Catalogue of Mites of India.* Indira Aca. Pub. House: Ludhiana, 1974.

Pritchard, A. E.; Baker, E. W. A Revision of the Spider Mite Family Tetranychidae. *Mem. Ser. Vol.* 2, *San Francisco Pacif Coast Ent. Soc.,* USA, 1955.

Pritchard, A. E.; Baker, E. W. The False Spider Mites (Acarina: Tenuipalpidae). *Univ. Calif. Publ. Ent.* **1958,** *14,* 175–274.

Putatunda, B. N.; Tagore, A. Effect of Temperature, Relative Humidity and Sunshine Hours on Mite Population. In *Mites, Their Identification and Management*; Yadav, P. R., Chauhan, R., Putatunda, B. N., Chhillar, B. S., Eds; CCS HAU: Hisar, India, 2003, pp. 23–28.

Puttarudriah, M. Field Control of the Leaf Mite of 101a *(Andropogon sorghum). Mysore Agr. J.* **1947,** *26,* 17–19.

Puttarudriah, M; ChannaBasavanna, G. P. A Preliminary Account of Phytophagous Mite of Mysore. *Proc. 1st All Indian Congr. Zool., Part 1.***1959,** *2,* 530–539.

Raja, J.; Rajendran, B.; Rajsekaran, S. ; Jhon Ambrose John, H. Efficacy of Certain Chemicals in the Control of Sheath Mite of Sugarcane. *Pestology* **1995,** *19,* 19–21.

Rajagopalan, K. First Record of Spider Mite, *Tetranychus ludeni* Zacher Transmitting *Dolichus* Enation Mosaic Virus. *Curr. Sci.* **1974,** *43,* 488–489.

Ramakrishnan, K.; Kandaswamy. *Investigations on Viral Diseases of Pulse Crops.* Final Technical Report, TNAU, Coimbatore, India, 1972.

Rao, J.; Prakash, A. Infestation of Tarsonemid Mite *Steneotarsonemus spinki* Smiley in Rice in Orissa. *J. Appl. Zool. Res.***1992,** *3,* 103.

Rao, P. R. M.; Bhavani, B.; Rao, T. R. M.; Reddy, P. R. Spikelet Sterlity/Grain Discolouration in Rice in Andhra Pradesh, India. *Int. Rice. Res. Newslett.* **2000,** *25,* 40.

Rao, Y. S.; Das, P. K. A New Mite Pest of Rice in India. *Int. Rice. Res. Newslett.* **1977,** *2,* 8.

Rasmy, A. H. Prospects of Biological Control in Integrated Management of Mite Pests. In *Progress in Acarology*; ChannaBasavanna, G. P.; Viraktamath, C. A., Eds; Oxford & IBH: New Delhi, 1989, pp. 337–341.

Rathi, Y. P. S. *Studies on Sterility Mosaic Disease on Pigeonpea (Cajanus cajan* (L.) Millsp.). Technical Report, G. B. Pant University of Agriculture and Tech., Pantnagar, India, 1983.

Ray, R.l Rai, L. Biology and Control of *Tetranychus neocaledonicus* (Acari: Tetranychidae) on Lady's Finger at Varanasi. In *Contribution to Acarology in India*; ChannaBasavanna, G. P., Ed.; Acarological Society of India, Bangalore, India, 1981, pp. 41–56.

Razmjou, J.; Tavakkoli, H.; Nemati, M. Life History Traits of *Tetranychus urticae* Koch on Three Legumes (Acari: Tetranychidae). *Munis Entomol. Zoo.* **2009,** *4* (1), 204–211.

Reddy, M. V.; Raju, T. N. Some Clues to Increased Incidence and Seasonal Variation of Pigeonpea Sterility Mosaic in Peninsular India. *Int. Pigeonpea Newslett.* **1993,** *18,* 22–24.

Reddy, M. V.; Sheila, V. K.; Nene, Y. L. *Cajanus scarabaeoids,* An Alternative Host of Pigeonpea Sterility Mosaic Pathogen and Its Vector *Aceria cajani. Int. Pigeonpea Newslett.* **1993,** *18,* 24–27.

Reddy, S. Y.; Reddy, M. Y.; Ghanekar, A. M.; Nene, Y. L.; Amort, K. S. Annual Recurrence of Pigeonpea in India, Sterility Mosaic in Eastern Uttar Pradesh. *Int. Pigeonpea Newslett.* **1988,** *7,* 30–31.

Rijn, P. C. J. van; Houten, Y. M. van. Life History of *Amblyseius cucumeris* and *A. barkaeri* (Acarina: Phytoseiidae) on a Diet of Pollen. In *Modern Acarology*; Dusbabek, F.; Bukva, Y.; Academia: Prague, 1991, pp. 645–670.

Roh, H. S.; Lim, E. G.; Kim, J.; Park, C. G. Acaricidal and Oviposition Deterring Effects of Santalol identified in Sandalwood Oil Against the Two-Spotted Spider Mite, *Tetranychus urticae* Koch (Acari: Tetranychidae). *J. Pest. Sci.* **2011,** *84,* 495–501.

Sabelis, M. W. Sampling Techniques. In *World Crop Pests, Spider Mites, Their Biology, Natural Enemies and Control*; Helle, W.; Sabelis, M. W.; Elsevier: Amsterdam, 1985, pp. 337–348.

Sadana, G. L.; Kumari, M. Seasonal History of *Brevipalpus phoenicis* on *Psidium guajava* cv. Seedless Guava. In Proceedings of First National Seminar on Acarology, Kalyani, West Bengal, 1987, p. 18.

Sadana, G. L. *Plant Feeding Mites of India.* Kalyani Publishers: Daryaganj, New Delhi, India, 1985.

Sadana, G. L.; Kumari, M. Influence of Temperature and Relative Humidity on the Development of *Brevipalpus phoenicis* (Geijskes) (Acari: Tenuipalpidae). In *Contribution to Acarological Researches in India*; Mukherjee, A. B.; Somchoudhary, A. K.; Sarkar, P. K., Eds; Kalyani, West Bengal, 1991, pp. 61–72.

Sadana, G. L.; Sharma, N. K. Biology of Predatory Mite, *Amblyseius finlandicus* (Oud.) (Phytoseiidae: Acari). *J. Assam Sci. Soc.* **1988,** *30,* 47–60.

Sadana, G. L.; Sidhu, R. New Species and New Host Record of Tenuipalpidae Mites from Punjab (Acari: Tenuipalpidae). *Acarologia* **1990,** *31,* 357–360.

Sandhu, G. S.; Singh, B.; Dhooria, M. S. Effect of Rain in Population of *Oligonychus indicus* (Hirst) (Acarina: Tetranychidae) on Different Varieties of Maize (*Zea mays* L.) in Punjab, India. *Int. J. Acarol.* **1975,** *1,* 10–13.

Sandhu, M. S.; Gatoria, G. S.; Sandhu, S. S.; Singh, S. Chemical Control of Tetranychid Mite *Eutetranychus banksi* Infesting Cotton in Punjab, India. *J. Res. Punjab Agric. Univ.* **1982,** *19,* 127–129.

Sandhu, S.; Chander, P.; Singh, J.; Sidhu, A. S. Effect of Insecticidal Sprays on the Plant and Secondary Pest Inductions in *Hirsutam* Cotton in Punjab. *Agri. Ecosystems Environ.* **1987,** *19,* 169–176.

Sarkar, P. K.; Somchoudhary, A. K. Influence of Major Abiotic Factors on the Seasonal Incidence of *Raoiella indica* and *Tetranychus tijiensis* on Coconut. In *Progress in Acarology*, ChannaBasavanna, G. P.; Viraktamath, C. A., Eds; Oxford & IBH: New Delhi, India, 1989, 67–72.

Sattar, A. Diseases of the Mango in Punjab. *Punjab Fruit J.* **1946,** *10,* 56–58.

Saxena, P. Role of Demographic Data in Monitoring Status of Women and Recent Fertility Transition. In Population Planning in India, Bose A.; Desai, P. B., Edsl B. R. Publishing Corporation: Delhi, India, 1989.

Seth, M. L. Transmission of Pigeonpea Sterility by an Eriophyid Mite. *Ind. PhyJopath.* **1962,** *15,* 225–227.

Sharma, A.; Kushwaha, K. S. Susceptibility of Different Varieties of *Ziziphus mauritinna* to *Larvacarus transitans* (Acari: Tenuipalpidae). In *Progress in Acarology*, ChannaBasavanna, G. P.; Viraktamath, C. A., Eds; Oxford & IBH: New Delhi, 1989, pp. 91–93.

Sharma, D. D. Occurrence of *Cephaleuros virescens,* a New Record of Leaf Curls in Litchi *(Litchi chinensis). Ind. J. Agril. Sci.* **1991,** *61,* 446–448.

Sharma, H. S.; Pande, Y. D. Seasonal Incidence of *Tetranychus* spp. on Four Improved Varieties of Brinjal. In *Contribution to Acarology in India*, ChannaBasavanna, G. P., Ed.; Acarological Society of India, Bangalore, 1981.

Sharmila, B. C.; Umamaheshwari, T.; Kanagarajan, R.; Ariudainami, S.; Swlvanarayanan, V. Feeding Site Preference of Okra Red Spider Mite. *J. Acarol.* **1999,** *14* (1–2), 80–81.

Sheila, Y. K.; Manohar, S. K.; Nene, Y. L. Biology and Morphology of *Aceria cajani. Int. Pigeonpea Newslett.* **1988,** *7,* 28.

Singh, J.; Mukherjee, I. N. Pest Status of Phytophagous Mites in Some Northern States of India. In *The First Asia-Pacific Conference of Entomology* (APCE), The Entomology and Zoology Association of Thailand, 1989, 192–203.

Singh, J.; Ray, R. *Stethorus* sp. (Coleoptera: *Coccinellidae),* A Predator of *Tetranychus neocaledonicus* Andre' on Okra at Varanasi. *Acarology Newsletter* **1977,** *4,* 5.

Singh, J.; Singh, R. K. Studies on Mango Bud Mite *(Aceria mangiferae* Sayed.). *Pestology* **1979,** *3,* 20–26.

Singh, J.; Gatoria, G. S.; Sidhu, A. S. Spider Mite *Tetranychus cinnabarinus* (Boisd.)—A Secondary Induced Pest of Cotton in Punjab. *Pesticides* **1987,** *21,* 43–44.

Singh, J.; Mukherjee, I. N.; Singh, R. K. Managing Gall Forming False Spider Mite in Ber. *Ind. Horticulture* **1996,** *41,* 30–31.

Singh, J.; Singh, R. K.; Mukherjee, I. N.; Singh, R. N.; Agarwal, L. Mites of Agricultural Importance and Their Management in India. In *Recent Advances in Entomology*; Mathur, Y. K.; Bhattacharya, A. K.; Pandey, N. D.; Upadhyaya, K. D.; Srivastava, J. P., Eds; Gopal Prakashan: Kanpur, 1987, pp. 169–185.

Singh, P.; Singh, R. N. Interaction of Environmental Factors with *Tetranychus neocaledonicus* Andre and Its Predatory Mite in Brinjal Ecosystem. *Annals Plant Protection Sci.* **2015,** *23* (1), 23–26.

Singh, P.; Singh, R. N. Interactions of Abiotic Factors with *Tetranychus neocaledonicus* andre and Its Management by Newer Acaricides in Brinjal Ecosystem. *Ecoscan* **2014,** *6,* 355–359.

Singh, P.; Somchoudhury, A. K.; Mukherjee, A. B. The Influence of Natural Enemy Complex on the Population of *Aceria litchii* (Acari: Eriophyidae). In *Progress in Acarology*; Channabasavanna, G. P., Viraktamath, C. A., Eds; Oxford & IBH: New Delhi, 1989, pp. 361–367.

Singh, R. N. Mite Menace on Vegetable Crops in India. *Shashpa* **1994,** *1,* 59–74.

Singh, R. N. Mites of Deciduous Fruits and Vegetables of Eastern Part of India and Their Economic Status. *Adv. Agric. Res. India* **1995,** *3,* 179–193.

Singh, R. N.; Singh, J. Incidence of *Tetranychus cinnabarinus* in Relation to Weather Factors in Varanasi. *Pestology* **1993,** *17* (8), 18–23.

Singh, R. N.; Singh, J. Evaluation of Some Pesticides Against Carmine Mite, *Tetranychus cinnabarinus* (Boisd.) on Lady's Finger. *Pestology* **1992,** *16,* 20–23.

Singh, R. N.; Singh, J. Incidence of *Tetranychus cinnabarinus* (Boisd.) (Acari: Tetranychidae) in Relation to Weather Factors in Varanasi. *Pestology* **1993,** *17* (8), 18–23.

Singh, S. P. *Technology for Production of Natural Enemies.* Project Directorate of Biological Control, Bangalore, 1994.

Singh, S. P. Production and Use of Phytoseiid Mites. First Acarology Workshop, G.A.U., Navasari, 1988, pp. 1–6.

Singh, V.S.; Bhatia, S. K. Reduction in Yield of Some Barley Varieties Due To Brown Wheat Mite Infestation. *Ind. J. Ent.* **1983**, *45*, 190–193.

Souliotis, P. P. The Present State of Biological Control of Mites in Protected Vegetable Crops in Greece. *Integrated Pest Management in Protected Vegetable crops*, Cavallaro, R.; Pelerents, C., Eds, 1990, pp. 107–110.

Spooner Hart, R. The Use of Predatory Mite, *Phytoseiulus persimilis* for the Control of Two Spotted Mite *Tetranychus urticae* in Horticultural Crops. Proceedings of the 1st National Conference, Australian Society for Horticultural Science, Macquarie University, Sydney, 1991, pp. 329–336.

Srinivasa, N.; Prabhakara, H.; Malik, B. Rice Sheath Mite, *Steneotarsonemus spinki* Smiley (Acari:Taronemidae). Status paper. AINP-Agricultural Acarology (ICAR), 2004, pp. 1–24.

Srivastava, R. P.; Bhutani, D. K. La 'Malifnrmation' du Manguier. *Fruits* (Paris) **1973**, *28*, 389.

Steiner, M. Y.; Spohr, L. J.; Goodwin, S. Impact of Two Formulations of the Acaricide Bifenazate on the Spider Mite Predator *Phytoseiulus persimilis* Athias-Henriot (Acari: Phytoseiidae). *Aust. J. Entomol.* **2011**, *50* (1), 99–105.

Sternlight, M.; Goldenberg, S. Mango Eriophyid Mites in Relation to Inflorescence. *Phytoparasitica* **1976**, *4*, 45–50.

Strebler, G. Les Me'diateurs chimiques: leur incidence sur la bioe'cologie des animaux. Lavoisier, Paris, 1989.

Summanwar, A. S.; Raychaudhuri, S. P.; Phatak, S. C. Association of the Fungus *Fusarium moniliforme* Sheld with Malformation in Mango *(Mangifera indica)*. *Ind. Phytopathol.* **1966**, *19*, 227.

Sunita Bionomics and Control of Mites on Okra (*Abelmoschus esculentus* Linn.) Ph.D. Thesis, CCS HAU, Hisar, p.96.

Tapondjou, A. L.; Adler, C.; Fontem, D. A.; Bouda, H.; Reichmuth, C. Bioactivities of Cymol and Essential Oils of Cupressus sempervirens and Eucalyptus Saligna Against *Sitophilus zeamais*. *J. Stored Products Res.* **2005**, *41* (1), 91–102.

Tewary, D. K.; Bhardwaj, A.; Shanker, A. Pesticidal Activities in Five Medicinal Plants Collected from Mid Hills of Western Himalayas. *Ind. Crop Prod.* **2005**, *22*, 241–247.

Thakur, A. P.; Sharma, D. D. Influence of Weather Factors and Predators on the Populations of *Aceria litchii* Keifer. *Ind. J. Plant Protection* **1990**, *18*, 104–112.

Thwaite, W. G. Acaricide Resistance and Its Management in Australian Deciduous Fruit Orchards. Spider Mites of Tree Fruit Crops. Meeting at Portland, Oregon, U.S.A, 1990.

Thwaite, W. G.; Phimister, S. R.; Edge, V. E. Resistance Management Strategies for Miticides. Proceedings of the Symposium on Mite Control in Horticulture Crops, Orange, Department of Agriculture, New South Wales, Australia, 1987, pp. 99–101.

Tirello, P.; Pozzebon, A.; Cassanelli, S.; Van Leeuwen, T.; Duso, C.) Resistance to Acaricides in Italian Strains of *Tetranychus urticae*: Toxicological and Enzymatic Assays. *Exp. Appl. Acarol.* **2012**, *57*, 53–64.

Van de Vrie, M. Studies on Prey–Predator Interactions Between *Panonychus ulmi* and *Typhlodromus (A.) potentillae* (Acarina: Tetranychidae: Phytoseiidae) on Apple in The Netherlands. Proceedings of FAD Conference on Ecological Relations Plant Pest Control, Rome, 1973, pp. 145–160.

Van de vrie, M.; McMurthy, J. A.; Huffaker, C. B. Ecology Oftetranychid Mites and Their Natural Enemies: A Review III. Biology, Ecology and Pest Status and Host–Plant Relation of Tetranychids. *Hilgardia* **1972**, *14*, 342–432.

Van den Boom, C. E. M.; Van Beek, T. A.; Dicke, M. Differences Among Plant Species in Acceptance by the Spider Mite *Tetranychus urticae* Koch. *J. Appl. Entomol.* **2003**, *127*, 177–183.

Varma, A. Mango Malformation. In *Exotic Plant Quarantine Pests and Procedures for Introduction of Plant Materials,* 1983, pp. 173–188.

Varma, A.; Lele, V. C.; Majumder, P. K.; Ram, A.; Sachchidananda, J.; Shukla, U. S.; Singh, G. C.; Yadav, T. D.; Raychaudhuri, S. P. *Mango Malformation.* ICAR Workshop Fruit Research, Ludhiana, 1969.

Varma, A.; Lele, V. C.; Raychoudhary, S. P.; Ram, A.; Singh, A. Mango Malformation: A Fungal Disease. *J. Phytopathol.* **1974**, *79*, 254.

Wafa, A. K.; Osman, A. A. Control of the Bud Mite, *Aceria mangiferae* (Sayed) on Mango Trees in Egypt. *Bull. Ent. Soc., Egypt. Econ. Ser.* **1973**, *7*, 265–273.

Walsh, D. B.; Zalom, F. G.; Shaw, D. V. Interaction of the two spotted spider mite (Acari: Tetranychidae) with Yield of Day-Neutral Strawberries in California. *J. Econo.. Entomol.* **1998**, *91*, 678–685.

Watt, G. The Pests and Blights of Tea Plant. Government Printing Press: Calcutta, India, 1898, pp. 400–408.

WAU. Integrated Pest Management in the Tropics, Needs and Constraints of Information and Documentation: A Feasibility Study. Wageningen Agricultural University, Dept. of Entomology, The Hague, The Netherlands, Ministry of Housing, Physical Planning and Environment, 1987.

Wermelinger, B., Oertli, J. J. and Baumgärtner, J. (1991). Environmental factors affecting the life-tables of *Tetranychus urticae* (Acari: Tetranychidae) III. Host-plant nutrition. *Exp. App. Acarol.*, 12: 259–274.

Wilson, L. J.; Bauer, L. R.; Lally, D. A. Effect of Early Season Insecticide Use on Predators and Outbreaks of Spider Mites (Acari: Tetranychidae) in Cotton. *Bull. Entomol. Res.* **1998**, *88*, 477–788.

Zote, K. K.; Mali, V. R.; Mayee, C. D.; Kulkarni, S. Y.; Mote, T. S. Outbreak of Sterility Mosaic of Pigeonpea in Marathwada Region, Maharashtra, India. *Int. Pigeonpea Newsletter* **1991**, 19–21.

Zwich, R.W.; Fields, G. I. Field and Laboratory Evaluation of Fenvalerate Against Several Insect and Mite Pests of Apple and Pear in Oregon. *J. Econ. Entomol.* **1978**, *71*, 793–796.

CHAPTER 6

MITE PROBLEMS IN SPICE CROPS AND THEIR MANAGEMENT

MRINALINI KUMARI[1,*] and DINESH PRASAD GOND[2]

[1]*Mandan Bharti Agriculture College, Agwanpur, Saharsa, Bihar Agricultural University, Sabour 852201, India*

[2]*Research Scholar Department of Endocrinology, Institute of Medical Science, Banaras Hindu University, Varanasi 221005, Uttar Pradesh, India*

Corresponding author. E-mail: mmrinalini35@gmail.com

ABSTRACT

Mites are the most diverse representatives of an ancient lineage in phylum-Arthropoda, subphylum-Chelicerata, subclass Acari. Their body plan is strikingly different to that of other arthropods in not having a separate head, instead, an anterior region, the cephalothorax, combines the functions of sensing, feeding, and locomotion. Antennae, mandibles and maxillae are also absent; rather, a pair of often pincer-like mouthparts are present, the so-called chelicerae. Mites constitute one of the most heterogeneous cheliceran groups, since they are extremely diverse in their morphology, biology and ethnology, enabling them to colonize different environments. Their remarkable diversity in acarine morphology is reflected in the variety of ecological and behavioral patterns that mites have adopted. Thus, species can inhabit soil, litter (i.e., Cryptostigmata, Mesostigmata, and Prostigmata), water (Hydrachnidia) or plants (Prostigmata or Mesostigmata). Those members of the subclass Acari, which feed on plants, are known as phytophagous mites. Phytophagy is widespread enough among the Trombidiform, Acariformes so as to suggest that there was an early evolution commitment to plant feeding by several primitive predaceous and saprophagous trombidiform lineages. Some Prostigmatan mites, chiefly spider mites, false spider mites and eriophoid mites, use their specialized mouthparts to feed on the vascular tissues of higher plants and with their activity they can cause losses to field and protected crops becoming economically important pests. In recent years such phytophagous mites have attracted the attention of acarologists and intensive farmers from all over the world.

6.1 INTRODUCTION

6.1.1 SPICES

The spice constitute an important part of horticultural production in India as well as the world, most of the people used spice as vegetable products or mixture thereof, free from extraneous matter, flavoring, seasoning, and imparting aroma in foods. India is well known as center of origin of spice crops and leading the area, production, productivity of several spice, for example, black pepper, cardamom (both large and small), ginger, garlic, turmeric, chilies, cinnamon, clove, coriander, cumin fennel fenugreek, nutmeg, tamarind, and number of tree and seed spice. In our country almost all the states grow one or more spice crops and major spice producing states are Andhra Pradesh, Kerala, Gujarat, Rajasthan, Maharashtra, West Bengal, Karnataka, Tamil Nadu, Odhisa, and Madhya Pradesh. The more than 90% of the spice and spice products consumed domestically and rest exported as raw or value added products in our country. Spices played a crucial role in the history of human civilization. It was defined by Christopher Morley as the plural of spouse (George, 1989), as per Webster, spices are specifically any of various aromatic vegetable production (Khan, 1990). According to the International spices group, "Spices are any of the flavored or aromatic substances of vegetable origin obtained from tropical or other plants commonly used as condiments or employed for other purposes on account of fragrance, preservative or medical qualities." Many spices are used for purposes of medicine and religious rituals in Asia and in cosmetics, perfumery, and liquorices in other parts of the world. Besides helping digestion, spices add color, flavor and zing to food.

Since immemorial times, India is known as the "Land of Spices" to the world. India has a unique position in the world as the largest producer of spice with 44% share in output and 36% in global spice trade (Bhattacharya, 1998). Ravindram and Manoj (2001) reported the importance in foreign trade and internal marketing spices are categorized into major and minor spices. Major spices such as turmeric, ginger, black pepper, chillies, cardamom, and so on. are export oriented whereas tree spices and other minor spices are consumed internally. Based on plants parts used, spices are classified as: fruit and seed spices (e.g. pepper, cardamom, aniseed, etc.), bud and flower spices (e.g. cloves, etc.), bark spices (e.g. cinnamon, etc.), root spices (ginger, turmeric, garlic, onion, etc.), and leaf spices (bay leaves, curry leaves, etc.).

The annual growth rate in area and production of spices is estimated to be 3.6 and 5.6%, respectively. India produces spices in 2 million hectares area (www.agriinfo.in). The annual production of spices in the country is around 6 million tones, out of which about 12–14% is exported while the rest is available for domestic consumption. India alone contributes 25–30% of the total world trade in spices. Nine spices namely pepper, ginger, clove, cinnamon, cassia, mace, nutmeg, pimento (all spice), and cardamom alone contribute as much as 90% of the total world trade. In India, the area and production is more in Rajasthan (556000.4 ha and 520000.6 MT) followed by Andhra Pradesh (317000.8 ha and 1235000.2 MT), Kerala (312000.3 ha and 159000.3 MT), Gujarat (299000.8 ha and 356000.8 MIT) (www.agriinfo.in).

The major factor responsible for low productivity of spice crops in India is due to infestation by pests. The insect pests of spice crops are well explored (Kotikal and Kulkarni, 2000; Paul, 2007; Kumar, 2011; Singh, 2014). Besides the insect pests, the number of noninsect pests is increasing vigorously, infesting the spice crops throughout their developmental stages, hence significantly reducing the crop yield. Among the noninsect pests, in recent years, mites are attracting greater attention from acarologists and farmers from all around the world. Some stray reports are only available informing mite infestation on spice crops (Gupta, 1985, 2002, 2003). Due to the lack of consolidated documents, the chapter was planned to explore the mite infestation in spice crops.

6.1.2 MITES

Mites are the most diverse representatives of an ancient lineage in phylum-Arthropoda, subphylum-Chelicerata, subclass Acari. Their body plan is strikingly different to that of other arthropods in not having a separate head, instead, an anterior region, the cephalothorax, combines the functions of sensing, feeding, and locomotion (Walter & Proctor, 1999). Antennae, mandibles and maxillae are also absent; rather, a pair of often pincer-like mouthparts are present, the so-called chelicerae. Mites constitute one of the most heterogeneous cheliceran groups, since they are extremely diverse in their morphology, biology, and ethnology, enabling them to colonize different environments. Their remarkable diversity in acarine morphology is reflected in the variety of ecological and behavioral patterns that mites have adopted (Krantz, 2009). Thus, species can inhabit soil, litter (i.e., Cryptostigmata, Mesostigmata, and Prostigmata), water (Hydrachnidia) or plants (Prostigmata or Mesostigmata). Those members of the subclass Acari, which feed on plants, are known as phytophagous mites. Phytophagy is widespread enough among the Trombidiform, Acariformes so as to suggest that there was an early evolution commitment to plant feeding by several primitive predaceous and saprophagous trombidiform lineages (Krantz, 2009). Some Prostigmatan mites, chiefly spider mites, false spider mites, and eriophoid mites, use their specialized mouthparts to feed on the vascular tissues of higher plants and with their activity they can cause losses to field and protected crops (Evans, 1992), becoming economically important pests. In recent years, such phytophagous mites have attracted the attention of acarologists and intensive farmers from all over the world.

The plant pests belong to the orders Tetranybhidae (spider mites, so called because some of them weave webs on the plants), Tenuipalpidae (false spider mites), tarsonemeidae (tarsonemids), Eriophydae (blister or gall mites), and Eupodidae (eupodids). Of these, the spider mites are most important and prevalent. Some of these mites (e.g., *Tetranychus urticae*) produce considerable amounts of fine silken webbing on the plant surface but other produce little or no webbing.

The presence of mite pests may go unnoticed because of their minute size. However, symptoms of attack are often apparent before the mites themselves can be seen. For control, it is important to be able to recognize mite injury at an early stage. Mites normally feed at the undersurface of the leaves but the symptoms are more easily seen on the upper surface. Two kinds of symptoms are produced in mites. While feeding carried out by tetranychids produce blotching on the leaf surface, that by tarsonemids and eriophyids produce distortion, puckering, or stunting of leaves and other parts of the plants. Some species of eriophyids produce distinct galls or blisters.

In India, 17 species of mites are considered as major pests and 30 as minor pests on different crops (Gupta, 1991; Singh and Mukherjee, 1989). The major plant feeding mites belong to families of spider mite (Tetranychidae), false spider mite (Tenuipalpidae), gall mite (Eriophyidae), and yellow or broad mite (Tarsonemidae).

The tetranychid mites, also known as spider mites, are widely spread and polyphagous in nature. They are serious pests of agricultural crops, vegetables, ornamentals, and fruits. The pest status of many potential mites has been described (Meyer, 1981; Singh and Mukherjee, 1989; Mori, et. al. 1990; Gupta, 1991). Indian Tetranychidae consists of a total of 101 species with 2 subfamilies, 6 tribes, and 10 genera (Gupta and Gupta, 1994). Recently, many species of tetranychid mites have assumed the status of major pest, which may be host-specific or infest a wide variety of plants. Mostly tetranychid mites feed on both surfaces of leaves. Pillai and Palaniswamy (1985) reported that on cassava the red mites, *Tetranychus cinnabarinus* and *T. neocaledonicus* usually feed on the lower surface whereas *Eutetranychus orientalis* and *Oligonychus biharensis* are always found on the upper surface of the leaves. While feeding, mites damage the epidermis with their needle like chelicerae to suck cell sap. This activity reduces chlorophyll content and leads to formation of numerous empty cells at a site resulting in formation of yellow or brown specks or stipples. Extensive feeding by large number of mites causes the leaves to appear yellow or brown. Upon severe injury, the leaves shrivel, die, and eventually drop off from the plant. Some species of tetranychid mites produce copious webs. Overcrowding of mites may also occur on leaf tip, fruit tip, and then they might migrate to other fields. The swarming phenomenon is characterized by hot spot of ballooning or foci. The male and female sexes are common in most of the species of tetranychid mites. Males are very less in ratio. The unfertilized eggs produce only male offsprings whereas fertilized eggs produce female only. Van de vrie et. al. (1972) and Jeppson et. al. (1975) have reported that the mated females produce both females and males because every egg has not received a spermatozoon.

The development of tetranychid mites takes place through egg, larva, protonymph, deutonymph, and adult stages.

The tenuipalpid mites are known as false spider mites. They are widely distributed and are more numerous in warmer zones of the globe (Baker and Tuttle, 1987). Meyer and Smith (1979) reported 504 species in 21 genera for the world fauna. Jeppson, et al. (1975) has described many mites of this group as pests of economic plants, such as ornamentals and fruit trees. These mites look like spider mites but do not spin silken web. They are flat, pear shaped, mostly reddish bright colored, size 0.3–0.35 mm, slow moving, and are normally found on the undersurface of the leaves near midrib or veins, twigs, and fruits. Most specialized members of this family form plant galls within which they feed (Jeppson, et al., 1975). The genera *Brevipalpus* and *Tenuipalpus* are of great economic importance in India. The first observable symptom on plant appears on the leaves as silvery areas that frequently become shrunken and brown. Seriously infested leaves become yellow and drop off from the plant.

Eriophyidae mites are commonly known as gall, rust, bud, blister, and erineum or simply eriophyid mites. They are exclusively plant feeders and host specific are of economic importance. Members of this family have two body regions, the gnathosoma and idiosoma. Body is elongated wormlike with two pairs of legs and size varying from 0.1 to 0.5 mm. Life cycle is simple but certain species may have complicated cycle that includes alternation of generations. There are two forms, one is an over-wintering female called deutogynes and the other consists of both sexes known as protogynes. The symptoms caused by eriophyid mites vary from simple rusting to coupled leaf galls, bud galls, or erineum formation. Witches broom, which is cluster of brush like eriophyid mites also cause growth of stunted twigs or branches on trees. Eriophyid mites also cause nongall abnormalities such as leaf folding and twist blisters. Ghosh et al. (1989) described the gall mites causing injuries such as big buds, curling/crinkling, and shrinkage of leaves, discoloration, erineum, rusting, and galls. These mites also induce rusting and silvering of leaves. Kiefer et al. (1982) described various symptoms of injuries caused by eriophyid mites on buds, shoots, stems, twigs, flowers, and fruits. Large number of eriophyid mites is confined to one species of host plant. Some eriophyid mites also act as vector of plant viral diseases transmitting pigeon pea sterility mosaic disease, fig mosaic disease, wheat streak mosaic disease, and sugarcane streak mosaic virus. About 1859 eriophyide mites belonging to 156 genera have been reported from different parts of the world (Boczek, et al., 1989; Davis, et al., 1982; Keifer, et al., 1982).

Tarsonemid mites are very small, 100–300μm in length. Mature forms have a relatively hard and shiny integument just like glistering white or shiny silver piece. The body and posterior legs are sparsely beset with setae. The terminal segment of the anterior pair of legs are more densely clothed with setae and are often equipped with specialized sensory setae of various configurations and sizes. These mites inhabit on the undersurface of leaves as well as in association of fungus. Males often carry female deutonymphs. Some of these mites are important crop pests (Gupta, 1985; Meyer, 1981; Singh, et al., 1987).

The outbreaks of such phytophagous mites on several crops have attracted the attention of agricultural scientists. Besides this, very limited documents explore the mite infestation on spice crops. Since the knowledge of spice crops infested by mites is necessary if the yield is affected due to lack of pest management, therefore, in the line of that the chapter was aimed to give a detailed account and elucidate some of the major mites infesting the spice crops and their management. Brief description of symptoms is also included to assist in the identification of specific mite pests.

6.2 IMPORTANT MITES OF SPICE CROPS

Yellow mite, Chilli mite, *Polyphagotarsonemus latus* Banks

Class	: Arachnida
Subclass	: Acari
Order	: Trombidiformes
Family	: Tarsonemidae
Genus	: *Polyphagotarsonemus*
Species	: *latus*

Distribution

Polyphagotarsonemus latus has worldwide distribution. It was first described by Banks in 1904 as *Tarsonemus latus* from the terminal buds of mango in a greenhouse in Washington, D.C., USA. It is known by a number of common names. In India and Sri Lanka it is known as yellow tea mite while in Bangladesh it is called as yellow jute mite. European countries call it as broad spider while in some parts of South America it is called as tropical mite or broad rust mite.

Marks of identification

Females are about 0.2 mm long, oval in outline, swollen body, light yellow to amber, or green in color with indistinct light median stripe forking near the back end of the body. Males are similar in color but lack stripe, 0.11 mm long faster moving than females.

Biology

Adult female lays colorless, elliptical 30–76 eggs (average 5 per day) on the underside of the leaves and in the depressions over an 8–13 day period and the die. Adult males may survive for 5–9 days. Mated female usually lays four female eggs for every male egg while unmated females lay eggs that become male. Eggs hatch in 2–3 days and slow moving larvae emerges out to feed. After 2–3 days, larvae develop into quiescent larval (nymph) stage. Quiescent female larvae are attracted by males and carried by them to the new foliage where they mate with them immediately after becoming adult.

Host plants

Chilli, peppers, seed spice, citrus, apple, avocado, cantaloupe, castor, coffee, grapes, guava, pear, potato, jute, mango, papaya, pole beans, tea, tomato, and sesame.

Nature of damage and symptoms of infestation

Adults and nymphs of these pests suck sap from the leaves and growing shoots. Affected leaves curl upward and downward resulting in damage called chilli leaf curl or chilli murda complex. Infestation is a characterized by elongation of leaf petiole and clustering of tender leaves at the tip of branches.

Red spider mite, Two-spotted spider mite, Glasshouse spider mite, Carmine spider mite, *Tetranychus urticae* Koch

Class	:	Arachnida
Subclass	:	Acari
Order	:	Trombidiformes
Family	:	Tetranychidae
Genus	:	*Tetranychus*
Species	:	*urticae*

Distribution

Tuttle and Baker in 1968 reported this species on the deciduous fruit trees in northern regions of United States and Europe. It is generally considered as a temperate zone species but is also found in the subtropical regions.

Marks of identification

Body oval-shaped, about 1/50″ long, green, greenish yellow, or translucent colored polyphagous mites. Females are about 0.4 mm in length with elliptical body bearing 12 pairs of dorsal setae. Males are elliptical with caudal end tapering and smaller than females.

Biology

Adult female lives 2–4 weeks, primarily on the underside of the leaves and is capable of laying several hundred eggs during her life. The eggs are attached to fine silk webbing and hatch in approximately 3 days. The life cycle is composed of the egg, the larva, two nymphal stages (protonymph and deutonymph) and the adult. The length of time from egg to adult varies greatly depending on the temperature. Under optimum conditions, spider mites complete their development in 5–25 days with many generations per year. The spider mites prefer hot dry weather of summer but may occur anytime throughout the year.

Host plants

Garlic, onion, pepper, turmeric, cardamom, curry leaves, wild mustard, vine, bean, cucumber, hop, cotton, clover, sunflower, and fruit trees.

Nature of damage and symptoms of infestation

Mites feed the plant tissues, remove the sap and the mesophyll tissue collapses forming a small chlorotic spot at each feeding site and feeding causes graying or yellowing of leaves. Continued feeding may cause stippled-bleached effect. Necrotic spots occur in the advanced stages of the leaf damage. Complete defoliation may occur in case of severe infestation.

The southern red mite, *Oligonychus ilicis* McGregor

Class	:	Arachnida
Subclass	:	Acari
Order	:	Trombidiformes
Family	:	Tetranychidae

Genus   : *Oligonychus*
Species : *ilicis*

Distribution

Pritchard and Baker in 1955 suggested its origin in the Far East. The general distribution includes the Northern Hemisphere and South America. Published distribution records includes: Brazil, Italy, Japan, Korea, the Netherlands, Paraguay, and the United States.

Marks of identification

The adult female is approximately 385μm (1/50") in length with a rotund-elliptical body. The adult male is approximately 300μm in length, much less rotund, and narrowed posteriorly. Both the sexes are reddish brown, and darker than most red spider mites. This species is more translucent towards the front end of the body.

Biology

Many generations occur each year, but population densities are highest during the cooler months of spring and fall when prolonged periods of high humidity occur. The species overwinters as red eggs on the undersides of leaves. However, darker summer eggs can be abundant on preferred hosts when infestations are not controlled. While some mites may be active during the summer months, most of the population is dormant (in aestivation) during the summer heat.

Host plants

Bay leaf, curry leaf, cinnamon, American sycamore, boxwood, camphor tree, coffee, meadow grass, walnut, eucalyptus, guava, hibiscus, juniper, sweet pepper bush, tea, and pyracantha.

Nature of damage and symptom of infestation

This mite feeds primarily on foliage. It usually attacks the lower leaf surface and as the population increases it moves toward the upper surface of the leaves and on to small succulent stems. It injures the leaves causing a graying, stippling or mesophyll collapse, "firing," and defoliation. The leaves may be distorted if infestation occurs when they are young and expanding.

Bulb mite, *Rhizoglyphus sps.*

Class	:	Arachnida
Subclass	:	Acari
Order	:	Sarcoptiformes
Family	:	Acaridae
Genus	:	*Rhizoglypus*

Distribution

Rhizoglyphus sps. were discovered in Europe but are now been found throughout the United States, Canada, Japan, China, Cuba, France, Argentina, Chile, Egypt, Taiwan, Turkey, former USSR, and the Bermuda islands. These mites are easily transported in shipment of infested bulbs.

Marks of identification

Adult mites are shiny, white bodied, somewhat transparent, smooth, 0.5–0.9 mm long, and have four pair of legs. They have reddish brown appendages.

Biology

Bulb mites appear in large colonies. All the stages of this mite can be found throughout the year. Development occurs in five or six stages (a hypopal stage is sometimes produced). In those forms with six stages, the life cycle proceeds from egg to larva to protonymph to hypopus to tritonymph to adult. Females lay up to 700 white, tanslucent, ellipsoidal, 0.12 mm long eggs each depending on the host. *R. echinopus* forms large colonies on a greater range of bulb crops. The mites can survive by feeding on paper and other sources of organic matter. Eggs mature in 5.1–27 days. The total life cycle from egg to adult could be as short as 13.9 days (at 25°C) depending on the host bulb, temperature, and relative humidity. Adults live longer at lower temperatures (up to 121 days) and males tend to live twice as long as females. The mites can survive at 35°C, but they cannot lay eggs at that temperature. On the other hand these mites cannot develop at temperatures below 11.8°C. The length of development is greatly dependent on temperature, relative humidity (100% is best), and available food. Hypopi form when the population becomes crowded, or the substrate becomes too polluted by decay. The hypopal stage attaches to insects visiting the bulbs and may be carried to other bulbs. Hypopi do not feed (no head), and they are resistant to starvation and desiccation during adverse conditions. The ratio of males to females varies from 1 to 1, to 1.9 to 1, depending on relative humidity, diet, and perhaps other factors. Besides their direct feeding, bulb mites are a threat because they carry pathogenic fungi.

Host Plants

Onion, garlic, leek, shallot, forced iris, lily, Narcissus, *Gloriosa*, *Hippeastrum*, *Eucharis*, orchid, hyacinth, tulip bulbs, dahlia tubers, freesia, and gladiolus corms.

Nature of damage and symptom of infestation

Bulb mites attack healthy new roots and corms, especially in greenhouses. The mites may penetrate into stems which become brittle. Stunted and distorted growth; rotted bulb.

Dry bulb mite, Wheat curl mite, Garlic mite, Onion mite *Aceria tulipae* **Keifer**

Class	:	Arachnida
Subclass	:	Acari
Order	:	Prostigmata
Family	:	Eriophyidae
Genus	:	*Aceria*
Species	:	*tulipae*

Distribution

Africa, Central America, Canada, Europe, Fiji, Asia, New Zealand, UK, USA, former USSR, present NSW, South Africa, Tasmania, and Victoria

Marks of identification

Body is 1/100 inch long, minute, elongated, spindle-shaped with two pairs of leg.

Biology

Adult mites actively move to young and newly emerged leaves, on the upper surface of the leaves. They lay their eggs along the veins of the leaf. Young ones feed on the leaves and move to top of the plant. When the host plant matures and dries out, mites move back to the newly emerged.

Host plants

Onion, garlic, shallot, leek, spring onion, cocksfoot

Nature of damage and symptom of infestation

Members of *Aceria tulipae* feed on the young leaves and between the layers in bulbs of onion, garlic, and so on.

Infested plants are dwarf and leaves are abnormal and wilted. Feeding causes stunting, twisting, curling, and discoloration of foliage and scarification and drying of bulb tissue. Green-yellow color streaks formation, abnormal, and wilted leaves

Eriophyid mite, the Gall mite, *Aceria doctersi* Nalepa

Class	:	Arachnida
Subclass	:	Acari
Order	:	Prostigmata
Family	:	Eriophyidae
Genus	:	*Aceria*
Species	:	*doctersi*

Distribution

India, Nepal, Bangladesh, China, Sri Lanka, USA, South Africa, Afghanistan, Pakistan, Russia, Turkey

Marks of identification

Tiny, microscopic, worm-like mites, yellow to pinkish white to purplish in color. They have only two pairs of legs and commonly cause galls or other damage to the plant tissues and hence known as gall mites.

Biology

The life cycle is composed of the egg, the larva, two nymphal stages (protonymph and deutonymph) and the adult. The adult female is capable of laying several hundred eggs during her life. Eggs incubate in 2–4 days. The eggs are attached to fine silk webbing and hatch in approximately 3 days. Protonymphal and deutonymphal period is 2–3 days, each. Adult mites survive about 8 days. The body of adult mites is about 200–250 μm in length and 36–52 μm in width. Total life cycle is 7–9 days. The length of time from egg to adult varies greatly depending on temperature.

Host plants

Cinnamon, onion

Nature of damage and symptom of infestation

Both adults and immature feed on the young leaves and between the layers in bulbs of onion. Feeding of adults and young ones causes stunting, twisting, curling and discoloration of foliage and scarification, and drying of bulb tissue in onion. Large irregular, conical shaped, and hairless galls are formed on the under surface of leaves, on leaf stalks, or on new stems in cinnamon. Individual galls become fused to form complex, irregular, massive structures, covering the entire leaf lamina including the midrib, veins, and vein lets. These further result in severe distortion and subsequent drying up of leaves due to hindrance in the photosynthetic efficiency of the plant (Nasareen et al., 2014).

Brown wheat mite, Legume mite, Oxalis spider mite, *Petrobia latens* Muller

Class	:	Arachnida
Subclass	:	Acari
Order	:	Prostigmata
Family	:	Tetranychidae
Genus	:	*Petrobia*
Species	:	*lateens*

Distribution

The genus *Petrobia* was first reported by Murray in 1877 and *Petrobia latens* Muller was designated as type species. The mite has cosmopolitan distribution.

Marks of identification

Body is about 0.5 mm long, red-brown to black, legs long and yellow, forelegs distinctly longer than the other legs. Claws pad-like; empodium hooked and with 2 pairs of tenent hair. Two types of eggs are found, bright-red nondiapause eggs and the white, capped diapause egg.

Biology

This parthenogenetic mite may complete a life cycle within a fortnight, raising three annual generations. The mite lays about 50–90 nondiapaused brick red eggs or 30 diapause eggs on straws and the dried leaves in winters and autumn. The larva hatches in following spring. The newly hatched larva are colorless with three pairs of legs but later changes to brown color. Newly molted brown protonymph bears four pairs of legs. The deutonymph resembles the adult in appearance except in size. The matured females are metallic in color with 13 pairs of dorsal setae and 6 segmented legs. The incubation period lasts 10.6 days. The mean larval, protonymphal and deutonymphal periods lasts for 1.4, 1.5, and 2.5 days, respectively. The mean duration taken from egg to egg stage is 25.5 days.

Host plants

Onion, garlic, coriander, cumin, fenugreek, fennel, wheat, barley, rice, oat, melon, cucumber, carrot, strawberry, cotton, lettuce, apple, sword lily.

Nature of damage and symptom of infestation

The larva, nymph, and adult cause webbing on the underside of the leaves and feed within the web. The mite first pierces cells on and near the leaf surface with its stylets and then sucks the sap from the leaf.

Feeding causes fine whitish or silverish mottling or stippling of the leaf. Prolonged feeding brings increased mottling and stippling as chlorophyll is withdrawn and the leaf assumes an overall pale yellow to bronze color. Eventually the damaged leaf turns brown and dies. Most infected plants exhibit moderate to severe dwarfing. Brown streaks appear on the leaves of extremely diseased plants and in some cases the plants die.

6.3   MANAGEMENT OF PHYTOPHAGOUS MITES IN SPICE CROPS

Cultural control

a.   **Regular Monitoring**. Brown wheat mites, *Petrobia latens* Muller, feed during the day, and the best time to scout for them is in midafternoon. Being dark-red, they are easily seen and accurate counting can be done by tapping plants over white paper and counting the dislodged mites. The presence of 30 or more mites/plant calls for control measures. In the Eriophyid mite, The Gall mite, *Aceria Doctersi* Nalepa, monitoring the presence of gall mites by periodic examination of the undersides of leaves; in large trees and removing infested leaves and branches or whole plant is effectively practiced.

b. Monoculture cropping should be avoided.

c. Intercropping using nonhost plants should be done.

d. Crop rotation and deep ploughing to destroy the diapause eggs in the soil and selection of a planting date so that the crop develops during weather conditions least favorable to the Brown wheat mites, *Petrobia latens* Muller, may reduce the probability for mite infestation.

e. Removal of host weeds to minimize source of re-infestation. The destruction of weeds or the clearing of fence rows reduces infestation. To control the Red spider mites, *Tetranychus urticae* Koch, destruction of weeds around the field in fall or early spring reduces the overwintering population.

f. Rough handling of bulbs such as onion, garlic, leek, shallot, and so on. should be avoided to prevent an injury which might afford an entry point for dry bulb mites, *Aceria tulipae Keifer, and* Bulb mites, *Rhizoglyphus sps.*

g. Spider mites are distributed over a field in two ways: (1) migration of females from a heavily infested area to a lightly infested one, and (2) natural or mechanical transportation of mites (by wind, mammals, and man or his machinery) from an infested area to an uninfested one. Man often spreads the mite inadvertently by walking from an infested area to an uninfested area, carrying mites on his pants legs. Therefore, known "hot spots" should be investigated last.

Mechanical control:

a. Striking the foliage sharply to fall off the mites after detection through regular monitoring is effectively practiced to control mites attack.

Physical control:

a. High humidity reduces the reproductive potential of tetranychids whose optimal environment is provided by hot and dry air (Sabelis, 1986; Duso et al., 2004).

b. Steam sterilization and methyl bromide at low concentrations eliminates the bulb mites from soil.

c. Lorenzato (1984) found that soap mixed with wet sulfur or mineral oil was the most effective of several disinfesting solutions tested against *Aceria tulipae* in stored garlic.

d. Almaguel et al. (1986) evaluated a range of chemical sterilants and water treatments for control of *Aceria tulipae* in garlic bulbs. Water treatments of 10, 15, and 20 minutes duration were most effective at 55°C, whereas dips of 10 and 15 minutes were most effective at 60°C. Although some water treatments reduced mite infestation, their use is not recommended because of the reduction in the clove sprouting capacity.

e. Application of horticultural soaps and oils at proper timing offer mite control and suppression.

f. Application of a layer kaolin clay covering foliage (top and bottom) may be effective to control the Gall mites, *Aceria Doctersi* Nalepa.

g. Full coverage with sulphur may control the gall mites, *Aceria Doctersi* Nalepa but its application should be avoided during periods of high humidity.

h. Host plant nutrition may have a positive effect on the reduction of the mite population by varying the fertilizer regime applied to the crops (Markkula and Tiittanen, 1969), large amount of nitrogen or a deficiency of potassium can increase the amount of soluble nitrogen available in the plant resulting sharp increases in the populations of mites (Sabelis, 1986). Since the plant response to such extreme regimes is economically unviable, so variations in host nutrition have not been used for pest control (Helle and Sabelis, 1985).

Chemical control:

1. Seed treatment with Imidacloprid 600 FS at 5 ml per kg seed is effective against yellow mites.

2. Spraying the acaricides such as Dicofol at 5 ml per liter or wettable sulphur 3 g per liter or diafenthiuron at 1 g per liter or Vertemic at 0.5 ml per liter is used for the management of yellow mites. Diafenthiuron (300 g a.i./ha) is reported to show the best efficacy

against motile stages followed by milbemectin (3.5 g a.i./ha) and propargite (1000 g a.i./ha) for the management of *Polyphagotarsonemus latus* in chilli (Chakrabarty and Sarkar, 2014).

3. Utilization of indigenous materials have confirmed that garlic chilli kerosene extract [GCK at 0.5%] along with Nimbecidine (2.5 ml/lit) can effectively combat the problem with yellow mites on spice crops.

4. Several available acaricides/miticides may provide effective chemical control against red spider mites. Since recommended miticides do not act as ovicides, a second application is often advised 5–7 days after the first.

5. Bulb mites are very tolerant to a number of synthetic pesticides apparently due to active oxidases, esterases, and transferases that detoxify such chemicals. Soaking bulbs in a miticide before planting has been shown to prevent bulb mite injury.

6. Bulbs treated with citral at 100 ppm, and a mixture of citral and a miticide gives significantly better control of bulb mites than the miticide by itself.

7. Sulfur dusting of plants in the field has been recommended, and dipping bulbs in pesticides just prior to planting is also effective against dry bulb mites (Doreste, 1963, 1965).

8. Over 90% mortality in dry bulb mite is achieved when infested garlic is fumigated with phosphine for 72 h in airtight plastic boxes (Na Seung Yong et al., 1998).

9. Any insecticides and miticides with active ingredients such as Abamectin benzoate, Bifenthrin, Carbaryl, Deltamethrin, Imidacloprid, Permethrin, and Pyrethrin could be used on severe infestation with gall mites.

Organophosphates and some fungicides have provided good control in the field against attack of *Petrobia latens*. However, decisions on controlling this pest are difficult because of its, sporadic, weather-dependent occurrence and uneven rates of damage.

Fortunately, the acaricides resistance has not developed so far due to their limited use in India. However, many reports are available on pesticides resistance in spider mites (Dittrich, 1975; Jeppson, et al., 1975; Van de vrie, 1973). The spider mites are potential danger for many crops because of their ability to develop resistance to chemicals, which initially give effective control (Herne et al., 1979). Cranham and Helle

(1985) discussed the pesticides resistance in Tetranychidae. Goodwin et al. (1991) reported about the miticide resistance in two spotted mite which had caused serious difficulties in maintaining product quality of flowers in Australia. Hoy et al. (1980) and Hoy and Knop (1981) reported increased tolerance in phytoseiid mites against pyrethroids. Carolyn and Spooner-Hart (1991) reported tolerance in *Phytoseilus persimilis* against improved organophosphate.

Biological control

Spices are high value and export oriented in nature and pesticide residues are most important nontariff barriers on trade of these products, therefore, biological control is more relevant in these crops. Biological control provides an environmentally safe, cost effective, and energy-efficient means of pest control either alone or as a component of IPM (Hoy, et al., 1983). Rasmy (1989) recommended biocontrol using predacious mites in mite pest management programme and suggested to follow IPM approach: (1) by utilizing naturally occurring predacious mites, (2) by releasing predatory mites, and (3) by utilizing pesticide resistant strains of predacious mites. Phytoseiid mites are well accepted bioagents for phytophagous mite control. McMurtry (1983) reported that most of the efficient predators of phytophagous mites in orchards belong to family Phytoseiidae.

Utilization of predatory mites

1. The predatory mites, *Amblyseius, Meraseiulus,* and *Phytoseiulus* regulate the spider mite population. Five species of predator mites are commercially available in the United States, *Phytoseiulus persimilis, Mesoseiulus longipes, Neoseiulus californicus, Galendromous occidentalis,* and *Amblyseius fallicus.*
2. For hot arid regions, *Galendromus occidentalis* (a western predatory mite) is effectively used to biologically control the gall mites.
3. Baier (1991) successfully used *Amblysieus barkeri* and *Amblysieus cucumeris* for the control of spider mite in Germany. Lo et al. (1990) reported more than 50 species of spider mites in Taiwan and recorded 30 species of native predators as natural enemies of spider mite.
4. Mori et al. (1990) described the use of native predator *Amblyseius longispinosus* and introduced predator *Phytoseiulus persimilis* against spider mite and concluded that combined action of both

native and introduced predators gave better suppression of tetranychid population under experimental conditions.

5. Osmelak and MacFarlane (1991) described the importance of *Phytoseiulus persimilis* for IPM of two-spotted mite in roses and carnations.

6. Jose et. al. (1989) found predatory mite, *Amblyseius alstoniae*, predatory thrips, *Scolothrips indicus* and beetle, *Stethorus paupers-culus* feeding on all stages of spider mites.

7. Toledo et al. (2013) has established the predatory potentiality of *Euseius alatus* (phytoseiidae) on all the life stages of *Oligonychus ilicis* (Tetranychidae) on coffee leaves.

8. Four predatory mites, *Amblyseius swirskii, Neoseiulus califor-nicus, Neoseiulus fallacies, and Euseius alatus* are used as biolog-ical agents to control the southern red mites, preferring the larvae over the adults.

9. A predaceous mite, *Cosmolaelaps claviger*, feeds readily on *Rhizoglyphus spp.* and other soil organisms.

Biological control through insects

Some of the insects which are voracious feeders on mites are as follows:

i. *Stethourus pauperculus* (order Coleoptera: Coccinellidea), some of the coccinellid beetles cannot *survive without* mite as food

ii. *Scolothrips indicus* (order Thysanoptera: Thripidae) and

iii. *Chrysopa* sp. (order Neuroptera).

Jose et. al. (1989) found predatory thrips, *Scolothrips indicus,* and beetle, *Stethorus pauperculus* feeding on all stages of spider mites.

Singh and Ray (1977) identified *Stethorus* beetle as a predator of *Tetranychus neocaledonicus.*

Host Plant Resistance

Certain wheat and barley cultivars have shown variable suscepti-bility to Wheat curl mite, *Petrobia latens* attack in India. Fewer mites occurred on these tolerant partially resistant cultivars in the field. However, susceptibility on spice crops to this mite attacks remains to be elucidated.

Genetical control

Genetic control is a form of biological control which exploits the pests mating ability by introducing genetic abnormalities, hence sterilizing it or through genetic manipulation of the sex ratio. Dicofol is an effective sterilant to control the dry bulb mites; its activity enhances when added in combination with mineral oil.

Repellents and Pheromones:

Rhizoglyphus spp. has an alarm pheromone, citral, which although not toxic to the mites, is definitely a repellent. This alarm pheromone makes the mites more active and increases their contact with the pesticide.

Botanical control

1. Neem oil (Azadirachtin) may be effective for control of gall mites.
2. Contact acaricidal activity of plant extracts have been well documented against several species of mites (Isman 2000, 2001; Gokce et al., 2011), including *T. urticae* (Roh et al., 2011).

KEYWORDS

- **integrated mite management**
- **spice crops**
- ***Tetranychus urticae***
- ***Oligonychus ilicis***
- ***Rhizoglypus spp.***

REFERENCES

Baier, B. Relative Humidity-a Decisive Factor for the Use of Oligophagous Predatory Mites in Pest Control. *In Modent Acarology;* Dusbabek, F.; Bubva, V., Eds.; Academia: Prague, 1991; Vol. 2, pp 661–665.

Baker, E. W.; Tuttle, D. M. The False Spider Mites of Mexico (Tenuipalpidae: Acari). *U.S. Dep. Agric. Tech. Bull.* **1987**, *No. 1706.*

Bakkali, F.; Averbeck, S.; Averbeck, D.; Idaomar, M. Biological Effects of Essential Oils—a Review. *Food Chem. Toxicol.* **2008**, *46*, 446–475.

Bhattacharya, H. K. Mite Spices: Present Scene and Prospects. *Yojana* **1998**, *2* 16–20.

Boczek, J. H.; Shevtchenko, V. G.; Davis, R. Genetic Key to World Fauna of Eriophyid Mites (Acarina: Eriophyoidea). Warsaw Agric. Uni. Press: Warsaw, Poland, 1989.

Carolyn, D.; Spooner-Hart, R. Improved Organophosphate Tolerance in *Phytoseiulus persimilis* Athian-Henriot for Integrated Mite Control in Horticultural Crops. In *Proceedings 1st National Conference,* Australian Society of Horticultural Science, Macquarie University, Sydney, 1991, 487–496.

Chakrabarty, S.; Sarkar, P. K. Studies on Efficacy of Some Acaricidal Molecules for the Management of *Polyphagotarsonemus latus* (Banks) (Acari: Tarsonemidae) Infesting Chilli (*Capsicum annuum* L.) in West Bengal. *Current Biotica.* **2014**, *7*, 299–305.

Cranham, J. E.; Helle, W. Pesticide Resistance in Tetranychidae. In *World Crop Pests, Spider Mites, their Biology, Natural Enemies and Control;* Helle, W.; Sabelis, M. W., Eds.; Elsevier: Amsterdam, 1985; pp 405–432.

Davis, R.; Flechtmann, C. H. W.; Boczek, J. H.; Barke, H. E.; Catalogue of *Eriophyid* Mites (Acari: Eriophyoidea) In *1955 "Mite" (Oligonychus), as a Causal Agent for "Ratada" Disease of Sorghum;* Desai, M. K.; Chavda, D. E., (Eds.) *Poona Agri. Coll. Mag.* 1982; Vol. 45, pp 138–141.

Dittrich, V. Acaricide Resistance in Mites. *Z. Angew. Entomol.* **1975**, *78*, 28–45.

Duso, C.; Malagnini, V.; Pozzebon, A.; Castagnoli, M.; Liguori, M.; Simoni, S. Comparative Toxicity of Botanical and Reduced-Risk Insecticides to Mediterranean Populations of *Tetranychus urticae* and *Phytoseiulus persimilis* (Acari Tetranychidae, Phytoseiidae). *Biol. Control* **2008**, *47*, 16–21.

Evans, G. O. *Principles of Acarology*; CAB International: Wallindford, UK, 1992; pp 563.

Ghosh, N. K.; Das, B.; Chakrabarti, S. Injury by Gall Mites (Acari: Eriophoidea) to Plants in North East India. In *Progress in Acarology;* ChannaBasavanna, G. P.; Viraktamath, C. A., Eds.; Oxford & IBH: New Delhi, 1989; Vol. 2, pp 135–140.

Gokce, A.; Isaacs, R.; Whalon, M. E. Ovicidal, Larvicidal and Antiovipositional Activities of *Bifora Radians* and Other Plant Extracts on the Grape Berry Moth *Paralobesia viteana* (Clemens). *J. Pest. Sci.* **2011**, *84*, 487–493.

Goodwin, S.; Gough, N.; Herron, G. Miticide Resistance and Field Control in Two Spotted Mite, *Tetranychus urticae* Koch, Infesting Roses. In *Proceedings 1st National Conference,* Australian Society of Horticultural Science, Macquarie University, Sydney, 1991; pp 325–328.

Gupta, S. K. A Monograph on Plant Inhabiting Predatory Mites of India, Part I, Orders: Prostigmata, Astigmata and Cryptostimata. *Mem. Zool. Surv. India.* **2002**, *19*, 1–183.

Gupta, S. K. A Monograph on Plant Inhabiting Predatory Mites of India, Part II: Order: Mesostigmata *Mem. Zool. Surv. India.* **2003**, *20*, 1–185.

Gupta, S. K.; Gupta, Y. N. A Taxonomic Review of Indian Tetranychidae (Acari: Prostigmata) with Description of Known Species and Keys to Genera and Species. *Mem. Zool. Surv. India* **1994**, *18*, 196.

Gupta, S. K. Hand Book-Plant Mite of India. *Zool. Surv. India Calcutta* **1985**, 1–520.

Gupta, S. K. The Mites of Agricultural Importance in India with Remark on their Economic Status. In *Modern Acarology;* Dusbabek, F.; Bukva, V., Eds.; Academia: Prague. **1991**, *1*, 509–522.

Helle, W., Sabelis, M. W., Eds.; *World Crop Pests, Spider Mites, their Biology, Natural Enemies and Control;* Elsevier: Amsterdam, 1985.

Herne, D. H. C.; Cranham, J. E.; Easterbrook, M. A. New Acaricides to Control Resistant Mites. In *Recent Advances in Acarology*; Rodriguez, J. G., Ed.; Academic Press: New York, USA, 1979; Vol. 1.

Hoy, M. A.; Knop, N. F.; Joos, J. L. Pyrethroid Resistance Persists in Spider Mite Predator. *Calif. Agri.* **1980**, *34*, 11–12.

Hoy, M. A.; Knop, N. F. Selection for and Genetic Analysis of Permethrin Resistance in *Metaseiulus Occidentalis:* Genetic Improvement of a Biological Control Agent. *Entomol. Exp. Appl.* **1981**, *30*, 10–18.

Isman, M. B. Pesticides Based on Plant Essential Oils for Management of Plant Pests and Diseases. In *Proceeding of Symposium on Development of Natural Pesticides from Forest Resources,* Republic of Korea, Seoul. 2001, pp 1–9.

Isman, M. B. Plant Essential Oils for Pest and Disease Management. *Crop Prot.* **2000**, *19*, 603–608.

Jeppson, L. R.; Keifer, H. H.; Baker, E. W. Mites Injurious to Economic Plants. *Univ. Calf. Press*, 1975; pp 614.

Jose, V. T.; Shah, A. H.; Patel, C. B. Feeding Potentiality of Some Important Predators of the Spider Mite, *Tetranychus macferlanei,* a Pest of Cotton. In *Progress in Acarology;* ChannaBasavanna, G. P., Viraktamath, C. A., Eds.; Oxford & IBH: New Delhi. 1989; Vol. 2, 357–360.

Keifer, H. H.; Baker, E. W.; Kono, T.; Delfinado, M.; Styer, W. E. An Illustrated Guide to Plant Abnormalities Caused by Eriuphyid Mites in North America. *U.S.D.A Hand book No.* 573, **1982**.

Kotikal, Y. Y.; Kulkarni, K. A. Insect Pests Infesting Turmeric in Northern Karnataka. *Karnataka. J. Agri. Sci.* **2000**, *13*, 858–866.

Krantz, G. W. Habits and Habitats. In *A Manual of Acarology;* Krantz G. W.; Walter, D. E., Eds.; Texas Tech Uni. Press: Lubbock, Texas, U S A. 2009; pp 64–82.

Kumar, Y. Diagnostic Symptoms and Loss Assessment Due to Insect Pests in Spices. In *Advances in Diagnosis of Arthropod Pests Damage and Assessment of Losses;* Saini, R. K.; Mrig, K. K.; Sharma, S. S., Eds.; C C S Haryana Agric. Uni.: Hisar, 2011; pp 153–157.

Lo, Kang-Chen, Lee; Wen-Tai., Wu; Tze-Kann; Ho, Chyi-Chen. Use of Predators to Control Spider Mites (Acarina: Tetranychidae) in the Republic of China on Taiwan. In *The Use of Natural Enemies to Control Agricultural Pests;* FFFC Book, 1990, Vol. 40, pp 166–178.

Markkula, M.; Tiittanen, K. Effect of Fertilisers on the Reproduction of *Tetranychus telarius* (L.), *Myzus persicae* (lz) and *Acyrthosiphon pisum* Harris. *Ann. Agric. Fenn.* **1969**, *8*, 9–14.

Mc Murtry, J. A. The Use of Phytoseiid for Biological Control Progress and Future Prospects. In *Recent Advances in Knowledge of the Phytoseiidae;* Viraktamath, M. A., Ed. Univ. of California Publication: U S A. 1983; pp 23–48.

Meyer, M. K. P. Mite Pests of Crops in Southern Africa. *Sci. Bull. Dept. Agric. Fish Repub. S. Afr. No.397.* **1981**, 92.

Meyer, M. K. P. The Tenuipalpidae (Acari) of Africa, with Keys to the World Fauna. *Entomol. Memp. Dep. Agric. Tech. Servo.* **1979**, *50*, 135.

Mori, H.; Saito, Y.; Nakao, H. Use of Predatory Mites to Control Spider Mites (Acarina: Tetranychidae) in Japan. In *The Use of Natural Enemies to Control Agricultural Pests;* FFFC Book, 1990; Vol. 40, pp 142–156.

Nasareen, P. N. M.; Vibija, C. P.; Ramani, N. Alterations in The Photosynthetic Pigments of *Cinnamomum Verum* (Presl.) Due to Infestation by The Gall Mite, *Aceria Doctersi* (Nalepa, 1909) (Acari: Eriophyidae). *Indian J. Appl. Res.* **2014**, *7*, 528–530.

Osmelak, J. A.; MacFarlane, J. Integrated Pest Management of two Spotted Mite in Roses and Carnations. In *Proceedings 1st National conference,* Australian Society of horticultural Science, Macquarie University, Sydney. 1991; pp 337–344.

Paul, A.V. N. Insect Pest of Spices and Condiments. In *Insect Pests and their Management*, Biological Control Laboratory, Division of Entomology, Indian Agricultural Research Institute, New Delhi, 2007; pp 56–57.

Pillai, K. S.; Palaniswamy, M. S. Spider Mites of cassava. *Tech. Bull.* Sr.1. Central Tuber Crops Research Institute, Trivandrum, 1985.

Rasmy, A. H. Prospects of Biological Control in Integrated Management of Mite Pests. In *Progress in Acarology*; Channa, B. G. P.; Viraktamath, C. A., Eds.; Oxford & IBH, New Delhi, 1989, Vol. 2, pp 337–341.

Roh, H. S.; Lim, E. G.; Kim, J.; Park, C. G. Acaricidal and Oviposition Deterring Effects of Santalol Identified in Sandalwood Oil Against the Two-Spotted Spider Mite, *Tetranychus urticae* Koch (Acari: Tetranychidae*). J. Pest. Sci.* **2011**, *84*, 495–501.

Sabelis, M. W. Sampling Techniques. In *World Crop Pests, Spider Mites, their Biology, Natural Enemies and Control*; Helle, W.; Sabelis, M.W., Eds.; Elsevier: Amsterdam, 1985; pp 337–348.

Singh, J.; Ray, R.; *Stethorus* sp. (Coleoptera: *Coccinellidae), a* Predator of *Tetranychus neocaledonicus* Andre on okra at Varanasi. *Acarology Newsletter* **1977**, *4*, 5.

Singh, J.; Singh, R. K.; Mukherjee, I. N.; Singh, R. N.; Agarwal, L. Mites of Agricultural Importance and their Management in India. In *Recent Advances in Entomology;* Mathur, Y. K.; Bhattacharya, A. K.; Pandey, N. D.; Upadhyaya, K. D.; Srivastava, J. P., Eds.; Gopal Prakashan: Kanpur, 1987; pp 169–185.

Singh, J; Mukherjee, I. N. Pest Status of Phytophagous Mites in Some Northern States of India. In *The First Asia-Pacific conference of Entomology (APCE).* Entomol. Zool. Assoc.: Thailand, 1989; pp 192–203.

Singh, N. P. Manual on Diagnosis of Insect Pest and Diseases of Plantation and Spice Crops of Goa. ICAR Research Complex for Goa, 2014; pp 50–51.

Toledo, M. D.; Reis, P. R.; Silveira, E. D.; Marafeli, P. D. P.; Pimentel, G. D. S. Predatory Potentiality of *Euseius alatus* (Phytoseiidae) on Different Life Stages of *Oligonychus ilicis* (Tetranychidae) on Coffee Leaves Under Laboratory Conditions. *Neotrop. Entomol.* **2013**, *42*, 185–190.

Van de vrie, M.; McMurthy, J. A.; Huffaker, C. B. Ecology of Tetranychid Mites and their Natural Enemies: A Review III. Biology, Ecology and Pest Status and Host-Plant Relation of Tetranychids. *Hilgardia* **1972**, *4*, 342–432.

Van de vrie, M. Studies on Prey-Predator Interactions between *Panonychus ulmi* and *Typhlodromus (A.) potentillae* (Acarina: Tetranychidae: Phytoseiidae) on Apple in The Netherlands. In *Proceeding of FAD Conf. Ecol.* Relations Plant Pest Control: Rome, 1973; pp 145–160.

Walter, D. H. P. *Mites: Ecology, Evolution and Behavior*; CABI Publishing, University of New South Wales Press: Sydney, New South Wales, 1999; pp 322.

CHAPTER 7

PESTICIDE RESIDUES IN VEGETABLES

PRITAM GANGULI

Department of Soil Science, Jute Research Station,
Bihar Agricultural University, Sabour 852201, Bihar, India

E-mail: pritam0410@gmail.com

ABSTRACT

In the recent years, the consumption of pesticides has shown a downward trend from 75,000 MT in 1991–1992 to around 41,822 Metric Tons in 2009–2010 due to popularization of Integrated Pest Management (IPM) approach which includes cultural, physical and mechanical, biological and need based use of safest chemical pesticides including neem based bio-pesticides in harmonized manner as well as use of low dose new molecules, ban on the Heptachlor, Chlordane and BHC etc. and the cultivation of Bt Cotton etc. Per hectare consumption of pesticide in India is 381 g a.i. (active ingredient), which is much less compared to the world average of 500 g a.i. But the problem associated with pesticides lies in its application. Excessive and indiscriminate use of pesticides not only increases the cost of production but also createstoxicity problems to non-target organismas well as causes environmental pollution. There are several reports available on pesticide contamination in vegetables in India. In a study conducted with vegetable samples taken from local market basket of Lucknow, residues of HCH, permethrin-II, dichlorvos and chlorofenvinfos were found above MRL in the samples of radish, cucumber, cauliflower, cabbage and okra.Vegetable samples were collected for pesticide residue analysis from different agricultural fields of central Aravalli region, when they were ready for transportation to market. About 40.11% of total analyzed samples were contaminated with different pesticide residues, among which 35.62% of total contaminated samples were exceeded the maximum residual limit

(MRL) values. Vegetable samples collected at harvest from farmer's fields around Hyderabad and Guntur showed HCH &mancozeb residues above MRL.

7.1 INTRODUCTION

Pesticides are the substances or mixture of substances which are intended to control pests those are having enough potency to harm at any stage of crop production, storage, transport, or at the time of processing. A pesticide may be chemical or biological agent which can destroy, prevent, attract, or repel pests in aim to get assured production both in terms of quality and quantity. These are broadly classified by target organism (e.g., insecticides, herbicides, fungicides, rodenticides, etc.), chemical composition (e.g., organic, inorganic, synthetic, biopesticides, etc.), and physical state (e.g., liquid, solid, gaseous, etc.). In this chapter, we shall restrict our discussion to chemical pesticides although biopesticides are also performing well in managing these devastating pests and thus deserve one of the most important subjects for further study.

7.2 STATUS OF PESTICIDE USE IN VEGETABLES IN INDIA

In the recent years, the consumption of pesticides has shown a downward trend from 75,000 MT in 1991–1992 to around 41,822 MT in 2009–2010 due to popularization of integrated pest management (IPM) approach which includes cultural, physical, and mechanical, biological and need-based use of safest chemical pesticides including neem-based biopesticides in harmonized manner as well as use of low dose new molecules, ban on the heptachlor, chlordane, and BHC, and so on. and the cultivation of Bt Cotton and so on.[1] Per hectare consumption of pesticide in India is 381 g a.i. (active ingredient), which is much less compared to the world average of 500 g a.i.[2] But the problem associated with pesticides lies in its application. Excessive and indiscriminate use of pesticides not only increases the cost of production but also creates toxicity problems to nontarget organism as well as causes environmental pollution.

It is well observed that pesticides play an important role in sustaining vegetable production in keeping pest population below

economic threshold. It is estimated that around 13–14% of total pesticides used in the country are applied on vegetables, of which insecticides account for two-thirds of total pesticides used in vegetables.
Among different vegetable crops the maximum pesticide usage is in
chili, brinjal, colecrops, and okra. The global agrochemical consumption on the other hand, is dominated by fruits and vegetables, which
account for 25% of the total market.[2] The effect of chemical pesticide
use is more pronounced in vegetables as these crops usually experience
several rounds of sprays in comparatively short time and practically
not maintaining the safe waiting period before harvest. Thus, there is
need to search for safer molecules, effective methods of pest control,
biocontrol agents, and botanical pesticides. Their use, however, has
several limitations and chemical control continuous to be the preferred
strategy in practice.

7.3 PESTICIDE RESIDUE AND ITS OCCURRENCE

Pesticide residues are now major concern as it seems that the use of these
agrochemicals is almost inevitable. Pesticide residue, as we all know, is
the left over portion of the certain agrochemical after application in the
target segment. But, unfortunately, these residues are not at all confined
to its target area of application. These are found in various substrates of
environment such as soil, air, water bodies and more importantly accu-
mulated in food chain even good agriculture practice (GAP) is followed.
Generally, these are toxic in nature, can do lot of harm to the non-target
organism including human being if not properly handled. Moreover, the
parent molecules undergo several transformations in the environment and
produce metabolites which may be more toxic than their parent molecule
itself.

7.4 PESTICIDE RESIDUES IN VEGETABLES

Pesticides applied to vegetables in the field can leave potentially harmful
residues. Organochlorine (OC) pesticides such as DDT, HCH, and so on.
can persist in food for a considerable period. If crops are sprayed shortly
prior to harvest without an appropriate waiting period, even moderate
or less persistent molecules can leave toxic residues in vegetables. In a

pesticide monitoring program conducted at national level, residues of certain pesticides found above maximum residue limit (MRL) in vegetables (see Table 7.1).

TABLE 7.1   List of Pesticides Found Above Mrl in Vegetables.

Vegetable	Predominant pesticide residues found above MRL
Cauliflower	Chlorpyriphos, Cypermethrin
Okra	Monocrotophos, Chlorpyriphos, and Cypermethrin
Cabbage	Chlorpyriphos, Monocrotophos, and Cypermethrin
Brinjal	Aldrin, Heptachlor, and Cypermethrin

Source: Monitoring of pesticide residues at national level, April 2008–March 2009; Indian Agriculture Research Institute, New Delhi under All India Network Project on Pesticide Residues, Department of Agriculture and Cooperation, Ministry of Agriculture, Govt. of India

There are several reports available on pesticide contamination in vegetables in India. In a study conducted with vegetable samples taken from local market basket of Lucknow, residues of HCH, permethrin-II, dichlorvos, and chlorofenvinfos were found above MRL in the samples of radish, cucumber, cauliflower, cabbage, and okra.[3] Vegetable samples were collected for pesticide residue analysis from different agricultural fields of central Aravalli region, when they were ready for transportation to market. About 40.11% of total analyzed samples were contaminated with different pesticide residues, among which 35.62% of total contaminated samples were exceeded the MRL values.[4] Vegetable samples collected at harvest from farmer's fields around Hyderabad and Guntur showed HCH and mancozeb residues above MRL.[5]

In order to get good price of their harvest at the market when the demand is high, farmers generally use high doses of pesticides in vegetables in a very short interval. Sometimes, based on their perception, farmers apply overdose of pesticides when there is higher pest incidence as well as "ineffectiveness" of pesticides at right dose. This overuse of pesticides may cause serious toxicity issues in human being upon consumption of the harvested vegetables.

7.5   PESTICIDE REGULATION

In order to protect human health and environment as well as to promote reduced-risk pest management, pesticide regulation is very much important. In most of the countries, pesticides are essentially approved for sale and use by a government agency. In the United States, the Environmental Protection Agency (EPA) is responsible for regulating pesticides under the Federal Insecticide, Fungicide, and Rodenticide Act (FIFRA) and the Food Quality Protection Act (FQPA). Registration, Evaluation, Authorisation and Restriction of Chemical substances (REACH) is a European Community regulation to improve the protection of human health and the environment. The major objective of pesticide regulation is to establish the conditions in which the particular pesticide is safe to use and economically effective against the intended pest(s).

Various government agencies are engaged in the regulation of the pesticide products in India. The Ministry of Agriculture regulates the manufacture, sale, transport, and distribution, export, import, and use of pesticides through the Insecticides Act, 1968 and the rules framed there under. The Central Insecticides Board (CIB), advises the Central and State governments on technical matters. The approval of the use of pesticides and new formulations to tackle the pest problem in various crops is given by the Registration Committee (RC) while the Union Ministry of Health and Family Welfare monitors and regulates pesticides residue levels in food. It also sets MRL of pesticides on food commodities.

In India, the pesticides regulations are governed under the following Acts/Rules:

1. The Insecticides Act 1968 and Rules 1971
2. Prevention of Food Adulteration Act 1954
3. The Environment (Protection) Act 1986
4. The Factories Act 1948
5. Bureau of Indian Standards Act
6. Air (Prevention & Control of Pollution) Act 1981
7. Water (Prevention & Control of Pollution) Act 1974
8. Hazardous Waste (Management & Handling) Rules 1989

Functions of CIB

a) Advise the Central Government on the manufacture of insecticides under the Industries (Development and Regulation) Act, 1951 (65 of 1951).

b) Specify the uses of the classification of insecticides on the basis of their toxicity as well as their being suitable for aerial application.

c) Advise tolerance limits for insecticides residues and establishment of minimum intervals between the application of insecticides and harvest in respect of various commodities.

d) Specify the shelf life of insecticides.

e) Suggest colorization, including coloring matter which may be mixed with concentrates of insecticides, particularly those of highly toxic nature.

f) Carry out such other functions as are supplemental, incidental, or consequential to any of the functions conferred by the Act or the Rules.[6]

7.5.1 PESTICIDE TOXICITY

The injurious effects produced by pesticides are known as toxic effects. The toxicity of a pesticide is the inherent ability of the chemical to cause damage. The hazard, whereas is the possibility of the compound to cause injury. Certain pesticides can cause loss of weight and appetite, irritability, insomnia, behavioral disorder, and dermatological problems in human being.

To assess the acute and delayed effects of a pesticide, a series of toxicological tests are carried out on laboratory animals known as short-term and long-term studies to know all the possible effects from skin irritation to premature birth cancer, and so on. Based on the scientific data generated in those studies, extrapolation of the possible effects to man is worked out. This type of extrapolation of laboratory experimental data to human beings may not be fool-proof as there is no animal species that has similar biochemical metabolic pathways in man. However, based on the toxicity, pesticides are classified and labeled in the container by color mark which represents the toxicity of the material by a color code (see Table 7.2).

TABLE 7.2   Classification of Insecticides Under Insecticides Rules, 1971.

Serial no	Classification of insecticide	LD50 mg/kg body weight of test animals (acute toxicity)	LD50 mg/kg body weight of test animals (dermal toxicity)	Color of identification on the label
1	Extremely toxic	1–50	1–200	Bright red
2	Highly toxic	51–500	201–2000	Bright yellow
3	Moderately toxic	501–5000	2001–20,000	Bright blue
4	Slightly toxic	More than 5000	More than 20,000	Bright green

Source: Annexure–17, Know Your Pesticides, Directorate of Plant Protection, Quarantine & Storage, Department of Agriculture, Cooperation and Farmers Welfare, Ministry of Agriculture and Farmers Welfare, Government of India

7.6   PESTICIDE RESIDUE ANALYSIS

Pesticide residue analysis of certain commodities is generally required for registration of pesticides, checking compliance with MRLs in foods, and to provide data for protection of the environment as well as risk assessment[7]. Determination of pesticide residues in food items is generally followed in several steps. In case of data generation for registration of a particular pesticide molecule, field trial has to be conducted in different agro-climatic regions for multiple seasons. This is required to bring out the clear picture about the performance of that particular molecule under different situations. Pesticide monitoring is also a well-organized initiative to check out compliance of different pesticides with their existing MRLs in different commodities which may require random sampling. Following are the steps involved in evaluating pesticide residue from a supervised field trial.

7.6.1   DESIGN OF EXPERIMENT

Trials should be designed to cover a range of representative field conditions, typical seasons of the year, cropping, and farming practices which are commonly encountered. This is the most important task and should be carefully performed in determining residue content of a particular pesticide.

7.6.1.1   SELECTION OF SITES

Experiment should be carried out in major areas of cultivation and all the relevant representative conditions (climatic, seasonal, soil, cropping system, farming, etc.) likely to be met for the intended use of the pesticide, should be taken care of. Areas or sites where typical conditions occur and which are not representative should be avoided.

7.6.1.2   REPLICATION

It is very useful to have at least three or four replicates at one site to study experimental uniformity and determine the within-site variations. Randomized block design (RBD) is generally followed in this study. Individual plot size may vary from crop to crop but should be large enough to draw samples from it at regular interval for analysis. A sufficient buffer zone (lanes, guard rows, etc.) should be maintained between plots to prevent cross-contamination.

7.7   PESTICIDE APPLICATION

Pesticide should be uniformly applied and contamination of neighboring plots, either during or after application, should be avoided.

7.7.1   DOSAGE

At least two dosage rates should be included in a residue trial, that is, the maximum rate which is likely to be recommended and another rate, preferably double the recommended rate. In addition to this, an untreated control plot should always be included in any residue experiment carried out to provide sample known to be free from residues of the pesticide under investigation. These control samples shall facilitate in understanding the matrix effect during analysis.

7.7.2   METHOD OF APPLICATION

Spray equipment should be calibrated before use to ensure proper application. Proper spray nozzle should be used for particular pesticide. For example, flat fan/flood zet nozzle is generally used for herbicide application whereas hollow cone nozzle is used for insecticide application. Pesticide container should be shaken well before use to resist sedimentation in case of EC formulation.

7.8   SAMPLING

It has always been found that it is extremely difficult to obtain uniform application of a pesticide in the field. Thus, "in taking a sample for residue analysis, it is necessary to approach the task in an intelligent, realistic manner if the results of analysis are to be valid or useful for estimating maximum residue levels".[8] Improper sampling may lead to erroneous results. "A sample must be statistically representative of the population from which it is taken".[9] The following points are very useful for effective sampling from a supervised trial plot.

i)   Sample should be drawn in a zigzag manner and must be representative of the entire plot. Diseased or undersized crop parts or commodities should be avoided during sampling at harvest as they would not normally be harvested.

ii) Sample size can vary depending upon the type of sample to be drawn. For vegetable samples, at least 2–5 kg field sample should be taken for analysis.[8]

iii) Sample from untreated plot also has to be taken.

iv) It is important to avoid contamination of the field sample under study during sampling, transportation, or subsequent operations.

v)   Sample should be packed and labeled properly so that analyst can easily identify the treatment details if required during analysis.

vi) Sample should be analyzed as soon as possible after collection before physical and chemical changes occur. If prolonged storage is unavoidable, it is generally preferable to extract the sample, remove most or all of the solvent and store the extracts at a low temperature, preferably at or below $-20^{\circ}C$.[8]

vii) In practice, a valid field sample is often much larger than the sample required by the analyst and cannot be handled properly during analysis. In such cases, a reduction in the size of the field sample is desirable to make it laboratory sample. The taking of a representative laboratory sample is obviously the most difficult task. If solid or liquid samples are to be analyzed individually, they are separated using a sieve or filter and individually mixed for subsampling. When both solid and liquid phases are to be analyzed as a unit, it is suggested to blend or otherwise homogenize the two before subsampling.

7.9   EXTRACTION AND CLEAN UP

7.9.1   EXTRACTION

Extraction is the process of separating pesticide residues from the matrix by using solvent. "The extraction procedure should be such that it quantitatively removes pesticides form matrix (high efficiency), does not cause chemical change in pesticide and use inexpensive and easily cleaned apparatus".[9] The extraction method and solvent type (usually organic) determine the extraction efficiency from substrates.

7.9.1.1   EXTRACTION FROM LIQUID SUBSTRATES

a) Partitioning—Water sample, body fluids, juices, and so on are extracted by partitioning with water immiscible solvent. The addition of sodium chloride in aqueous samples generally improves the extraction efficiency by reducing the solubility of pesticide in water. It also helps to prevent the emulsion formation, which is frequently encountered during partitioning.

b) Use of absorbent—Pesticide residues from liquid samples can be extracted by passing the sample through solid adsorbents packed in glass column. The adsorbents have strong affinity for pesticide molecules, therefore, they are held up on the absorbent whereas water passes out. The solid adsorbents are then extracted with organic solvent. Examples of solid absorbents are activated charcoal, silica gel, florisil, magnesium sulphate, and so on.

7.9.1.2 EXTRACTION FROM SOLID SUBSTRATES

a) Fresh residues—Dipping, tumbling, and shaking method is usually followed for solid substrate when pesticide residues are present on the surface.

b) Weathered residues—After long time exposure of pesticide with the substrate, residues penetrate in the substrate matrix and exist in adsorbed form. The matrix needs to be broken down in fine particles before extraction with solvent. Techniques such as macerating/blending followed by column extraction, soxhlet extraction are usually followed to extract these adsorbed residues from substrate matrix.

7.9.1.3 CHOICE OF SOLVENTS

Selection of solvent for extraction depends on the (a) nature of the substrate and (b) the type of pesticide to be extracted. However, the solvent should have following features:

a) Should have high solubility for the pesticide molecule and least solubility for co-extractives.
b) Should not chemically react with the pesticide.
c) Economical
d) Low boiling.
e) Can be easily separated from the substrate.

The general thumb rule regarding solvents to solubilize substance is "like dissolves like." Thus, polar solvents can solubilize polar pesticide molecules while nonpolar pesticide can be solubilized by nonpolar solvents. Few examples of polar solvents are acetonitrile, methanol while hexane, cyclohexane are considered to be nonpolar solvents.

7.9.2 CLEAN UP

"Cleanup refers to a step or series of steps in the analytical procedure in which the bulk of the potentially interfering co-extractives are removed by physical or chemical methods".[9] The co-extractive generally extracted along with pesticide from various substrates are moisture, colored pigment like chlorophyll, xanthophylls, anthocyanins, and so on, color-less compounds like oil, fat, waxes, and so on. which can interfere in the procedure and may produce misleading results. Several chromatographic techniques that have been developed for cleanup are as follows:

a) Thin Layer Chromatography (TLC)—Silica gel plates are normally used but other adsorbents like alumina can also be used.

b) Ion Exchange Chromatography—Ion exchange resins can be used for cleanup of ionic pesticide molecules. For cationic pesticides like paraquat and diquat, cation exchange resins (H+) while for anionic pesticide like 2,4-D, anion exchange resins (Ch) are gener-ally used.

c) Gel Permeation Chromatography and Molecular Sieves—The separation in gel permeation chromatography and molecular sieves is based on the principle of size exclusion.

d) Adsorption Column Chromatography—This is the most common and widely used technique for clean up. Different type of adsor-bent or mixtures of adsorbent have been used in this purpose.

Liquid-liquid partitioning is also a popular technique followed by analyst where co-extractives from the extract are removed by partitioning the residues between two immiscible solvents. Cleanup can also be done through chemical treatment such as saponification of fats and oils, precip-itation, oxidation with sulphuric acid, and so on. Recently, solid phase extraction (SPE) cartridge is developed which serves the dual purpose of extraction and clean up. Advantages of SPE device over other conventional solvent extraction and cleanup of pesticides includes better reproducibility, less use of solvent, high speed, versatility, freedom from interferences, and field applications.

7.10   IDENTIFICATION AND QUANTIFICATION

It is very essential to identify the pesticide in samples as possible metab-olism, photo degradation, and break down of the molecule may further complicate the analysis. In that case, cleanup procedure may not be able to eliminate all the interfering co-extractives. Chromatographic instrument having mass spectrometer (MS) provides structural information of the molecules. Infrared spectrophotometer is generally used to detect functional group of a particular organic molecule. Nuclear magnetic resonance spectrophotometer is also a useful tool in structural identification of organic compounds. Chemical derivatization is occasionally done for positive identification of pesticides. Some commonly used procedures are oxidation, reduction, addition, rearrangement, and so on.

Several modern chromatographic instruments like gas liquid chromatography (GLC), high performance liquid chromatography (HPLC), and so on. are used to quantify the pesticide residues at very lower level.

7.10.1 GAS LIQUID CHROMATOGRAPHY

The basic principle involved in GLC is to separate volatile components that are distributed between a fixed stationary phase and a mobile inert gas. The stationary phase consists of a nonvolatile liquid distributed on a solid support. The sample mixture got vaporized in the injection port and carried away by the inert mobile gas into the packed column where partitioning of the components occurs between the phases. Then the individual components are carried away into the detectors for identification. Temperature programming is usually being done during analysis. The detectors generally used in GLC are electron capture detector (ECD), flame ionization detector (FID), thermal conductivity detector (TCD), flame photometric detector (FPD), and so on.

7.10.2 HIGH PERFORMANCE LIQUID CHROMATOGRAPHY

HPLC is a form of column chromatography that pumps a sample mixture or analyte in a solvent (known as the mobile phase) at high pressure through a column with chromatographic packing material (known as stationary phase). Several detectors being used in HPLC are ultra violet–visible (UV–VIS) detector, photo diode array (PDA) detector, fluorescence detector, and so on. Gradient elution may be necessary when series of compounds over wide range of polarities are being separated. The separation is usually carried out at room temperature. Thermo labile compounds are used to pass through the column without degradation. Both volatile and nonvolatile compounds can be analyzed in HPLC.

7.11   CONFIRMATION OF RESIDUE IDENTITY

After identification, it is essential to confirm the identified residues in the matrix. Primitive technique like TLC can be followed with reference standard for further confirmation. Mass spectrometry (MS) or combined Gas Chromatography–Mass Spectrometry (GC–MS) is recommended for confirmation of residues if other methods are not sufficient enough to do so.

7.12   METHOD VALIDATION

The analytical procedure must be validated with positive identification of pesticide residue for data generation. Recovery experiment is generally being followed for method validation which includes several steps. Known quantity of target pesticide is added in the control samples of the substrate which must be free from any pesticide contamination before fortification. Then, same steps for processing the samples like extraction, cleanup is followed and analyzed in the instrument. The expected recovery should be minimum 80–85% as per guideline and level of fortification at least includes the level of quantification (LOQ) of the instrument. Blank sample without spiking as well as blank reagent without sample should be run to check out if any significant interference of sample or reagent is found or not. The method would be widely acceptable on the basis of precision and accuracy if confirmed by different laboratories through repeat analysis.

7.13   LABEL CLAIM OF PESTICIDES IN DIFFERENT VEGETABLE CROPS IN INDIA

It is well observed that vegetable crop faces huge loss caused by various insect pests, disease causing pathogens as well as harmful weeds. A number of pesticides have been registered against major pests on vegetables by CIB, Govt. of India with dosage recommendation (see Table 7.3). Each and every pesticide container should have the label in which particular information of its dosage should be clearly mentioned.

TABLE 7.3 Label Claim of Pesticides in Different Vegetables.

Product	Crop	Pest	Dosage/ha			Waiting period (days)
			a.i. (g)	Formulation (g/ml)	Dilution in water (l)	
Insecticides						
Acetamiprid 20% SP	Cabbage	Aphids	15	75	500–600	7
	Okra	Aphids	15	75	500–600	3
	Chili	Thrips	10–20	50–100	500–600	3
Buprofezin 25% SC	Chili	Yellow mite	75–150	300–600	500–750	5
Buprofezin 70% DF	Okra	Jassid	200	286	500	5
Carbaryl 5% DP	Okra	Jassid	1000	20,000	–	8
	Cabbage	Cabbage borer	600	20,000	–	8
	Cauliflower	Cabbage borer	600	12,000	–	8
Carbaryl 10% DP	Okra	Fruit borer, jassids	2500	25,000	–	–
	Cabbage	Diamond back moth, Armyworm	2500	25,000	–	–
Carbaryl 50% WP	Chili	Thrips	1000	2000	500–1000	–
	Brinjal	Fruit borer, jassids	1000	2000	500–1000	5
	Okra	Fruit borer	1000	2000	500–1000	3
	Cauliflower	Cabbage borer	800	1600	500–1000	8
	Cabbage	Cabbage borer	800	1600	500–1000	5
Carbofuran3%CG	Tomato	Whitefly	1200	40,000	–	–
	Potato	Aphid Jassids	500	16,600		–
			1000	33,300		
	Okra	Jassids	1000	33,300	–	–
	Chili	Aphid, thrips	1000	33,300	–	–

TABLE 7.3 *(Continued)*

Product	Crop	Pest	a.i. (g)	Formulation (g/ml)	Dilution in water (l)	Waiting period (days)
				Dosage/ha		
	Cabbage	Nematode	1000	50,000	—	—
	Brinjal	Root knot nematode	2000	66,600	—	
		Reni form nematode	2000	66,600	—	
	French bean	White grubs	750	23,300		—
		Grey and stem weevil	1000	33,300		
Carbosulfan 25% EC	Chili	Leaf folder,	200–250	800–1000	500–1000	14
		White aphid	200–250	800–1000	500–1000	8
Chlorantraniliprole 18.5% SC	Cabbage	Diamond back moth	10	50	500	3
	Tomato	Fruit borer	30	150	500	3
	Chili	Fruit borer	30	150	500	3
	Brinjal	Fruit and shoot borer	40	200	500–750	22
	Bitter gourd	Fruit borer, caterpillar	20–25	100–125	500	7
	Okra	Fruit borer	25	125	500	5
Chlorfenapyr 10% SC	Cabbage	Diamond back moth	75–100	750–1000	500	7
	Chili	Mites	75–100	750–1000	500	5
Chlorfluazuron5.4% EC	Cabbage	Diamond back moth, Tobacco leaf eating caterpillar	75	1500	500	7
Chlorpyrifos 20% EC	Brinjal	Fruit and shoot borer	200	1000	500–1000	—
	Cabbage	Diamond back moth	400	2000	500–1000	—
	Onion	Root grub	1000	5000	500–1000	—

TABLE 7.3 (*Continued*)

Product	Crop	Pest	Dosage/ha			Waiting
			a.i. (g)	Formulation (g/ml)	Dilution in water (l)	period (days)
Cyantraniliprole 10.26% OD	Cabbage	Cabbage aphid, mustard aphid, diamond back moth, tobacco caterpillar	60	600	500	5
	Chili	Thrips, fruit borer, tobacco caterpillar	60	600	500	3
	Tomato	Leaf miner, aphids, thrips, white fly, fruit borer	90	900	500	3
	Gherkins	Leaf miner, red pumpkin beetle, aphids, thrips, white fly, pumpkin caterpillar, fruit fly	90	900	500	5
Cypermethrin 0.25% DP	Brinjal	Fruit and shoot borer	50–60	20,000–24,000	–	3
Cypermethrin 10% EC	Cabbage	Diamondback moth	60–70	650–760	100–400	7
	Okra	Fruit borer	50–70	550–760	150–400	3
	Brinjal	Fruit and shoot borer	50–70	550–760	150–400	3
Cypermethrin 25% EC	Okra	Fruit and shoot borer, jassids	37–50 37–50	150–200 150–200	500 500	3 3
	Brinjal	Fruit and shoot borer jassids, epilachna grub	37–50	150–200	500	1

TABLE 7.3 *(Continued)*

Product	Crop	Pest	Dosage/ha a.i. (g)	Dosage/ha Formulation (g/ml)	Dosage/ha Dilution in water (l)	Waiting period (days)
Deltamethrin 2.8% EC	Okra	Fruit and shoot borer, jassid	10–15	400–600	400–600	1
			10	400	400–600	1
	Chili	Fruit borer	10–12.5	400–500	400–600	5
	Brinjal	Fruit and shoot borer	10–12.5	400–500	500	3
Dichlorvos 76% EC	Cucurbit	Red pumpkin beetle	500	627	500–1000	–
Dicofol 18.5% EC	Okra	Red spider mite	250–500	1350–2700	500–1000	15–20
	Brinjal	Yellow mite	500–1000	2700–5400	500–1000	15–20
	Bottle and bitter gourd	Red spider mite	250–500	1350–2700	500–1000	15–20
Diafenthiuron 50%WP	Cabbage	Diamondback moth	300	600	500–750	7
	Chili	Mites	300	600	500–750	3
	Brinjal	Whitefly	300	600	500–750	3
Dimethoate 30% EC	Okra	Aphid	700	2310	500–1000	–
		Leaf hopper, jassid	600	1980	500–1000	–
	Brinjal	Shoot borer	200	660	500–1000	–
	Cabbage and cauliflower	Aphid, painted bug Mustard aphid	200	660	500–1000	–
	Chili	Mite	300	990	500–1000	–
	Onion	Thrips	200	660	500–1000	–
	Potato	Thrips	200	660	500–1000	–
	Tomato	Aphids	200	660	500–1000	–
		Whitefly	300	990	500–1000	–

TABLE 7.3 *(Continued)*

Product	Crop	Pest	Dosage/ha			Waiting period (days)
			a.i. (g)	Formulation (g/ml)	Dilution in water (l)	
Emamectin benzoate 5% SG	Okra	Fruit and shoot borer	6.75–8.5	135–170	500	5
	Cabbage	Diamond back moth	7.5–10	150–200	500	3
	Chili	Fruit borer, thrips, and mites	10	200	500	3
	Brinjal	Fruit and shoot borer	10	200	500	3
Ethion 50% EC	Chili	Mites, thrips	750–1000	1500–2000	500–1000	5
Etoxazole 10% SC	Brinjal	Red spider mite	40	400	400–500	5
Fenazaquin 10% EC	Chili	Yellow mite	125	1250	400–600	10
	Okra	Red spider mite	125	1250	500	7
	Brinjal	Red spider mite	125	1250	500	7
	Tomato	Two spotted spider mite	125	1250	500	7
Fenpropathrin 30% EC	Chili	Thrips, whitefly, mites	75–100	250–340	750–1000	7
	Brinjal	Whitefly, fruit and shoot borer, mites	75–100	250–340	750–1000	10
	Okra	Whitefly, fruit and shoot borer, mites	75–100	250–340	750–1000	7
Fenpyroximate 5% EC	Chili	Yellow mite	15–30	300–600	300–500	7
Fenvalerate 20% EC	Cauliflower	Diamond back moth, American boll worm, aphids, jassids	60–75	300–375	600–750	7

TABLE 7.3 *(Continued)*

Product	Crop	Pest	Dosage/ha			Waiting period (days)
			a.i. (g)	Formulation (g/ml)	Dilution in water (l)	
Fipronil 5% SC	Brinjal	Fruit and shoot borer, aphids	75–100	375–500	600–800	5
			75–100	375–500	600–800	5
	Okra	Fruit and shoot borer, jassids	60–75	300–375	600–750	7
			60–75	300–375	600–750	7
	Cabbage	Diamond back moth	40–50	800–1000	500	7
	Chilies	Thrips, aphids, fruit borers	40–50	800–1000	500	7
Flubendiamide 20% WG	Tomato	Fruit borer	48	100	375–500	5
	Cabbage	Diamond back moth	18.24	37.5–50	375–500	7
Flubendiamide 39.35% SC	Chili	Fruit borer	48–60	100–125	500	7
	Tomato	Fruit borer	48	100	375–500	5
	Cabbage	Diamond moth back	18.24	37.5–50	375–500	7
Flumite 20% SC/ Flufenzine 20%SC	Brinjal	Mite	80–100	400–500	500–1000	5
Hexythiazox 5.45% EC	Chili	Yellow mites	15–25	300–500	625	3
Imidaclopride 70% WG	Okra	Jassids, aphids, thrips	21–24.5	30–35	375–500	3
	Cucumber	Aphids, jassids	24.5	35.0	500	5
Imidacloprid 48% FS	Okra	Jassid, aphid	300–540 g/100 kg seed	500–900	—	—

TABLE 7.3 *(Continued)*

Product	Crop	Pest	Dosage/ha			Waiting period (days)
			a.i. (g)	Formulation (g/ml)	Dilution in water (l)	
Imidacloprid 70% WS	Okra	Jassid, aphid	350–700 g/100 kg seed	500–1000	—	—
	Chili	Jassid, aphid, thrips	700–1050 g/100 kg seed	1000–1500	—	—
Imidacloprid 17.8% SL	Chili	Jassid, aphid, thrips	25–50	125–250	500–700	40
	Okra	Aphid, jassid, thrips	20	100	500	3
	Tomato	Whitefly	30–35	150–175	500	3
Indoxacarb 14.5% SC	Cabbage	Diamond back moth	30–40	200–266	400–750	7
	Chili	Fruit borer	50–60	333–400	300–600	5
	Tomato	Fruit borer	60–75	400–500	300–600	5
Indoxacarb 15.8% EC	Cabbage	Diamond back moth	40	266	500–1000	5
Lambda-Cyhalothrin 4.9% CS	Brinjal	Fruit and shoot borer	15	300	500	5
	Okra	Fruit borer	15	300	500	5
	Tomato	Fruit borer	15	300	500	5
	Chili	Thrips, pod borer	25	500	500	5
Lambda-Cyhalothrin 5% EC	Brinjal	Fruit and shoot borer	15	300	400–600	4
	Tomato	Fruit borer	15	300	400–600	4
	Chili	Thrips, mite, pod borer	15	300	400–600	5

TABLE 7.3 (Continued)

Product	Crop	Pest	Dosage/ha			Waiting period (days)
			a.i. (g)	Formulation (g/ml)	Dilution in water (l)	
Lufenuron 5.4% EC	Onion	Thrips	15	300	300–400	5
	Okra	Jassids, Shoot borer	15	300	300–400	4
	Cabbage	Diamondback moth	30	600	500	14
	Cauliflower	Diamondback moth	30	600	500	5
	Chili	Fruit borer	30	600	500	5
Malathion 50% EC	Okra	Aphid	500	1000	500–1000	–
		Jassids	625	1250	500–1000	
		Spotted boll worm	750	1500	500–1000	
	Brinjal	Mites	750	1500	500–1000	–
	Cabbage	Mustard aphid	750	1500	500–1000	–
	Cauliflower	Head borer	750	1500	500–1000	–
	Radish	Stem borer	750	1500	500–1000	–
	Turnip	Tobacco caterpillar	600	1200	500–1000	–
	Tomato	Whitefly	750	1500	500–1000	–
Metaflumizone 22% SC	Cabbage	Diamond back moth	165–220	750–1000	500	3
Methomyl 40% SP	Tomato	Pod borers	300–450	750–1125	500–1000	5/6
	Chili	Pod borers, thrips	300–400	750–1125	500–1000	5/6
Milbemectin 1% EC	Chili	Yellow/white mite	3.25	325	500	7
Novaluron 10% EC	Cabbage	Diamond back moth	75	750	500–1000	5
	Tomato	Fruit borer	75	750	500–1000	1–3
	Chili	Fruit borer, tobacco caterpillar	33.5	375	500	3

TABLE 7.3 *(Continued)*

Product	Crop	Pest	Dosage/ha a.i. (g)	Formulation (g/ml)	Dilution in water (l)	Waiting period (days)
Oxydemeton-Methyl 25% EC	Okra	Whitefly	250	1000	500–1000	—
		Jassid/leaf beetle	400	1600	500–1000	
	Chili	Aphid	400	1600	500–1000	—
		Mites	500	2000	500–1000	
		Thrips	250	1000	500–1000	
	Onion	Thrips	300	1200	500–1000	—
	Tomato	Whitefly	250	1000	500–1000	—
	Potato	Aphids	250	1000	500–1000	—
Phorate 10% CG	Brinjal	Aphid, jassids, Lace wing bug, Red spider mite	1500	15,000	—	—
		Thrips	1000	10,000		
	Cauliflower	Aphid	2000	20,000	—	—
	Chili	Aphid, mite, thrips	1000	10,000	—	—
	Potato	Aphid	1000	10,000	—	—
	Tomato	Whitefly	1500	15,000	—	—
Phosalone 35% EC	Brinjal	Fruit borer	500	1428	500–1000	—
	Cabbage	Aphid	500	1428	500–1000	—
	Tomato	Fruit borer	450	1285	500–1000	—
Phosphamidon 40% SL	Brinjal	Jassid, aphid, whitefly	250–300	625–750	500	10
Propargite 57% EC	Chili	Mite	850	1500	500–625	7

TABLE 7.3 *(Continued)*

Product	Crop	Pest	Dosage/ha			Waiting period (days)
			a.i. (g)	Formulation (g/ml)	Dilution in water (l)	
Pyridalyl 10% EC	Brinjal	Two spotted spider mite	570	1000	400	6
	Okra	Fruit and shoot borer	50–75	500–750	500–750	3
	Cabbage	Diamond back moth	50–75	500–750	500–750	3
Quinalphos 25% GEL	Chili	Aphid	250	1000	500–1000	–
Quinalphos 20% AF	Okra	Fruit and shoot borer	250–300	1250–1500	750–1000	7
	Tomato	Fruit borer	300–350	1500–1750	750–1000	7
Quinalphos 25% EC	Okra	Fruit borer	200	800	500–1000	–
		Leaf hopper	250	1000	500–1000	
		Mite	250	1000	500–1000	
	Cauliflower	Stem borer	500	2000	500–1000	–
	Chili	Aphid mite	250	1000	500–1000	–
			375	1500	500–1000	
	Tomato	Fruit borer	250	1000	500–1000	–
Quinalphos 1.5% DP	French bean	Stem fly	300	20,000	–	–
	Chili	Aphid	300	20,000	–	–
Spinosad 45.0% SC	Chili	Fruit borer, thrips	73	160	500	3
Spinosad 2.5% SC	Cabbage and cauliflower	Diamond back moth	15.0–17.5	600–700	500	3
Spiromesifen 22.9% SC	Brinjal	Red spider mite	96	400	500	5
	Chili	Chili yellow mite	96	400	500–750	7
	Okra	Red spider mite	96–120	400–500	500	3
	Tomato	Whiteflies, Mites	150	625	500	3

TABLE 7.3 (Continued)

Product	Crop	Pest	Dosage/ha a.i. (g)	Formulation (g/ml)	Dilution in water (l)	Waiting period (days)
Thiacloprid 21.7% SC	Chili	Thrips	54–72	225–300	500	5
	Brinjal	Fruit and shoot borer	180	750	500	5
Thiodicarb 75% WP	Cabbage	Diamond back moth	750–1000	1000–1330	500	7
	Brinjal	Fruit and shoot borer	470–750	625–1000	500	6
	Chili	Fruit borer	470–750	626–1000	500	6
Thiamethoxam 30% FS	Chili	Thrips	2.1	7.0	–	–
	Okra	Jassids	1.7	5.7	–	–
Thiamethoxam 70% WS	Okra	Aphids, jassids	200	286	–	–
	Tomato	Aphids, thrips	420	600	–	–
Thiamethoxam 25% WG	Okra	Jassid, aphid, whitefly	25	100	500–1000	5
	Tomato	Whitefly	50	200	500	5
	Potato	Aphids -Foliar application	25	100	500	77
		-Soild rench	50	200	400–500	77
	Brinjal	Whitefly	50	200	500	3
Thiometon 25% EC	Brinjal	Aphid, jassid fruit, and shoot borer	250	1000	750–1000	–
Tolfenpyrad 15% EC	Cabbage	Diamond back moth, aphids	150	1000	500	5
	Okra	Aphids, jassids, thrips, and white fly	150	1000	500	3

TABLE 7.3 (Continued)

Product	Crop	Pest	Dosage/ha			Waiting period (days)
			a.i. (g)	Formulation (g/ml)	Dilution in water (l)	
Trichlorfon 5% GR	Brinjal, cabbage, cauliflower, cucurbits, tomato	Fruit and shoot borer Diamond back moth Tobacco caterpillar Red pumpkin beetle	500 500 750 500 (a.i. %)	—	—	—
Trichlorfon 5% Dust	Brinjal, cabbage, cauliflower, cucurbits, tomato	Fruit and shoot borer Diamond back moth Tobacco caterpillar Red pumpkin beetle	500 500 750 500	—	—	—
Trichlorfon 50% EC	Brinjal, cabbage, cauliflower, cucurbits, tomato	Fruit and shoot borer Diamond back moth Tobacco caterpillar Red pumpkin beetle	500 500 750 500	—	—	—
Betacyfluthrin 8.49% + Imidacloprid 19.81% OD	Brinjal	Aphids, jassids, fruit and shoot borer	15.75 + 36.75 – 18 +42	175–200	500	7
Cypermethrin 3% + Quinalphos 20% EC	Brinjal	Fruit and shoot borer	—	350–400	500–600	7
Flubendiamide 19.92% + Thiacloprid 19.92%	Chili	Thrips, fruit borer	48 + 48 – 60 + 60	200–250	500	5
Indoxacarb 14.5% + Acetamiprid 7.7% SC	Chili	Thrips, fruit borer	88.8–111	400–500	500	5

TABLE 7.3 *(Continued)*

Product	Crop	Pest	Dosage/ha			Waiting period (days)
			a.i. (g)	Formulation (g/ml)	Dilution in water (l)	
Novaluron 5.25% + Indoxacarb 4.5% SC	Tomato	Fruit borer, leaf eating caterpillar	43.31 + 37.13 – 45.94 + 39.38	825–875	500	5
Pyriproxyfen 5% EC + Fenpropathrin 15% EC	Brinjal	Whitefly, shoot, and fruit borer	25 + 75 – 37.5 + 112.5	500–750	500–750	7
	Okra	Whitefly, fruit borer	25 + 75 – 37.5 + 112.5	500–750	500–750	7
	Chili	Whitefly, Fruit borer	25 + 75 – 37.5 + 112.5	500–750	500–750	7
Fungicides						
Azoxystrobin 23% SC					500–750	
					500	
					500	
					500	
Benomyl 50 % WP					600	
					600	
					600	
					600	
					600	

TABLE 7.3 *(Continued)*

Product	Crop	Pest	Dosage/ha			Waiting period (days)
			a.i. (g)	Formulation (g/ml)	Dilution in water (l)	
Captan 50% WG					500	
					500	
Captan 50% WP					750–1000	
					750–1000	
Captan 75% WP					1000 soild rench in the nursery	
					1000	
					1000	
					1000	
					1000	
Captan 75% WS					1	
					1	
					1	
Carbendazim 50% WP					750	
					400	
					600	
					600	
Chlorothalonil 75% WP					600–800	
Copper oxy chloride 50% WP					750–1000	
					750–1000	
					750–1000	

TABLE 7.3 *(Continued)*

Product	Crop	Pest	Dosage/ha			Waiting period (days)
			a.i. (g)	Formulation (g/ml)	Dilution in water (l)	
Copper Hydroxide 53.8% DF					62	
Cyazafamid 34.5% SC					500	
					500	
Difenoconazole 25% EC					500	
					500	
Dimethomorph 50% WP					750	
Flusilazole 40% EC					500	
Hexaconazole 2% SC					500	
					500	
Iprodione 50% WP					500	
Kitazin 48% EC					As required depending upon crop stage and plant protection equipment used	
					As required depending upon crop stage and plant protection equipment used	
					As required depending upon crop stage and plant protection equipment used	

TABLE 7.3 (Continued)

Product	Crop	Pest	Dosage/ha			Waiting period (days)
			a.i. (g)	Formulation (g/ml)	Dilution in water (l)	
					As required depending upon crop stage and plant protection equipment used	
Mancozeb 35% SC					500 l water or as required depending upon crop stage and equipment used	
Mancozeb 75% WG					500	
Mancozeb 75% WP					750	
					750	
					1	
					750	
					750	
					1	
					750	
Mandipropamid 23.4% SC					500	
					500–750	
Metalaxyl-M 31.8% ES					—	
					—	
Metiram 70% WG					500–750	

TABLE 7.3 *(Continued)*

Product	Crop	Pest	Dosage/ha a.i. (g)	Formulation (g/ml)	Dilution in water (l)	Waiting period (days)
M.E.M.C. 6% FS					100 ml	
Myclobutanil 10% WP					500	
Propineb 70% WP					As required depending upon crop stage and plant protection equipment used	
					As required depending upon crop stage and plant protection equipment used	
					As required depending upon crop stage and plant protection equipment used	
Pyraclostrobin 20% WG					500	
Sulphur 52% SC					400	
Sulphur 80% WP					750–1000	
Tebuconazole 25.9% EC					500	
Tebuconazole 25% WG					500	
Thiophanate Methyl 70% WP					750–1000	
Thiram 75% WS					1000	

TABLE 7.3 *(Continued)*

Product	Crop	Pest	Dosage/ha a.i. (g)	Formulation (g/ml)	Dilution in water (l)	Waiting period (days)
Zineb 75% WP					750–1000	
					750–1000	
					750–1000	
					750–1000	
					750–1000	
					750–1000	
					750–1000	
Ziram 80% WP					750–1000	
					750–1000	
Ametoctradin + Dimethomorph 20.27% SC					500	
Azoxytrobin 11% + Tebuconazole 18.3% SC					500–750	
Azoxystrobin 18.2% + Difenoconazole 11.4% SC					500	
Captan 70% + Hexaconazole 5% WP					500.	
Carbendazim 12% + Mancozeb 63% WP					—	

TABLE 7.3 *(Continued)*

Product	Crop	Pest	Dosage/ha a.i. (g)	Formulation (g/ml)	Dilution in water (l)	Waiting period (days)
Carbendazim 25% + Mancozeb 50% WS					2	
Cymoxanil 8% + Mancozeb 64% WP					500–600	
					500–600	
					500–600	
Famoxadone 16.6% + Cymoxanil 22.1% SC					500	
					500	
Fenamidone 10% + Mancozeb 50% WG					500	
Metalaxyl M 4% + Mancozeb 64% WP					500–1000	
					2.0 l/m²	
Metalaxyl M 3.3% + Chlorothalonil 33.1% SC					500	
Metalaxyl 8% + Mancozeb 64% WP					1000	
Metiram 55% + Pyraclostrobin 5% WG					500	
					500	
					750	
					750	
Tebuconazole 50% + Trifloxystrobin 25% WG					500	

TABLE 7.3 *(Continued)*

| Product | Crop | Pest | Dosage/ha | | | Waiting period (days) |
			a.i. (g)	Formulation (g/ml)	Dilution in water (l)	
Thiophanate Methyl 450g/l + Pyraclostrobin 50g/l FS					500	
Herbicides						
2,4-D Dimethyl amine salt 58% SL						
Oxy flour fen 23.5% EC						
Paraquat dichloride 24% SL						
Pendimethalin 38.7% CS						
Propaquizafop 10% EC						
Quizalofop ethyl 5% EC						

Source: Major uses of pesticides, Central Insecticides Board & Registration Committee, Directorate of Plant Protection, Quarantine & Storage, Department of Agriculture, Cooperation and Farmers Welfare, Ministry of Agriculture and Farmers Welfare, Government of India

7.14 MAXIMUM RESIDUAL LIMITS FOR VEGETABLE CROPS

Pesticide residues are now major concern in fresh vegetables and their products for domestic consumption and export. Technically, pesticide residue in food commodity is regulated using a set of quantitative standards called MRL. A MRL is the highest level of a pesticide residue that is legally tolerated in or on food or feed when pesticides are applied following GAP. MRLs are often misunderstood as toxicological safety limits. These are safe limits that define the maximum expected level of a pesticide on a food commodity after its safe and authorized use.[10]

In India, The Ministry of Health and Family Welfare regulates MRLs of pesticides and agrochemicals in food products through the amended Prevention of Food Adulteration Act (PFA), 1955. However, with the implementation of Food Safety and Standards Act (FSSA), 2006, the PFA rules are being integrated into the Food Safety and Standards Regulations, 2010. The new act authorizes the Food Safety and Standards Authority of India (FSSAI) to specify the limits for use of food additives, crop contaminants, pesticide residues, and residues of veterinary drugs, heavy metals, processing aids, mycotoxins, antibiotics, pharmacological active substances, and irradiation of food.[2] Like other crops, the amount of pesticide on the vegetables shall not exceed the tolerance limit prescribed in the standard regulation (see table 7.4).

TABLE 7.4   Tolerance Limit of Pesticides in Vegetables.

Product	Crop	Tolerance limit (mg/kg)
Insecticides		
Aldrin, dieldrin (the limits apply to aldrin and dieldrin singly or in any combination and are expressed as dieldrin)	Vegetables	0.1
	Okra, leafy vegetables	10.0
Carbaryl	Potato	0.2
	Chili, other vegetables	5.0
Chlordane (residue to be measured as cis plus trans chlordane)	Sugar beet	0.3
	Other vegetables	0.2
D.D.T. (the limits apply to D.D.T., D.D.D. and D.D.E. singly or in any combination)	Vegetables	3.5
Diazinon	Vegetables	0.5

TABLE 7.4   *(Continued)*

Product	Crop	Tolerance limit (mg/kg)
Dichlorvos [content of di-chloroacetaldehyde (D.C.A.) be reported where possible]	Vegetables	0.15
Dicofol	Vegetables	5.0
Dimethoate (residue to be determined as fruits and dimethoate and expressed as dimethoate)	Vegetables	2.0
	Chili	0.5
Endosulfan (residues are measured and reported as total of endosulfan A and B and endosulfan-sulphate)	Vegetables	2.0
	Chili	1.0
Fenitrothion	Vegetables	0.3
Heptachlor (combined residues of heptachlor and its epoxide to be determined and expressed milled as heptachlor)	Vegetables	0.05
Hexachloro cycle hexane and its isomers		
(a) Alfa (α) isomer	Vegetables	1.0
(b) Beta (β) isomer	Vegetables	1.0
(c) Gamma (γ) isomer (known as lindane)	Vegetables	1.0
(d) Delta (δ) isomer	Vegetables	1.0
Malathion (malathion to be determined and expressed as combined residues of malathion and malaoxon)	Vegetables	3.0
Parathion (combined residues of parathion and paraoxon to be determined and expressed as parathion)	Vegetables	0.5
Parathion methyl (combined residues of parathion methyl and its oxygen analogue to be determined and expressed as parathion methyl)	Vegetables	1.0
Phosphamidon residues (expressed as the sum of phosphamidon and its desethyl derivative)	Vegetables	0.2
Pyrethrins (sum of pyrethrins I & II and other structurally related insecticide ingredients of milled pyrethrum)	Vegetables	1.0
Chlorienvinphos (residues to be measured as alpha and beta isomers of chlorienvinphos)	Vegetables	0.05
Chlorpyrifos	Potato, onion, cauliflower, cabbage	0.01
	Other vegetables	0.2

TABLE 7.4 *(Continued)*

Product	Crop	Tolerance limit (mg/kg)
Ethion (residues to be determined as ethion and its oxygen analogue and expressed as ethion)	Cucumber, squash	0.5
	Other vegetables	1.0
Formothion (determined as dimethoate and its oxygen analogue and expressed as dimethoate except in case of citrus fruits where it is to be determined as formothion)	Vegetable	2.0
	Tomato	1.0
Monocrotophos	Carrot, potato, sugar beet	0.05
	Onion	0.1
	Chili, other vegetables	0.2
Phosalone	Potato	0.1
	Other vegetables	1.0
Trichlorfon	Sugar beet	0.05
	Other vegetables	0.1
Thiometon (residues determined as thiometon its sulfoxide expressed as thiometon)	Potato, carrot, sugar beet	0.05
	Other vegetables	0.5
Aldicarb (sum of aldicarb its sulphoxide and sulphone, expressed as aldicarb)	Potato	0.5
Carbofuran (sum of carbofuran and 3-hydroxy carbofuran expressed as carbofuran)	Vegetables	0.1
Cypermethrin (sum of isomers) (fat soluble residue)	Brinjal, okra	0.2
	Cabbage	2.0
	Onion, bean	0.1
Fenthion (sum of fenthion, its oxygen analogue and their sulphoxides and sulphones expressed as fenthion)	Potato	0.05
	Tomato	0.5
	Other vegetables	1.0
Fenvalerate (fat soluble residue)	Cauliflower, brinjal, okra	2.0
Phorate (sum of phorate, its oxygen analogue and their sulphoxides and sulphones, expressed as phorate)	Tomato	0.1
	Other vegetables	0.05
Permethrin	Cucumber	0.5
Quinalphos	Chili	0.2
Triazophos	Chili	0.2

TABLE 7.4   *(Continued)*

Product	Crop	Tolerance limit (mg/kg)
Spinosad	Cauliflower, cabbage	0.02
Chlorfenapyr	Cabbage	0.05
Indoxacarb	Cabbage	0.1
Lufenuron	Cabbage	0.3
Novaluron	Tomato, cabbage	0.01
Fungicides		
Carbendazim	Sugar beet	
	Other vegetables	
Benomyl	Sugar beet	
	Other vegetables	
Captan	Vegetables	
Copper oxychloride (determined as copper)	Potato	
	Other vegetables	
Dithiocarbamates (expressed as mg/CS_2/kg and refer separately to the residues arising from any or each group of dithiocarbamates) (a) Dimethyl dithiocarbamates residue resulting from the use of ferbam or ziram, and	Tomato	
(b) Ethylene bis-dithiocarbamates resulting from the use of mancozeb, maneb, or zineb (including zineb derived from nabam plus zinc sulphate) © Mancozeb	Potato	
	Chili	
Captafol	Tomato	
Chlorothalonil	Potato	
Dimethomorph	Potato	
Iprodione	Tomato	
Metiram	Tomato	
Propineb	Potato	
	Green chili	
Herbicides		
2,4-D	Potato	
Paraquat dichloride (determined as paraquat cations)	Potato	
	Other vegetables	

Source: Food safety and standards (contaminants, toxins, and residues) regulation, 2011, Food Safety and Standards Authority of India, Ministry of Health and Family Welfare, Government of India

7.15   MINIMIZATION OF PESTICIDE RESIDUE IN VEGETABLES

The following ideas can be implemented to minimize pesticide residues in vegetables:

i)   Pesticide should be applied on the crop when it is absolutely necessary and no other alternative is available.

ii) Pesticide should be applied at recommended dose. Amount of pesticide needed to apply in spray tank should be carefully determined to avoid over dose.

iii) Highly persistent pesticide molecules like OC compounds can be avoided.

iv) Matured vegetables must be plucked before pesticide application in case of multiple picking. Crops are only be harvested after the recommended waiting period of the particular pesticide is over.

v)   Use of biopesticides and nonchemical control can be encouraged.

7.16   PESTICIDE DOSAGE CALCULATION

Successful pesticide application is very much dependent on executing right dose of the product. Calibrating spray equipment is very essential in this regard. After calibration, one must determine how much chemical is required to put into spray tank to apply the recommended dosage. Therefore, the following calculation method can be very much useful in determining the quantity of the product required for application.

a)   For emulsifiable concentrate (EC) and wettable powder (WP) formulations:

$$C_1 V_1 = C_2 V_2$$

C_1 = concentration of given formulation (%)

V_1 = volume/amount of formulation required (ml or g)

C_2 = concentration of spray fluid required (%)

V_2 = volume/amount of spry fluid required (ml or g)

Example 1. How much spray fluid of 2.5% concentration can be prepared from 500 g of mancozeb 75 WP?

Answer:

C_1 = 75%
C_2 = 2.5%
V_1 = 500 g
V_2 = ?
V_2 = $(C_1 V_1)/C_2$
        = 15,000 ml
        = 15 l

Example 2. How much quantity of quinalphos 25 EC is required for spraying 0.025% spray fluid at the rate of 250 L/ha for controlling the fruit borer on tomato over an area of one-twenty-fifth of a hectare?

Answer:

C_1 = 25%
V_1 = ?
C_2 = 0.025%
Rate of application= 250 L/ha
For one-twenty-fifth of a hectare = $(250 \times 1/25)$ L/ha
V_2 = 10 L = 10,000 mL
$C_1 V_1$ = $C_2 V_2$
V_1 = $(C_2 V_2)/C_1$
        = $(0.025 \times 10,000)/25$
        = 10 mL

b) For granules and dust formulations:

$C_1 V_1$ = 100 RA
C_1 = Concentration of formulation available (%)
V_1 = Amount of formulation required (g or kg)
R   = Recommended rate of pesticide application [g or kg ingredient (a.i.)/ha]
A   = Area to be treated (ha)

Example 3. Calculate the amount of 5% carbofuran granules applied @ 0.20 kg a.i. to 1 ha area.

C_1 = 5%
V_1 = ?

$$R = 0.20 \text{ kg}$$
$$A = 1 \text{ ha}$$
$$C_1 V_1 = 100 \text{ RA}$$
$$V_1 = (100 \text{ RA})/C_1$$
$$= (100 \times 0.20 \times 1)/5$$
$$= 4 \text{ kg}.$$

Any pesticide product necessarily provides detailed information about its active ingredients along with its percentage in the particular formulation. Using this information one can work out the quantity of formulated product required for spraying for one hectare of land by using the following formula:

$$\text{Formulated product (g/ha)} = \frac{\text{Does in g.a.i./ha} \times 100}{\% \text{ a.i. in the kg/l formulation}}$$

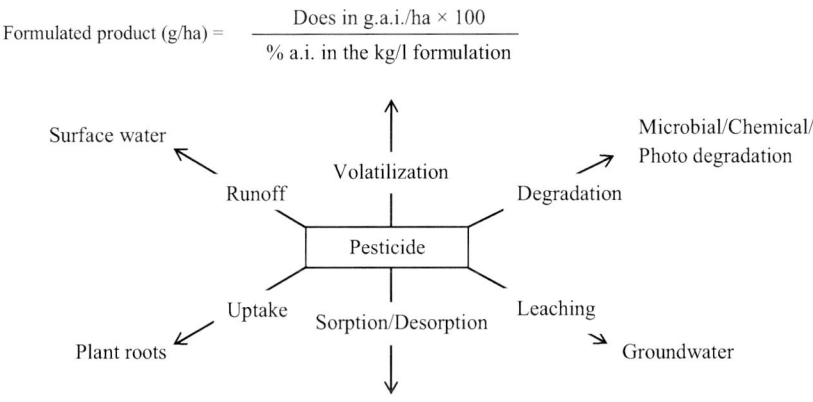

FIGURE 7.1   Fate of pesticides in environment.

KEYWORDS

- **pesticide residues**
- **vegetable crops**
- **good agricultural practices**
- **MRL**
- **ADI**

REFERENCES

1. Directorate of Plant Protection, Quarantine & Storage, Department of Agriculture, Cooperation and Farmers Welfare, Ministry of Agriculture and Farmers Welfare, Government of India. http://ppqs.gov.in/PMD.htm#consumption (accessed March 01, 2016).
2. Kodandaram, M. H.; Sujoy, Saha; Rai, A. B.; Prakash S., Naik. Compendium on Pesticide use in Vegetables. *IIVR Extension Bulletin* No. 50, IIVR, Varanasi, 2013; pp 133.
3. Srivastava, K. A.; Trivedi, P.; Srivastava, M. K.; Lohani, M.; Srivastava, L. P. Monitoring of Pesticide Residues in Market Basket Samples of Vegetable From Lucknow City, India: Quechers Method. *Environ. Monit. Assess.* **2011,** *176,* 465–472.
4. Charan, P. D.; Ali, S. F.; Kachhawa, Y.; Sharma, K. C. Monitoring of Pesticide Residues in Farmgate Vegetables of Central Aravalli Region of Western India. *Am.-Eurasian J. Agric. Environ. Sci.* **2010,** *7,* 255–258.
5. Reddy, J.; Narsimha, B. R.; Sultan, M. A.; Narsimha, K. R. Pesticide Residues in Farmgate Vegetables. *J. Res. ANGRAU* **1998,** *26,* 6–10.
6. Central Insecticides Board & Registration Committee, Directorate of Plant Protection, Quarantine & Storage, Department of Agriculture, Cooperation and Farmers Welfare, Ministry of Agriculture and Farmers Welfare, Government of India. http://www.cibrc.nic.in/ (accessed April 27, 2016).
7. Perihan, Y. O.; Dilek, B.; Árpad, A.; Artemis, K.; Samim, S. An Overview on Steps of Pesticide Residue Analysis and Contribution of the Individual Steps to the Measurement Uncertainty. *Food Anal. Method* **2012,** *5,* 1469–1480.
8. Food and Agriculture Organization of The United Nations, 1986. Guidelines on Pesticide Residue Trials to Provide Data for the Registration of Pesticides and the Establishment of Maximum Residue Limits, pp 10.
9. Food Safety and Standards Authority of India, 2015. Manual of Methods of Analysis of Foods – Pesticide Residues, pp 18.
10. Dureja, P., Singh, S. B.; Parmar, B. S. Pesticide Maximum Residue Limit (MRL): Background, Indian Scenario. *Pestic. Res. J.* **2015,** *27,* 4–22.

CHAPTER 8

INSECT PESTS OF TOMATO AND THEIR MANAGEMENT

ATANU SENI

Regional Research and Technology Transfer Station (OUAT), Chiplima 768025, Sambalpur, Odisha, India

E-mail: atanupau@gmail.com

ABSTRACT

Tomato is a one of the important vegetable crop grown in almost every country of the world—in outdoor fields, greenhouses and net houses. In India, it is grown over an area of 0.86 m ha with an estimated production of 16.82 mt and productivity is 19 t/ha. They have many nutritional properties as they are rich in vitamins A, C, K, Potassium, Iron and lycopene. Vitamin A helps our body in many ways such as takes part in bone growth, cell division and differentiation, helping in the regulation of immune system and maintaining surface linings of eyes, respiratory, urinary and intestinal tracts. Vitamin C helps in forming collagen, a protein that gives structures to bones, cartilage, muscle and blood vessels. It also helps maintain capillaries, bones and teeth and aids in the absorption of iron. Lycopene is a very powerful antioxidant which can help prevent the development of many forms of cancer. Besides their medical properties and one of the popular vegetable, their production is reduced by the attack of large number of insect pests from the time of emergence in the seed bed until the harvest of the crop. There are numbers of insect pests like, fruit borer (*Helicoverpa armigra*), aphids (*Aphis gossypii, Myzus persicae*), jassids (*Amarasca biguttula*), American serpentine leaf miner, (*Liriomyza trifolii*), root knot nematodes (*Meloidogyne* sp.) etc have been recorded as serious problem. Further this, infestations by sucking insect pests not only affect the crop but also hamper the plant vitality by transmitting different pathogenic viruses.

8.1   INTRODUCTION

Tomato is a one of the important vegetable crops grown in almost every country of the world—in outdoor fields, greenhouses, and net houses. In India, it is grown over an area of 0.86 million hectare with an estimated production of 16.82 mt and productivity is 19 t/ha (Choudhary and Kundal, 2015). They have many nutritional properties as they are rich in vitamins A, C, K, potassium, iron, and lycopene. Vitamin A helps our body in many ways, such as it aids in bone growth, cell division, and differentiation; helps in the regulation of immune system; and in maintaining surface linings of eyes, respiratory, urinary, and intestinal tracts. Vitamin C helps in forming collagen, a protein that gives structures to bones, cartilage, muscle, and blood vessels. It also helps maintain capillaries, bones, and teeth and aids in the absorption of iron. Lycopene is a very powerful antioxidant, which can help prevent the development of many forms of cancer. Inspite of their medical properties and being one of the popular vegetable, their production is reduced by the attack of large number of insect pests from the time of emergence in the seed bed until the harvest of the crop. There are a large number of insect pests, like fruit borer (*Helicoverpa armigra*), aphids (*Aphis gossypii, Myzus persicae*), jassids (*Amarasca biguttula*), American serpentine leaf miner, (*Liriomyza trifolii*), root knot nematodes (*Meloidogyne* sp.), etc. that has recorded as serious problem for tomatoes. Further, infestations by sucking insect pests not only affect the crop, but also hamper the plant vitality by transmitting different pathogenic viruses (Butani and Verma, 1976).

8.2   CHEWING INSECT PESTS ATTACKING TOMATO

Different chewing insect pests such as tomato fruit borer, tobacco caterpillar, serpentine leaf miner, tomato pinworm, and cut worm causes enormous damage to tomato plants by eating both foliage and fruits at different growth stages of the plant.

8.2.1 TOMATO FRUIT BORER; HELICOVERPA ARMIGERA (HbBNER) NOCTUIDAE: LEPIDOPTERA

Distribution: It is a widely distributed pest occurring throughout Africa, Southern Europe, Asia, Australia, New Zealand, and many eastern Pacific Islands.

Host Range: It is a polyphagous pest and attacks more than 182 plant species (Gowda, 2005; Seni et al., 2010) including several important crops such as tomato, cotton, corn, chickpea, pigeonpea, sorghum, sunflower, soybean, and groundnut (Fitt, 1989).

Identification: Adult males are normally yellowish-brown, light yellow, or light brown in colour and females are orange-brown in colour. Forewings have a black or dark brown kidney-shaped marking near the center. Hind wings are creamy white with a brown or dark gray band on outer margin. Newly emerged first instars larvae are translucent and yellowish-white in color. The full-grown larvae are long and ventrally flattened but convex dorsally; and they are brownish or pale green with brown lateral stripes with a distinct dorsal stripe.

Life Cycle: Female lays 500–1000 eggs, even may go up to a maximum of 3000 spherical, yellowish eggs singly on tender parts, buds, and flowers of plants. The egg period lasts for 2–4 days. The larval period lasts for 18–25 days. When full grown, they measure 3.7–5 cm in length. They pupate in the soil in an earthen cell and emerge in 16–21 days. They completed 11 generations in optimal tropical conditions as compared to 2–5 generations in subtropical and temperate regions (Tripathi and Singh, 1991). Warm weather conditions followed by light rains and dry spells are congenial for their multiplication.

Damage: After hatching newly emerged larva starts feeding on the leaves for some time and then attacks fruits. Internal tissues are eaten severely and completely hollowed out. While feeding, the caterpillar thrust its head inside leaving the rest of the body outside. They make round holes on infested fruit. It causes 24–73% yield losses in tomato (Rai et al., 2014).

Management:

- Field should be kept clean
- Ecological engineering of tomato with growing intercrops such as cowpea, onion, maize, coriander, urdbean, etc. (Satyagopal et al., 2014)
- Grow four rows of maize/sorghum/bajra around the field as a gourd guard/barrier crop
- Collection and destruction of eggs and early stages of larvae
- Deep ploughing after harvesting the crop to expose the pupa for natural killing affords
- Intercropping tomato with African marigold, as a trap crop. Planting of marigold should be done in such a way that tomato flowering coincides with the tight-bud stage of the marigolds
- Use light trap @ 1/acre and operate between 6 pm and 10 pm for monitoring the adult population or Install pheromone traps @ 4–5/ acre for monitoring adult moths activity (replace the lures with fresh lures after every 2–3 weeks)
- Erecting of bird perches @ 20/acre for encouraging predatory birds such as King crow, common mynah, etc.
- Conserve parasitoids; such as *Trichogramma* spp., *Tetrastichus* spp., *Chelonus* spp., *Telenomus* spp., *Bracon* spp., *Ichneumon* spp., *Carcelia* spp., *Campoletis* spp. etc.; and Predators such as Lacewing, ladybird beetle, spider, red ant, dragon fly, robber fly, reduviid bug, praying mantis, black drongo (King crow), wasp, common mynah, big-eyed bug (*Geocoris* sp), earwig, ground beetle, and pentatomid bug (*Eocanthecona furcellata*) (Satyagopal et al., 2014)
- Inundative release of *Trichogramma pretiosum* @ 40,000/acre, 4–5 times from flower initiation stage at weekly intervals
- Spray *Bacillus thuringiensis* var *gallariae* @ 400–600 g in 200 L of water/acre
- Apply entomopathogenic nematodes (EPNs) @ 20–120 crore infective juveniles of *Steinernema feltiae*/acre (Satyagopal et al., 2014)
- Spray with indoxacarb 14.5% SC @ 160–200 mL in 120–240 L of water/acre or flubendiamide 20% WG @ 40 g in 150–200 L of water/acre or flubendiamide 39.35% M/M SC @ 40 mL in 150–200 L of water/acre or novaluron 10% EC @ 300 mL in 200–400 L of water/acre or chlorantraniliprole 18.5% SC @ 60 mL in 200 L of

water/acre or lambda cyhalothrin 5% EC @ 120 mL in 160–200 L of water/acre or methomyl 40% SP @ 300–450 mL in 200–400 L of water/acre or quinalohos 25% EC @ 400 mL in 200–400 L of water/acre (Satyagopal et al., 2014)

8.2.2   TOBACCO CATERPILLAR

Spodoptera Litura **Fabricius**

Noctuidae: Lepidoptera

Distribution: It is one of the most important insect pests of agricultural crops in the Asian tropics and widely distributed throughout tropical and temperate Asia, Australasia, and the Pacific Islands (Anon, 2016).

Host Range: It is polyphagous in nature and infests almost 120 species including different important crop plants, such as cotton, castor, groundnut, tomato, soybean, jute, groundnut, chilli, various cruciferous vegetable crops, and other ornamental plants (Anon, 2016). It has caused 12–23% damage to kharif season grown tomatoes and 9–24% damage in the winter season grown tomatoes (Patnaik, 1998).

Identification: Moth is medium sized and stout bodied with forewings pale grey to dark brown in color having wavy white crisscross markings. Hind wings are whitish with brown patches along the margin of wing. Larva are velvety, black with yellowish green dorsal stripes and lateral white bands with incomplete ring—like dark band on anterior and posterior end of the body and measures about 35–40 mm in length, when full grown.

Life Cycle: Female lays about 300 eggs in clusters but it may vary from 2000 to 2600 eggs. The eggs are covered over by brown hairs and they hatch in about 3–5 days. It takes 15–30 days to complete larval stage by passing through six instars (the larval period required 303° days above a 10°C threshold) (Anon, 2016). Pupation takes place inside the soil; pupal stage lasts for 7–15 days (almost 155° days above a 10°C threshold) (Anon, 2016). Adults live for 7–10 days. Total life cycle takes 32–60 days. Moths are very active at night and they breed throughout the year. There are almost eight generations in a year. Warm weather conditions and rainy conditions are favorable for their multiplication.

Damage: After emergence, caterpillars scrape the chlorophyll content of leaf lamina giving it a papery white appearance. Later they become voracious feeders making irregular holes on the leaves and finally leaving only veins and petioles. They feed the inner content of the fruit and irregular bored holes are found from outside.

Management:

- Collection and destruction of the infested plant parts
- Plucking of leaves harboring egg masses/gregarious larvae and destroying them
- Setting up light traps for adults
- Setting up of pheromone traps @ 12/ha
- Conserve parasitoids; such as *Trichogramma* spp., *Tetrastichus* spp., *Chelonus* spp., *Telenomus* spp., *Bracon* spp., *Ichneumon* spp., *Carcelia* spp., *Campoletis* spp. etc.,; and Predators such as Lacewing, ladybird beetle, spider, red ant, dragon fly, robber fly, reduviid bug, praying mantis, black drongo (King crow), wasp, common mynah, big-eyed bug (*Geocoris* sp), earwig, ground beetle, pentatomid bug (*Eocanthecona furcellata*), etc.
- Spraying NSKE 5% against eggs and first instar larva (Satyagopal et al., 2014)
- Spraying NPV @ 200LE/ac in combination with jaggery 1 kg, sandovit 100 mL or Robin Blue 50 g thrice at 10–15 days interval on observing the eggs or first instar larvae in the evening hours
- Release of egg parasitoid *Trichogramma* @ 50,000/ha/week four times
- ETL: one egg mass/10 plants
- Foliar spraying with endosulfan 2 mL/L or thiodicarb 1.5 g/L or quinalphos 2.5 mL/L; in severe incidence, novaluran 1 mL/L or lufenuron 1 mL/L.
- Baiting with rice bran 10kg + jaggery 2 kg+ chlorpyriphos 750 mL or thiodicarb 300 g in sufficient quantity of water in form of small balls and broadcasting in evening hours in one acre.

8.2.3 *SERPENTINE LEAF MINER:* LIRIOMYZA TRIFOLII (BURGESS) (AGROMYZIDAE: DIPTERA)

Distribution: It is native of Florida and believed to be accidentally introduced into India. It is widely distributed in Andhra Pradesh, Tamil Nadu, Karnataka, and Maharastra causing damage to fiber crops, pulses, ornamentals, vegetables, fodder, etc. (Nath, 2007).

Host Range: It is an invasive pest and attacks a large number of plants belonging to 16 families, but seems to favor families of Cucurbitaceae, Leguminosae, and Solanaceae plants. Among them, it is a serious pest on tomato, cotton, ridge gourd, brinjal, cucumber, potato, and cowpea (Nath, 2007).

Identification: Adults are small, measuring less than 2 mm in length, with a wing expanse of 1.25–1.9 mm. The head is yellow with red eyes. The thorax and abdomen are mostly gray and black, whereas ventral surface of the body and legs are yellow. The wings are more or less transparent. Females are usually larger in size and more robust than the males, and have an elongated abdomen. Initially the larvae are nearly colorless, becoming greenish and then yellowish as they mature. Mouthparts are apparently black in all instars.

Life Cycle: Female lays 110–140 oval-shaped eggs on the ventral surface of the leaf and inserted just below the epidermis and hatch in about 2–8 days. The Larval period ranges from 4 to 25 days. The mature larva cuts a semicircular slit on the upper surface of the leaf just prior to formation of the puparium. The larva usually emerges from the mine, drops from the leaf, and burrows into the soil to a depth of a few centimeters to form a puparium. After about 7–15 days of pupation, the adult emerges out from the puparium, mainly in the early morning hours. Total life cycle takes about 3 weeks. Warm weather conditions are congenial for their multiplication.

Damage: Maggot mines the leaves by scraping the chlorophyll content between the layers, which results in destruction of mesophyll cells. The mine becomes noticeable after 3–4 days after oviposition and becomes larger in size as the maggot matures. Extensive mining also causes premature leaf drop, which can result in lack of shading and sun scalding of fruits. In Karnataka, it causes 35% yield losses in tomato (Rai et al., 2014).

Management:

- Collection and destruction of the infested plant parts
- Avoid excessive use of nitrogenous fertilizer
- Foliar spray of neem oil 5 mL/L can minimize the incidence
- Ecological engineering of tomato with beans as intercrop reduces leaf miner infestation
- Conserve the parasitoids such as *Chrysocharis pentheus, Diglyphus isaea, Gronotoma micromorpha,* etc. and predators such as Lacewing, ladybird beetle, spider, red ant, etc.
- In severe infestation, spraying of Cyantraniliprole 10.26% OD @ 360 mL in 200 L water/acre is helpful to suppress the pest population (Satyagopal et al., 2014).

8.2.4 TOMATO PINWORM OR TOMATO LEAF MINER

Tuta Absoluta (Meyrick) (Lepidoptera: Gelechiidae)

The occurrence of tomato pinworm is reported as an invasive pest on tomato and potato in India in 2014.

Distribution: Since the 1960s, it has become one of the key pests of tomato in South America (Garcia and Espul, 1982). Now, it is present in Morocco, Tunisia, France, Italy, Netherlands, Albania, Portugal, Bulgaria, Cyprus, Spain, Germany, Israel, Hungary, Greece, Bahrain, Iraq, Isreal, Japan, Jordan, Kuwait, Qatar, Saudi Arabia, Syria, Turkey, Yemen, Ukraine, India, and other countries (CABI, 2014; Sridhar et al., 2014).

Host Range: It is an oligophagous pest infesting many solanaceous crops, such as tomato, potato, pepper, and tobacco, etc. (Pereyra and Sanchez, 2006).

Identification: Adult moths are about 5–7 mm long. They were brown or silver with black spots on the narrow wings. Early instar larvae are white or cream with a black head. As larvae grow older, they turn greenish to pink with a brown head (Sridhar et al., 2014).

Life Cycle: Female lays upto 260 cream-colored oval-cylindrical eggs, singly on the both sides of leaves, on buds, or on the calyxes of green fruit.

Incubation period is 4–5 days. Young neonate larvae start feeding outside the leaf, stem, or fruit for a short time before tunneling into the plant part. Larvae take almost 2 weeks to complete its four instars developmental stages. They pupate within the leaf or on the leaf surface or in the soil. Pupae are brown in color and takes 9–11 days for an adult moth to emerge.

Damage: After hatching, newly emerged larvae immediately mined into leaves, apical buds, stalks, or fruits. Feeding resulted in conspicuous wide mines (blotches) and galleries on leaves causing severe loss of photosynthetic activity and pinhole-sized holes on fruits from the stalk end generally covered with the frass. Several mines were observed on a single leaf. The mines have dark frass (excrement) visible inside and the mined areas turn brown and dried over time (Sridhar et al., 2014). In severe cases, it causes defoliation and death of the plant. Tuta damage symptoms are different from the mines made by the leafminer fly (*Liriomyza* spp), which are narrower and more circuitous whereas former is thicker and less circuitous (Anon, 2010).

Management:

- Collection and destruction of the infested plant parts
- Avoid excessive use of nitrogenous fertilizer
- Allow a minimum of 6 weeks fallow lands to prevent carry-over of the pest from previous crop
- Destroy the weeds to prevent multiplication in alternative weed host (especially *Solanum, Datura, Nicotiana*)
- Foliar spray of neem oil 5 mL/L can minimize the incidence
- Black sticky traps can be used for monitoring purpose. Traps should be placed at 15–20 cm above the ground level
- Conserve the parasitoid *Trichogramma achaeae* and predator *Nesidiocoris tenuis*

8.2.5   CUT WORM: AGROTIS IPSILON HUFNAGEL, LEPIDOPTERA: NOCTUIDAE

Although several species of cutworms may injure tomato seedlings and newly set plants in the field, *A. ipsilon* is one of the most destructive cutworms. It is also known as dark sword-grass or black cutworm.

Distribution: Cutworms are polyphagous insects and cosmopolitan in distribution and occurs in the United States, Europe, Canada, Japan, New Zealand, South Africa, South America, and the Pacific. This pest was first found on Oahu in 1879 and is now present on all islands. They are nocturnal in habit and attack the young seedling of the plants at night (Chandel, 2007).

Host Range: It attacks the seedlings of most crops. Crops attacked include beans, broccoli, cabbage, carrot, Chinese broccoli, Chinese cabbage, Chinese spinach, corn, eggplant, flowering white cabbage, green beans, head cabbage, lettuce, mustard cabbage, potato, spinach, sugarcane, sweet potato, tomato, turnip, as well as many other plants (Chandel, 2007).

Identification: Moths are medium sized (22–26 mm longer), stout with greyish brown wavy lines and spots on fore wings and hind wings have a thin dark line along the margins. The moths are active at dusk and are attracted by light. Full-grown caterpillars are 40–50 mm long dirty black in color and have habit of coiling at slightest touch.

Life Cycle: The female starts laying eggs 4–6 days after emergence and lays 300–450 eggs in 4–11 days. Eggs are laid during night either singly or in batches of 30–50 eggs on the underside of leaves or on shoots. The larvae inhibit in the upper layer of soil and fourth instar and are a voracious feeder (Trivedy and Rajagopal, 1999; Chandel et al., 2007). Larval period is 22–30 days. Pupation takes place in soil and lasts for 12–15 days. Life cycle is completed in 39–53 days (Singh, 1987).

Damage: The larvae damage the crop by cutting off young plants near the ground and later on by feeding the shoots and leaves. They hide under clods and in cracks in the soil by day and appear at night.

Management:

- Monitor for cut worms by digging around the base of the damaged plants and sifting the soil for caterpillars. They are best found at dawn or at night.
- Heaps of green grasses may be kept at suitable interval for attracting the larvae in infested field during evening and destroy the collection with caterpillars in next day early morning.

- Deep ploughing of tomato fields during summer months is helpful for exposing the immature stages to high temperature and predatory birds.
- Clean cultivation and mechanical destruction of caterpillars also help in reducing pest infestation.
- Irrigation also brings them on the surface and birds shall predate them.
- Use well decomposed organic manure.
- Light traps installed in or around the potato fields attract the adults of cutworms, and helps in mass collection and destruction of the moths.
- Conserve biological control agents such as *Microgaster* sp, *Micropilitis dimilis, Bracon kitchener, Broscus punctatus*, and *Liogryllus bimaculatus* (predator) (Satyagopal et al., 2014).
- Spraying of *Bacillus thuringiensis* on the crop @10⁹ spores/mL gives a good control (Mishra et al., 1995; Chandla et al., 2011). Entomopathogenic fungus, *Metarrhizium anisopliae* is also shows good result for cutworm control (Mishra et al., 1995).
- In severe infestation, spraying the chemical Chlorpyriphos 20 EC 2.5 L/ha or Coragen 20 SC 300 mL/ha or Spinosad 48 SC 200 mL/ha or Indoxacarb 30 WG 130 g/ha is helpful for reducing the pest incidence.

8.3   SUCKING INSECT PESTS ATTACKING TOMATO

Among different sucking pests, whitefly, aphids, thrips, leafhopper, and mealybug; and among other arthropod pests, mites are important and causes damage to the tomato plants at different times.

8.3.1   *WHITEFLY,* BEMISIA TABACI *GENN.*

Aleyrodidae: Hemiptera

Distribution: It was first described in 1889 as a pest of tobacco in Greece and named as *Aleyrodes tabaci*, the tobacco whitefly (Gennadius, 1889). From 1926 to 1981, it was reported as sporadic pest and was the most important vector of plant viruses in subtropical, tropical, and temperate regions where winters are mild enough to permit year round survival (Cock, 1986).

Host Range: It attacks more than 500 species of plants from 63 plant families (Greathead, 1986). Beside okra, it infests cotton, radish, water melon, cucumber, chillies, brinjal, tomato, potato, tobacco, etc. It is also acts as a vector of leaf curl virus disease of tobacco, cotton, and vein-clearing disease of okra.

Identification: Adults are winged, they are 1.0–1.5 mm long and their yellowish bodies are slightly dusted with white waxy powder. They have two pairs of white wings and have prominent long hindwings.

Life Cycle: The females lay stalked eggs singly on the underside of the leaves, average 80–110 eggs per female. The eggs hatch in 3–5 days during summer, 5–33 days in winter. The nymphs feed on cell sap and grow into three stages to form the pupae within 9–14 days in summer and 17–81 days in winter. In 2–8 days, the pupae change into whiteflies. The total lifecycle completed in 14–100 days depending on weather conditions. A female may live up to 60 days, while the male lives for shorter period (9–17 days). They completed 11–15 generations in a year.

Damage:

- The milky white minute whiteflies and nymphs suck the cell sap from the leaves. It causes weakening and early wilting of the plant and hampers plant growth and yield (Berlinger, 1986). It may also cause leaf chlorosis, leaf withering, curl, vein thickening, leaf enations, leaf cupping, plant stunting, and premature dropping of leaves and plant death.
- They excrete the honeydew, which serves as a substrate for the growth of black sooty mold on leaves and fruit. The mold reduces photosynthesis and lessens the market value of the yields (Berlinger, 1986).
- They are also responsible for transmitting Tomato yellow leaf curl virus (TYLCV) and Bean golden mosaic virus (BGMV).

Management:

- Installation of Yellow sticky traps @ 5/ha is helpful to monitor this pest at early stage.

- Plant tall border crops like maize, sorghum, or pearl millet to reduce whitefly infestations (four rows).
- Cultivation of peppermint plants is helpful to repellent the whitefly.
- Many weed plants harbor whiteflies; removal of weed hosts found to reduce both the incidence of whiteflies and associated viral diseases.
- Protection of seedlings in nursery by nylon net (200 mesh) covering for 25–30 days (Shivalingaswamy et al., 2006) is able to reduce the whitefly incidence.
- Conservation of natural enemies like *Encarsia brevivena*, *Eretmocerus corni*, *Eretmocerus mundus* are some of the parasitoids, *Chrysoperla zastrowi sillemi*, *Mallada boninensis*, *Coccinella septempunctata* are the predators and *Beauveria bassiana* and *Paecilomyces farinosus* are the entomopathogens recorded from *B. tabaci* were found promising for the management of this pest.
- Release of *Chrysoperla carnea* @ 8000 larvae/acre.
- Spraying of *Beauveria bassiana* 1×10^4 cfu/g two times with a week interval reduce the whitefly nymphs and adult.
- Spray NSKE 5% or azadirachtin 0.03% (300 ppm) neem oil-based WSP @ 1000–2000 mL in 200–400 L of water/acre or azadirachtin 5% W/W neem-extract concentrate @ 80 mL in 160 L of water/acre.
- Seed treatment with imidacloprid 70WS @ 2.5 g/kg of seed provides protection for 25–30 days.
- In severe infestation, 4–5 foliar sprays of imidacloprid 17.8 SL (0.002%) or thiamethoxam 25% WG @ 40 g in 200–400 L of water/acre or pyriproxyfen 5% EC + fenpropathrin 15% EC @ 200–300 mL in 200–300 L of water/acre at an interval of 10 days effectively controls the whitefly population.

8.3.2 MEALY BUG, PHENACOCCUS SOLENOPSIS TINSLEY

Pseudococcidae: Hemiptera

Distribution: It was first described originally from the United States in 1898 and it remained known only in the United States, where it is widespread, until 1992. In 1992, it was reported in Central America, the Caribbean, and Ecuador. In 2002, it was recorded as a pest of *Solanum* *muricatum* in Chile and its first record from Brazil in 2005, India in 2006 (Vinobaba and Prishanthini, 2009).

Host Range: They are polyphagous in nature and infested 154 plant species of 53 families comprising 20 field and horticultural crops, 45 ornamentals, 64 weeds, and 25 bushes and trees (Arif et al., 2009).

Identification: They are pale yellowish in color and oblong in shape. A submarginal line of dark mark on thorax and abdomen were observed. There are 18 pairs of short-to-medium–sized waxy filaments around the body, anal filaments being larger. Ventral was with dark circulas, which secrete cottony fibers to form ovisacs. Dorsal body surface was covered with dense waxy dust. The female adults are wingless; whereas, male adults have one pair of wings.

Life Cycle: A female lays 300–700 eggs usually in an ovisac beneath her body. The eggs hatch in few minutes to 2 days. The newly emerged nymphs (crawlers) crawl out and start feeding on tender plant parts. The female mealy bug passes through three nymphal instars in 12–17 days. The male mealy bug passes through four nymphal instars in 14–18 days. The longevity of adult female is longer (14–19 days) as compared to male, which is shorter (1–3 days). The total life cycle of female is completed in 25–38 days, whereas of male in 17–24 days.

Damage:

- Both nymphs and adults suck the sap from leaves, flower buds, petioles, twigs, fruits, and even from the stem of the plants and cause curling and distortion of plant tissue.
- Infested buds may not open and fruits can be deformed and reduce the vitality of the plants.
- In severe cases, they produce copious amount of honeydew, which helps to develop sooty mould affecting the photosynthetic ability of the plants.

Management:

- To check the spread of mealy bug, remove regularly the weeds growing adjacent to road sides, pathways, water channels, and waste lands.

- The hymenopterous parasitoid, *Aenasius bambawalei* is effective against this mealy bug. Avoid spray of insecticides if parasitized mealy bug mummies are observed. Conserve the predators, that is, *Cryptolaemus* montrouzieri, *Brumoides suturalis*, and *Hyperaspis maindroni*.
- In case of severe infestation, spray 1.25 L of profenophos 50EC or 2.0 L of quinalphos 25EC or 625 g of thiodicarb 75WP or 250 g of thiamethoxam in 500 L of water.

8.3.3 LEAFHOPPER, AMRASCA BIGUTTULA BIGUTTULA ISHIDA, CICADELLIDAE: HEMIPTERA

Distribution: It is present throughout India, South, and Southeast Asia, and in the Marina Island (Singh, 2007).

Host Range: It feeds on brinjal, tomato, beans, cowpea, sunflower, sunhemp, cotton, etc.

Identification: Adults are small having wedge-shaped body measuring about 2–3 mm long and greenish yellow in color having a black spot on each forewing and a black spot on the vertex.

Life Cycle: The females lay about 15–30 yellowish, pear-shaped eggs on the underside of the leaves, embedding them into the leaf veins. The eggs hatch in 4–11 days and give rise to nymphs, which are wedge-shaped and are very active and move laterally in a peculiar fashion after disturbance. They suck cell-sap from the underside of the leaves and pass through six stages of growth in 7–21 days. On transformation into winged adults, they live for 5–8 weeks, feeding constantly on the plant sap. They render more economic loss of the crop if they infest within 45–75 days after planting (Schreiner, 2000).

Damage:

- Both nymphs and adults suck the sap from the ventral side of the leaves and damage phloem tubes. While feeding, they also inject toxic saliva into plant tissues, which causes the leaves become yellowish and curl upward along the margins and have a burnt look like symptoms called hopperburn, which later extend over the entire leaf area.

- Severe infestation leads to cupping of leaves and hampered photo-synthesis and stunted growth of the plants (Shivalingaswamy and Satpathy, 2007).

Management:

- Destroy the alternate host plants.
- Cultivate tolerant varieties.
- Growing of okra as a trap crop save the tomato plant from leaf-hopper damage.
- The soil application of neem cake @ 250 kg/ha at sowing, and two repeated applications at 30–45 days interval reduces the incidence of hoppers (Moorthy and Kumar, 2004).
- Conserve the parasitoids, that is, *Lymaenon empoascae* (egg), *Anagrus flaveolus*, *Stethynium triclavatum*, and Predators, that is, Lady beetle, ants *Distina albino*, *Chrysoperla* spp., mired bug (*Dicyphus hesperus*), big-eyed bug, (*Geocoris* sp), etc.
- Spray NSKE 5% or azadirachtin 0.03% (300 ppm) neem oil-based WSP @ 1000–2000 mL in 200–400 L of water/acre or azadirachtin 5% W/W neem extract concentrate @ 80 mL in 160 L of water/acre.
- Seed treatment with imidacloprid 48% FS @ 500–900 mL/100 kg seed or imidacloprid 70% WS @ 500–1000 mL/100 kg seed.
- In severe infestation, spray thiamethoxam 25% WG @ 40 g in 200–400 L of water/acre or cypermethrin 25% EC @ 60–80 mL in 200 L of water/acre or fenvalerate 20% EC @ 120–150 mL in 240–300 L of water/acre or imidacloprid 70% WG @ 12–14 g in 150–200 L of water/acre or imidacloprid 17.8% SL @ 40 mL in 200 L of water/acre.

8.3.4   APHID, *APHIS GOSSYPII* GLOVER

Aphididae: Hemiptera

Distribution: It is found in tropical and temperate regions throughout the world, except extreme northern areas. It is common in North and South America, Central Asia, Africa, Australia, Brazil, East Indies, Mexico, and Hawaii, and in most of Europe (Berim, 2003).

Host Range: It has a very wide host range with at least 700 host plants worldwide including many important vegetables, such as watermelons, tomato, cucumbers, cantaloupes, squash, pumpkin, pepper, okra, brinjal, and asparagus.

Identification: The nymphs are small, greenish brown, and soft-bodied insects found in colonies on the tender parts of the plants and under surface of the leaves. The adults exist in both winged and wingless forms.

Life Cycle: The winged and wingless females multiply parthenogenetically and viviparously. In a day, a female may give birth to 8–20 nymphs. The nymphs moult four times to become adults in 7–10 days. Dry condition favors rapid increase of the pest population.

Damage:

- The damage is caused both by the nymphs and adults by sucking plant sap.
- Severe infestation results in curling of leaves, stunted growth, and gradual drying and death of young plants.
- Black sooty mould develops on the honey dew of the aphids, which falls on the leaves hamper photosynthesis process in plants and quality of fruit may be impaired.
- It also transmits potyviruses, such as, cucumber mosaic virus, watermelon mosaic virus 2, and zucchini yellow mosaic virus.

Management:

- Destroy the infested plant parts.
- Follow clean cultivation.
- Install yellow sticky trap @ 4–5/acre.
- Apply judicious nitrogenous fertilizers.
- Conserve the parasitoid, that is, *Aphidius colemani* and predators, that is, Anthocorid bugs/pirate bugs (*Orius* spp.), mirid bugs, syrphid/ hover flies, green lacewings (*Mallada basalis* and *Chrysoperla carnea*), predatory coccinellids (*Stethorus punctillum*), staphylinid beetle (*Oligota* spp.), cecidomyiid fly (*Aphidoletis aphidimyza*), gall midge (*Feltiella minuta*), earwigs, ground beetles, rove beetles, spiders, wasps, etc. Chemical control measures are same as in case of leafhopper.

- Spray azadirachtin 5% W/W neem extract concentrate @ 80 mL in 160 L of water/acre.
- Spraying of *Verticillium lecanii* @ 1×10^7 viable spores/mL reduce the aphid population (Nirmala et al., 2006).
- Seed treatment with imidacloprid 48% FS @ 500–900 mL/100 kg seed or imidacloprid 70% WS @ 500–1000 mL/100 kg seed.
- In severe infestation, spraying of acetamiprid 20% SP @ 30 g in 200–240 L of water/acre or imidacloprid 70% WG @ 12–14 g in 150–200 L of water/acre or imidacloprid 17.8% SL @ 40 mL in 200 L of water/acre or thiamethoxam 25% WG @ 40 g in 200–400 L of water/acre or dimethoate 30% EC @ 500 mL in 200–400 L of water/acre is helpful.

8.3.5   THRIPS, *THRIPS TABACI* LINDEMAN (THYSANOPTERA: THRIPIDAE)

Distribution: It is cosmopolitan in distribution. In India, *T. tabaci* is widely distributed throughout the country and is considered as most important pest of onion and garlic.

Host Range: It feeds on many cultivated crops as well as uncultivated plants in at least 25 families including onion, garlic, cauliflower, cabbage, cucumber, squash, melon, tomato, turnip, bean, beets, cotton, etc., which are seriously attacked and destroyed by this pest. The adult thrips and their nymphs lacerate the surface tissues of the foliage.

Identification: The adult measures only 1 mm in length. The body of adult is yellowish-brown in color, with slender thorax and abdomen. Abdomen tapers posterior. Female bears full wings but in male the wings are extremely reduced or absent. Wings are narrow, consisting of stiff portion bearing a fringe of hairs. The legs are adapted for running. Each leg terminates in two jointed tarsus and a peculiar vesicle or bladder without claws.

Life Cycle: This pest remains active throughout the year and intense breeding generally occurs between November and May month. The population of male is extremely low, it has been observed that a female most often reproduces without mating and the eggs develop parthenogenetically. Adult female lives for two to four weeks. During this span it produces 50–60 eggs at the rate of 4–6 eggs per day. The eggs are laid singly and are inserted in the leaves of host plant through the slit made by ovipositors of the female. The eggs are white and bean shaped. Nymph hatches out from the eggs within 4–9 days. The nymphs are very much similar to the adults. They start feeding immediately after their emergence by lacerating the epidermal tissues of the leaves and swallowing the sap of the host. Nymphal period lasts for 4–6 days, during which they pass through four instars. The last two instars are non-feeding stage; and during this period, they descend to the soil. Prepupal and pupal stages are spending at a depth of about 2.5–5.0 cm inside the soil. Prepupal stage lasts for one to two days and pupal stage for two to four days, after which adult emerges out. Life cycle is completed in 11–21 days. Shortest during of life cycle have been recorded during April (about 14 days) and longest in December (about 23 days). Population increase occurs at its peak during April–May.

Damage:

- They puncture the leaves of the host plant and suck the exuding sap. In the initial phase of attack by this pest, the leaves show spotted appearance and later on the leaf becomes blighted. Later, they feed on flowers and leaf buds.
- This pest is also acts as a vector of tomato spotted wilt virus (TSWV) and has been involved in transmitting the disease in pineapple, tomatoes, and certain other crops.

Management:

- Destroy the infested plant parts.
- Follow clean cultivation.
- Spray azadirachtin 5% W/W neem extract concentrate @ 80 mL in 160 L of water/acre.
- Seed treatment with imidacloprid 48% FS @ 500–900 mL/100 kg seed or imidacloprid 70% WS @ 500–1000 mL/100 kg seed.
- In severe infestation, Cyantraniliprole 10.26% OD @ 360 mL in 200 L water/acre (Satyagopal et al., 2014).

8.3.6   RED SPIDER MITE, *TETRANYCHUS URTICAE* KOCH (TETRANYCHIDAE: ARACHNIDA)

Distribution: *Tetranychus urticae*, commonly called red spider mite, is the most common and destructive species attacking okra all over India. Although its native place is Eurasia, it has acquired a cosmopolitan distribution (Raworth, 2002).

Host Range: It is highly polyphagous having a very wide range of host plants. These include many important vegetables such as peppers, tomatoes, potatoes, okra, beans, etc.

Identification: Eggs are translucent pearl like, about 0.1 mm in diameter. Larvae are about 0.2 mm in length and pinkish in color. Nymphs are greenish-red in color and about 3 mm in length. Larvae and nymphs look alike in shape but can be easily be distinguished as larvae have three pairs of legs while nymphs and adults have four pairs of legs. There are only two nymphal stages—protonymphal and deutonymphal. Adults are ovate in shape, reddish-brown in color, and 0.4 mm (male)–0.5 mm (female) in length with four pairs of legs.

Life Cycle: Female lays several 100 eggs during her life time. The eggs are attached to fine silk webbing and hatch in approximately 4–7 days. The life cycle is composed of the egg, the larva, two nymphal stages (protonymph and deutonymph), and the adult. Larval development takes 3–5 days; protonymphal and deutonymphal stages last for 3–4 days each. Longevity of adult males and females is 4–9 and 9–18 days, respectively. The length of time from egg to adult stage varies greatly according to temperature. Under optimum conditions, they complete their development in 5–20 days. The females that are active during summer in northern India become active with the onset of monsoon and lay eggs parthenogenetically. These unfertilized eggs give rise to males only, but the subsequent generations are sexual. The activity of these mites increases during the dry season and their activity declines with the drop of temperature (Puri and Mote, 2004).

Nature of Damage:

- They produce fine silken webs on ventral surface of leaves.
- Innumerable yellow spots appear on the dorsal surface of leaves.

- Affected leaves gradually start curling and finally get wrinkled and crumpled. This, in turn, affects the growth and fruit-formation capacity of the plants.

Management Strategies:

- Grow nurseries away from infested crops and avoid planting next to infested fields.
- Destruction of severely infested plant parts can reduce the mite population.
- Grow healthy crops; avoid water and nutrient stress.
- Apply mulch and incorporate organic matter into the soil to improve the water-holding capacity and reduce the evaporation rate.
- Grow perennial hedges, such as pigeon peas, as they encourage the predatory mites.
- Keep the field free of weeds.
- Conservation and release of predatory mite such as *Amblyseius* sp., *Phytoseiulus persimilis* can efficiently regulate the pest under field conditions.
- Botanicals like *Kochea* and *Calotropis* leaf extract show strong acaricidal action under field conditions (Rai et al., 2014).
- In case of severe infestation, spraying of dicofol 18.5 EC @ 2.5 mL/L or propargite @ 3 mL/L (Singh and Singh, 2004) or apply fenazaquin 10 EC @ 500 mL in 200 L of water/acre or spiromesifen 22.9 SC @ 250 mL in 200 L of water/acre should be done at 10 days interval at the initiation of symptoms. Dusting with sulphur dust or spray wettable sulphur 80WP @ 2g/L also suppress the pest population.

KEYWORDS

- **pest management**
- **tomato**
- **insect pests**
- ***Helicoverpa armigera***
- ***Spodoptera litura***
- **sucking insect pests**

REFERENCES

Anonymous. Real IPM Technical Brief *Tuta absoluta*. 2010, www.realipm.com.

Anonymous. *Spodoptera Litura* (Taro Caterpillar). 2016, www.cabi.org/isc/datasheet/44520.

Arif, M. I.; Rafiq, M; Ghaffar, A. Host Plants of Cotton Mealybug (*Phenacoccus solenopsis*): A New Menace to Cotton Agro-Ecosystem of Punjab. *Int. J. Agri. Biol.* **2009,** *11*, 163–167.

Berim, M. N. *Aphis gossypii* Glov. *Cotton Aphid.* **2003,** http://www.agroatlas.ru/en/content/pests/Aphis_gossypii/.

Berlinger, M. J. Host Plant Resistance to *Bemisia tabaci*. *Agri. Ecosyst. Environ.* **1986,** *17*, 69–82.

Butani, D. K.; Jotwani, M. G. *Insects in Vegetables*; Periodical Expert Book Agency: India, 1984; p 356.

CABI. Crop Protection Compendium. **2014,** http://www.cabicompendium.org.

Chandel, R. S.; Dhiman, K. R.; Chandla, V. K.; Kumar, V. Integrated Pest Management in Potato. In *Entomology: Novel Approaches*; Jain, P. C., Bhargava, M. C., Eds.; New India Publishing Agency: New Delhi, 2007; 377–398.

Chandla, V. K.; Kashyap, S.; Sharma, A. Eco-Friendly Management of Potato Pest. In *Potential Plant Protection Strategies*. Prasad, D., Sharma, R., Eds.; I. K. International Publishing House Pvt. Ltd.: New Delhi, 2011; 173–194.

Choudhary, K.; Kundal, R. A Study on Area, Production and Yield of Tomatoes in India from 2002 to 2011. *Int. J. Adv. Res. Comp. Sci. Manag. Stud.* **2015,** *3* (7), 90–94.

Cock, M. J. W. *Bemisia Tabaci*: A Literature Survey on the Cotton Whitefly with an Annotated Bibliography. FAO **1986,** p 121.

Fitt, G. P. The Ecology of *Heliothis Armigera* Species in Relation to Agro-Ecosystem. *Ann. Rev. Entomol.* **1989,** *34*, 17–52.

Garcia, M. F.; Espul, J. C. Bioecologia de la Polilla del Tomate (*Scrobipalpula Absoluta*) en Mendoza, Republica Argentina. *Rev. Invest. Agropecuarias INTA (Argentina)* **1982,** *XVII*, 135–146.

Gennadious, P. Disease of the Tobacco Plantation in the Triconia. The Aleurodid of Tobacco. *Ellenike Georgia* **1889,** *5*, 1–3.

Gowda, C. L. L. *Heliothis/Helicoverpa* Management. In *Emerging Trends and Strategies for Future Research*; Sharma, H. C., Ed.; H C ICRISAT, Hyderabad, Oxford and IBH Publishing Co Pvt Ltd.: New Delhi, 2005.

Greathead, A. H. Host Plants. In *Bemisia Tabaci—A Literature Survey on the Cotton Whitefly with an Annotated Bibliography*; Cock, M. J. W., Ed.; CAB International Institute of Biological Control, Ascot: UK, 1986; pp 17–25.

Misra, S. S.; Chandla, V. K.; Singh, A. K. Potato Pests and Their Management. *Tech. Bull. No.* 45, Central Potato Research Institute, Shimla, **1995,** p 57.

Moorthy, K. P. N.; Kumar, K. N. K. Integrated Pest Management in Vegetable Crops. In *Proceedings of the Integrated Pest Management in Indian Agriculture*; Pratap, S. B., Sharma, O. P., Eds.; NCIPM: New Delhi, 2004, pp 95–108.

Nath, P. Emerging Pest Problems in India and Critical Issues in Their Management. In *Entomology: Novel Approaches*. Jain, P. C., Bhargava, M. C., Eds.; New India Publishing Agency, New Delhi, 2007; pp 43–96.

Patnaik, H. P. In *Pheromone Trap Catches of Spodoptera Litura F. and Extent of Damage on Hybrid Tomato in Orissa. Advances in IPM for Horticultural Crops*, Proceedings of the First National Symposium on Pest Management in Horticultural Crops: Environmental Implications and Thrusts, Bangalore, India, October 15–17, 1997; pp 68–72.

Pereyra, P. C.; Sanchez, N. Effect of Two Solanaceous Plants on Developmental and Population Parameters of the Tomato Leafminer, *Tuta Absoluta* (Meyrick) (Lepidoptera: Gelechiidae). *Neotrop. Entomol.* **2006,** *35,* 671–676.

Puri, S. N.; Mote, U. N. Emerging Pests Problems in India and Critical Issues in Their Management: An Overview. In *Frontier Areas of Entomological Research*; Subrahamaniyam, B., Ramamurthy,V. V., Singh, V. S., Eds.; In *Proceedings on the National Symposium on Frontier Areas of Entomological Research*, November, 5–7, 2003, Division of Entomology, IARI, New Delhi-110012, India, 2004; pp 13–24.

Rai, A. B.; Halder, J.; Kodandaram, M. H. Emerging Insect Pest Problems in Vegetable Crops and Their Management in India: An Appraisal. *Pest Manag. Horticul. Ecosyst.* **2014,** *20* (2), 113–122.

Raworth, D. A.; Gillespie, D. R.; Roy, M.; Thistlewood, H. M. A. *Tetranychus Urticae Koch, Twospotted Spider Mite (Acari: Tetranychidae). In Biological Control Programmes in Canada, 1981–2000*; Mason, P. G., Huber, T. J.; CAB International, 2002; pp 259–265.

Satyagopal, K.; Sushil, S. N.; Jeyakumar, P.; Shankar, G.; Sharma, O. P.; Boina, D. R.; Varshney, R.; Sain, S. K.; Rao, N. S.; Sunanda, B. S.; Asre, R., Kapoor, K. S.; Arya, S.; Kumar, S.; Patni, C. S.; Chattopadhyay, C.; Krishnamurthy, A.; Devi, U.; Rao, K.; Vijaya, M.; Sireesha, K.; Madhavilatha, S.; Sreedharan, R. P.; Chandel, Y. S.; Kotikal, J. H.; Saha, S.; Sathyanarayana, N.; Latha, S. AESA-Based IPM Package for Tomato. Department of Agriculture and Cooperation, Ministry of Agriculture, Government of India, 2014; pp 50.

Schreiner, I. Okra Leafhopper (*Amrasca Biguttula Biguttula* Ishida). *Agric. Pests of the Pacific*, ADAP 2000-11, 2000.

Seni. A.; Dilawari, V. K.; Gupta, V. K. Behavioral Response of H*elicoverpa Armigera* (Hübner) Larvae to African Marigold. *J. Insect Sci.* **2010,** *23* (2), 200–203.

Shivalingaswamy, T. M.; Satpathy, S. Integrated Pest Management in Vegetable Crops. In *Entomology: Novel Approaches*; Jain, P. C., Bhargava, M. C., Eds.; New India Publishing Agency., New Delhi, 2007; pp 353–375.

Shivalingaswamy, T. M. S.; Satpathy, S.; Rai, A. B.; Rai, M. Insect Pests of Vegetable Crops: Identification and Management. *Tech. Bull. IIVR,* **2006,** *30,* 15.

Singh, R. Elements of Entomology; Deep and Deep Publication: New Delhi, 2007; pp 570.

Singh, S. P. Studies on Some Aspects of Biology-Ecology of Potato Cut Worms in India. *J. Soil. Biol. Ecol.* **1987,** *7,* 135–143.

Sridhar, V.; Chakravarthya, A. K.; Asokan, R.; Vinesh, L. S.; Rebijith, K. B.; Vennila, S. New Record of the Invasive South American Tomato Leaf Miner, *Tuta absoluta* (Meyrick) (Lepidoptera: Gelechiidae) in India. *Pest Manag. Horticul. Ecosyst.* **2014,** *20* (2), 148–154.

Tripathi, S.; Singh, R. Population Dynamics of *Helicoverpa Armigera* (Hubner) (Lepidoptera: Noctuidae). *Insect Sci. Appl.* **1991,** *12*, 367–374.

Trivedi, T. P.; Rajagopal, D. Integrated Pest Management in Potato. In *IPM System in Agriculture, Cash Crops*, Vol 6; Upadhyay, R. K., Mukherji, K. G., Dubey, O. P., Eds.; Aditya Books Pvt. Ltd.: New Delhi, 1999; 299–313.

Vinobaba, M. L; Prishanthini, M. A New Invasive Species of Mealy Bug from the East, http. //www.esn.ac.Ik/*zoology*, 2009, 1–5.

CHAPTER 9

INSECTS OF TUBER CROPS AND THEIR MANAGEMENT

P. CHAND[1*], AMIT SINGH[1], and R. VISHWAKARMA[2]

[1]Department of Agricultural Entomology, Bidhan Chandra Krishi Vishwavidyalya, Mohanpur, Nadia, West Bengal, India

[2]Department of Entomology, Bihar Agricultural University, Sabour, Bihar, India

*Corresponding author. E-mail: drocshastri82@gmail.com

ABSTRACT

Perhaps the greatest economic impact that *L. decemlineata* has had on agriculture has been since its development of resistance to insecticides. *L. decemlineata* became resistant to DDT (now banned) in the mid1950s and has since become resistant to >25 insecticides belonging to the traditional chemical classes. Recently, the cost of controlling *L. decemlineata* infestations in the eastern USA averages between $138 and 368 per ha. Because insecticide resistance in *L. decemlineata* populations is inevitable, agribusiness industries continue to invest millions of dollars into developing new insecticides and genetically modified crops that produce insecticidal toxins. Adults and larvae are easily seen because of their large size. *L. decemlineata* has a tendency to release its hold on plants that are shaken and this characteristic can be used to detect insects hidden among foliage. Visual sampling of potato fields was as efficient for estimating population density as the whole plant bag sampling method, and more efficient than sweep netting. Soil sampling at harvest for buried beetles in diapause provides reliable results in area surveys. A sequential sampling plan has been reported for estimating populations of Colorado potato beetle egg masses and of adults and larvae.

9.1 ORIGIN AND DISTRIBUTION

The beetle was first discovered by Thomas Nuttall in 1811 and was described in 1824. Suddenly, in 1859, the Colorado potato beetle began devastating potato crops 100 miles west of Omaha, Nebraska, USA (Pope and Madge, 1984), whether the attacks stemmed from a change in food first choice by the beetle. Over the next few years, the beetle caused crippling damage as it extended eastward to the Atlantic coast, which it reached in 1874 (Pope and Madge, 1984). *Leptinotarsa decemlineata* became well-known in Europe following its introduction from the USA to Bordeaux, France in 1922, after several unsuccessful attempts from 1876. The beetle spread rapidly in Europe despite intensive control operations to contain it. This was primary reported in Belgium and Spain in 1935, Luxembourg in 1936, the Netherlands and Switzerland in 1937, Austria in 1941, Hungary and the former Czechoslovakia in 1945, Poland and Romania in 1947 and Turkey in 1949. It is extensive in the European part of the former USSR, and has progressively spread eastwards to most potato-growing areas, reaching the Far Eastern provinces. *L. decemlineata* was reported in Xinjiang, China in 1993 (WenChao et al., 2010; Jian et al., 2012). A record of *L. decemlineata* in Zhejiang (CABI/EPPO, 2003) published in previous versions of the Compendium was based on a paper by Mo and Cheng (2003) which only refers to laboratory studies on the pest resistance to insecticides. *L. decemlineata* has not been practical in the field in Zhejiang. *L. decemlineata* has been detected from Denmark, Finland, Norway, Sweden and the UK, including the islands of Guernsey and Jersey (Thomas and Wood, 1980), but the beetle is not established in these countries.

Perhaps, the greatest economic impact that *L. decemlineata* has had on agriculture has been since its development of resistance to insecticides. *L. decemlineata* became resistant to DDT (now banned) in the mid 1950s and has since become resistant to >25 insecticides belonging to the traditional chemical classes (Forgash, 1985; Gauthier et al., 1981; Heim et al., 1990; Bishop and Grafius, 1991). Factors responsible for insecticide resistance development in this pest are described in detail in (Zehnder et al.,1994). Recently, the cost of controlling *L. decemlineata* infestations in the eastern USA averages between $138 and 368 per ha (Grafius, 1997). Because insecticide resistance in *L. decemlineata* populations is inevitable, agri-business industries continue to invest millions of dollars into developing new insecticides and genetically modified crops that produce insecticidal toxins.

9.1.1 ENVIRONMENTAL IMPACT

L. decemlineata invaded North America, and then Europe and Western Asia, in a classic pattern of regular geographical spread, hardly impeded by measures taken against it. However, this spread could not be called invasive because it occurred in an introduced crop planted over large areas as a monoculture. Though *L. decemlineata* attacks other Solanaceae, there are no indications that it affects wild plants in the natural environment to any significant extent. Control of the pest leads to the use of insecticides in potato crops, which most probably would not need such treatment in its absence. So to a certain degree, *L. decemlineata* is responsible for an increased pesticide load in the environment.

9.1.2 SOCIAL IMPACT

Though a serious pest of potato, *L. decemlineata* never caused such losses as those due to *Phytophthora infestans*, with their attendant social consequences. By the time of its introduction, peasant farmers were not so dependent on the single potato crop. In the early days, Colorado beetle could be partly controlled by hand removal and destruction of larvae and adults. Effective cheap insecticides fairly rapidly became available.

9.1.3 DETECTION AND INSPECTION

Adults and larvae are easily seen because of their large size. *L. decemlineata* has a tendency to release its hold on plants that are shaken and this characteristic can be used to detect insects hidden among foliage. Visual sampling of potato fields was as efficient for estimating population density as the whole plant bag sampling method, and more efficient than sweep netting (Senanayake and Holliday, 1988). Soil sampling at harvest for buried beetles in diapause provides reliable results in area surveys (Glez, 1983). A sequential sampling plan has been reported for estimating populations of Colorado potato beetle egg masses and of adults and larvae (Hamilton et al., 1997).

9.2   HOST PLANTS

L. decemlineata attacks potatoes and several other cultivated crops including tomato and egg plant. It attacks wild solanaceous plants as well, which occur widely and can act as a reservoir for infestation. The adults feed on the tubers of host plants as well as leaves, stems and growing points.

Plant name	Family
Hyoscyamus niger (black henbane)	Solanaceae
Nicotiana tabacum (tobacco)	Solanaceae
Solanaceae	Solanaceae
Solanum lycopersicum (tomato)	Solanaceae
Solanum melongena (aubergine)	Solanaceae
Solanum tuberosum (potato)	Solanaceae

9.3   NATURAL ENEMIES

L. decemlineata is attacked by many hymenopterous and dipterous parasitoids; Many predators including pentatomid, reduviid and mired bugs, beetles, including ground beetles and ladybird beetles; Lacewing larvae and parasitic nematodes, protozoans, fungi, and bacteria.

Two hymenopteran egg parasitoids, the Nearctic eulophid *Edovum puttleri* and the Palaearctic mymarid *Anaphes flavipes*, are reported from *L. decemlineata*, but the former is potentially more important with regard to biological control. The New World *E. puttleri* parasitizes eggs of *L. decemlineata* (Grissell, 1981; Jansson et al., 1987; Loganet al., 1987; Williams, 1987; Ruberson et al., 1987; Schroder and Athanas, 1989a), which become less suitable for parasitization and parasitoid development as they age (Hu et al., 1999). Even though it has been released for biological control of the potato beetle in the potato growing areas of the USA, *E. puttleri* has not been effective in potato due to it lacks a hibernal diapause and is badly adapted to the low temperatures that prevail during the period when first generation Colorado potato beetle eggs are present (Obrycki et al., 1985). It has, however, been used in inundative release programmes for biological control in egg plants (Lashomb et al., 1987).

Several Tachinidae are identified to parasitize the Colorado beetle. *Meigenia mutabilis* attacks the larvae of *L. decemlineata* in Russia (Bjegovic, 1967) and *Myiopharus doryphorae* (Geismer, 1920) and attack the larvae in the USA. *M. abberans* (Trouvelot, 1932) and *M. australis* [*M. americanus*] (Reinhard, 1935) attack adults in the USA. *M. doryphorae* is larviparous and first to lastinstar larvae of Colorado beetle are attacked (Tamaki et al., 1983a, b). Maximum parasitism rate in this host reached almost 75% in late August–September when the crop had already suffered damage. The effectiveness of the parasite was limited by its low large quantity during the first generation of *L.decemlineata*.

In Germany, the phorid *Megaselia rufipes* parasitized 0.219 per cent of *L. decemlineata* adults depending on location; Hibernation of the phorid occurs in the pupal state (Eisenschmidt, 1958). The pentatomid heteropteran *Podisus maculiventris* (the spined soldier bug) is an indigenous, generalist predator commonly found throughout North America, east of the Rocky Mountains (Baker and Lambdin, 1985). This bug preys on eggs and larvae of the Colorado beetle in Illinois, Missouri, Iowa (Anon., 1868) and Delaware (Heimpel and Hough Goldstein, 1992; Hough Goldstein and McPherson, 1996). *P. maculiventris* is fascinated to host plant volatiles produced in response to prey feeding (Dickens, 1999). In the field, *P. maculiventris* employs a search strategy that causes it to consume a constant number of preys, which is independent of prey density (O'Neil, 1997).

The pentatomid *Perillus bioculatus* preys on the adults, eggs and larvae of *L. decemlineata* (Chittenden, 1911; Tamaki and Butt, 1978; Drummond et al., 1984). Prey consumption by an individual of *P. bioculatus* during its development from egg to adult averages 285 eggs, or 3.7 fourth instar larvae or 5.1 adults of *L. decemlineata* (Franz and Szmidt, 1960). After hibernation, adults generally consume 0.70.8 adult potato beetles or 0.5 fourth instar *L. decemlineata* per day. Similar to *P. maculiventris, P. bioculatus* responds to host plant volatiles produced in response to feeding by Colorado potato beetle (Dickens, 1999). *P. bioculatus* contributes to beetle mortality but natural populations are generally ineffective in suppressing Colorado potato beetle densities, especially when beetle densities are high (Harcourt, 1971;Tamaki and Butt, 1978).

The generalist predator *Euthyrhynchus floridanus*, a pentatomid occurring widely in the southern states of the USA, Central and South America, was detected feeding on larvae of *L. decemlineata* in South Carolina

(Chittenden, 1911). *E. floridanus* apparently is of limited importance as a predator of Colorado potato beetle (Oetting and Yonke, 1975).

The potential for classical biological control of *L. decemlineata* using natural enemies is suggested by recent collections of natural enemies from a wide range of climates and habitats in Mexico (Logan et al., 1987). A total of 18 species of natural enemies were encountered feeding on the Colorado potato beetle. The densities of Mexican populations of the beetle were typically low, suggesting that natural enemies may be limiting factors. *Oplomus dichrous*, an egg/larval predator, was the most common predator. *O. dichrous* is native to Central America, Mexico and south western USA.

Drummond et al. (1987) attained 95% control in the field with a release ratio of about one *O. dichrous* adult to 4050 Colorado potato beetle eggs. However, preliminary studies of *O. dichrous* undertaken in Rhode Island, USA, indicated that the bug was inadequately adapted to temperate potato production regions because its populations were low synchronized with those of the Colorado potato beetle, its population growth rate was poorly under cool temperatures and it appeared to lack the ability to overwinter in the north eastern USA.

In addition, recent studies in the USA verified that the Neuropteran *Chrysoperla rufilabris* can effectively prey on the eggs and larvae of Colorado potato beetle in both laboratory and field cage tests. In field cage experiments, releases of *C. rufilabris* larvae at a rate of 80940/ha resulted in an 84% reduction of the Colorado potato beetle population (Nordlund et al., 1991). The carabid beetle Lebia grandis is thought to feed almost exclusively on the Colorado potato beetle (Lindroth, 1969). The adults forage nocturnally on potato foliage, where they prey on eggs and larvae (Hazzard et al., 1991). The larvae exhibit a parasitoid like lifestyle by developing to maturity on a single larva or pupa of *L. decemlineata*. This carabid may be the most important endemic predator of the Colorado potato beetle (Heimpel and Hough Goldstein, 1992).

The coccinellids *C. maculata* and *Hippodamia convergens* are polyphagous predators that feed on eggs and small larvae of Colorado potato beetle in North America (Riley, 1869; Franz, 1957; Groden et al., 1990; Cappaert et al., 1991). *C. maculata* is perhaps the most abundant predator of *L. decemlineata* in many potato production areas, and the only one consistently present wherever the beetle eggs were found in the field (Hazzard et al., 1991; Hilbeck and Kennedy 1996). It caused 40% mortality in the second generation in 1986, 38% in the first generation in 1987, and

58% in the second generation in 1987. In a sequence of greenhouse and field experiments, consumption of potato beetle eggs by adult *C. maculata* was inversely related to the population density of egg masses (Arpaia et al., 1997). However, under commercial potato production conditions the abundance of *C. maculata* and daily predation rate appeared to be independent of prey density (Hilbeck and Kennedy, 1996). The high mobility of *C. maculata*, relative to the insect herbivores on which it feeds, may contribute to its effectiveness as a biological control agent in agricultural ecosystems (Coll et al., 1994). Though, the arrangement of the agro-ecosystem is an important determinant of the contribution of *C. maculata* to biological control within specific crop in an agro-ecosystem (Groden et al., 1990; Nault and Kennedy, 2000). *L. decemlineata* is the natural host for the microsporidia *Nosema leptinotarsae* and *Endoreticulatus* (*Pleistophora*) *fidelis* (Brooks et al., 1988). *N. leptinotarsae* occurs in the haemolymph but infections of *E. fidelis* are limited to the epithelial cells of the midgut (Hostounsky and Weiser, 1975; Brooks et al., 1988). *E. fidelis* has been observed to cause a gradual decline in laboratory colonies of Colorado potato beetle. Most work has been conducted on *L. decemlineata* as an experimental host for other *Nosema* species. Experimental infection by *N. gastroideae* causes a reduction in growth rate, apparently due to inadequate food intake (Hostounsky and Weiser, 1973). A high death rate was reported in *L. decemlineata* infected with *N. gastroideae* and *N. polygrammae* (Hostounsky and Weiser, 1975). Chrysomelid microsporidians can be transmitted not only horizontal transmission (from one insect to another of the same species) but also vertical transmission (from one generation to another). Horizontal transmission is oral, with the hosts becoming infected after swallowing spores (Hostounsky and Weiser, 1973; Hostounsky, 1984). The fungi *Beauveria bassiana* and *Scopulariopsis* have been reported from *L. decemlineata* (Humber, 1992). Usual outbreaks of *B. bassiana* cause remarkable levels of mortality in the Colorado potato beetle in the absence of pesticides. In Poland, *B. bassiana* was the dominant fungal pathogen infecting 21% of diapausing Colorado potato beetle (Mietkiewski et al., 1996). *B. bassiana* does not infect Colorado potato beetleeggs, although all other stages are susceptible (Long et al., 1998). Another fungus, *Paecilomyces farinosus* parasitizes a wide range of insect hosts, often causing epizootics. The potential of *P. farinosus* as a microbial insect pathogen was evaluated using *L. decemlineata* (Samsina kova and Kalakova, 1978).

9.4   DAMAGE

Colorado potato beetle is the major defoliator of potato throughout most of North America, both adults and larvae feed on leaves and when foliage has been consumed they will gnaw on stem tissue, and even tubers. First instars are responsible for about 3% of total leaf consumption and second, third, and fourth instars for 5%, 15%, and 77.7, respectively. Total leaf consumption is projected at 33–45 sq cm for larvae and adults consume foliage at a rate of 7–10 sq cm per day. Potato is more susceptible to injury during bloom and shortly thereafter when tubers are rapidly expanding. Generally, defoliation should not exceed 10–25% during this period. Late season defoliation is much less damaging (Ferro et al., 1983). Tomato is readily damaged by early season defoliation, and mean population densities as low as 0.5 beetles per plant reduce tomato yields (Schalk and Stoner, 1979).

9.5   TAXONOMIC AND NOMENCLATURE

Domain	:	Eukaryota
Kingdom	:	Metazoa
Phylum	:	Arthropoda
Subphylum	:	Uniramia
Class	:	Insecta
Order	:	Coleoptera
Family	:	Chrysomelidae
Genus	:	Leptinotarsa
Species	:	*Leptinotarsa decemlineata*

9.6   MORPHOLOGY

Beetle is oval, convex, shiny, 812 mm in length. Elytra are light yellow, each one with 5 black longitudinal stripes. Other parts of body are brown yellow or red orange with black spots. Eggs are oval, shiny, about 1.5 mm in length; their number in a batch may reach 100, though it is usually 3040 eggs. The color of eggs and pupae varies from yellow to brick red and does not change over the course of their growth. Larva is shortened, convex dorsally, up to 1516 mm in length. Head and legs are black; Abdomen is red brown during fifth instars and pink orange or yellow at the end of IV instars (before pupation). The length of pupa is 812 mm and weight is 50,170 mg. Beetles and larvae live openly on plants and they feed on leaves, eating them completely. If the leaves are absent, the imagoes eat any plant part, including non harvested tubers. Females lay eggs on lower leaf surfaces. The fertility of female is usually 4,001,000, maximum 5000 eggs. Larvae pupate in soil at a depth of 510 cm. depending on the temperature, eggs ranges 517 days, larvae on most favorable food plants over 1030 days, and pre-pupae and pupae both over 820 days. Juvenile beetles of a new generation emerge from soil and feed intensively over 620 days, forming a fat body. Later, they either diapause and bury themselves in the soil or they couple and lay eggs of the next generation before diapause. Only imagoes are capable of wintering; they usually hibernate in soil at a depth of 2050 cm. Life span is 14 years due to the diversity of diapause types in this species.

9.7   LIFE CYCLE, REPRODUCTION, AND DISPERSION

The winter is passed in the adult stage, which emerges from the soil in the spring at the time potatoes are emerging. A complete generation can occur in about 30 days. The northern most portion fits range usually allows only a single generation per year, but two generations occur widely, even in Canada. Southern areas such as Maryland and Virginia can support three annual generations, but not all the beetles go on to form a third generation; the third generation often develops on weeds rather than potatoes, so it is ignored. Diapause is induced by a grouping of photo-period, temperature and host quality. Long day lengths normally promote continued reproduction, whereas short day lengths promote diapause, hut diapause induction varies widely among populations. For instance, populations in Washington and Utah do not reproduce when the photoperiod is less than 15 h, whereas beetles from Arizona reproduce at a 13h photo phase. Beetles from southern Texas and Mexico are relatively insensitive to photoperiod and reproduce at photoperiods of 10–18 h (Hsiao, 1988). The overwintering population may be comprised of individuals from more than one generation. In Massachusetts, egg hatch begins in late May and first generation larvae are present until July. First generation adults, the offspring of over-wintered beetles, begin to emerge in July. Those emerging before August first normally go on to reproduce, whereas those emerging later usually enter diapause (Ferro et al., 1991). Some beetles remain in diapause for more than a year.

9.7.1 EGG

Overwintering adults usually feed for 5–10 days before mating and producing eggs, while some beetles mate in the autumn and can oviposition without mating in the spring. The eggs are orange and elongate oval, measuring about 1.7–1.8 mm long and 0.8 mm wide. They are deposited on end in clusters of about 5–100 eggs, but 20–60 eggs per cluster is normal. The eggs are deposited on the lower surface of foliage and anchored with a small amount of yellowish glue. The eggs do not change markedly in appearance until about 12 h before hatching, when the embryo becomes visible. Indicate development time of eggs is 10.7, 6.2, 3.4, and 4.6 days when held at 15, 20, 24, and 30°C, respectively.

9.7.2 LARVA

The larvae are reddish and black, and easily observed and recognized. The larvae arc very plump, with the abdomen strongly convex in shape. The larvae bear a terminal pro-leg at the tip of the abdomen, in addition to three pairs of thoracic legs. Young larvae are dark red with a black head, thoracic plate, and legs. Two rows of black spots occur along each side of the abdomen. The larger larvae are lighter red, with the black coloration of the thoracic shield reduced to the posterior margin. There are four instars. I lead capsule widths are about 0.65, 1.09, 1.67, and 2.5 mm for instars 1–4 respectively. During these instars body length increases from 15–2.6, 2.8–5.3, 5.5–8.5, and 9–15 mm, respectively. Developmental thresholds vary geo-graphically, but 8–12 CC is common. Mean development time of first instars is about 6.1, 3.7, 2.1, and 1.4 days at 15, 20, 24, and 28°C, respectively. For the second instar, mean development times are about 5.0, 3.8, 2.2, and 1.6 days; for the third instar 2.8, 2.5, 2.3, and 1.7 days and for the fourth instars 9.5, 6.6, 3.3, and 2.4 days, respectively, when reared at 15, 20, 24, and 28°C.

9.7.3 PUPA

At maturity level, larvae drop to the soil and burrow to depths of 2–5 cm where they form a small cell. After two days they develop into pupae. The pupae are oval and orange in color. They measure about 9.2 mm long and 6.4 mm wide. The form of the adult beetle is identifiable in the pupal stage, though the wings and antennae are twisted ventrally. Point to development time of the pupal stage, exclusive of the period spent in the soil prior to actual pupation, is about 5.8 days. Ferro el al. (1985) report the below ground combined pre-pupal, pupal and post-pupal period to average about 22.3, 14.9, 11.7, and 8.8 days at 15°, 20°, 24°, and 28°C, respectively

9.7.4 ADULT

After promising from the pupal to the adult stage, the beetle remains in the soil for 3–4 days before digging to the surface. Adults are healthy in form, and oval in shape when viewed from above. The dorsal surface of adults is principally yellow, but each brewing is marked with five longitudinal black lines. The head bears a triangular black spot and the thorax is dotted with about ten irregular dark markings. The underside of the beetle and legs also arc mostly dark. Beetles produce eggs over a 4–10 week period, with most of the eggs produced during weeks 1–5. Fecundity of beetles under field conditions has been expected about 200–500, but this is likely an underestimate. Under laboratory conditions, mean fecundity was 3348 eggs perform ale when fed potato, and 2094 when fed hairy nights hade (Brown et al., 1980). In the autumn, newly emerged beetles feed for a time, and then dig into the soil, normally to a depth of 7–13 cm, to pass the winter. When beetles emerge in the spring they are not highly dispersive, mostly seeking out hosts by walking, though they are capable of flight.

9.7.5 ECOLOGY

The beetles overwinter as diapausing adults in the soil, typically at depths of 7.6 –12.7 cm (Lashomb et al., 1984). Overwintered adult beetles emerge from the ground over a period of quite a few weeks in spring or early summer, depending on the climate and their physiological condition (Hare, 1990). Following emergence in the spring overwintered adults disperse to find suitable host plants by walking and by flight. Without eating beetles display greater flight activity than those that locate a suitable host and begin feeding soon after emergence (Ferro et al., 1999). Host plants are located largely by chance, through random searching. However, potato plant odor is attractive to the beetles and potato plants damaged by feeding are more attractive than undamaged plants, under laboratory conditions (Bolter et al., 1997; Landolt et al., 1999). Maximum food consumption occurs at 25°C temperature.

Beetles typically mate before entering hibernal diapause, and mate repeatedly in the spring, often within 24 h of emergence from the soil as well (Ferro et al., 1999). Sperm precedence in multiply mated females is incomplete (Alyokhin and Ferro, 1999). Oviposition begins 510 days after emergence at 1530°C temperature. Eggs are laid in masses, containing 1030 eggs, on the lower surface of the leaf. Eggs laying usually continues over quite a lot of weeks, with each female laying up to 2000 eggs. The larvae hatch using egg bursters or oviruptors situated on the meso and metathorax and abdominal segment 1 (Cox, 1988). They hatch in 414 days. After freeing themselves from the chorion, the larvae partly or entirely consume the chorion before feeding on leaf tissue. Larvae moult four times, the last of which is the larval/pupal moult. Larval development requires as small as 8 days or as long as 28 days at average temperatures of 29 and 14°C, respectively. Mature fourth instars larvae burrow into the soil where they pupate. The pupal stage naturally lasts 8–18 days, depending on temperature. Developmental thresholds, which range from 8–12°C temperature, vary among populations and life stages. At stable temperatures, development is most rapid between 25 and 33°C; at higher temperatures larval growth is slowed and mortality increases (Hare, 1990).

The larvae are hardy and resistant to adverse weather, although heavy rain and strong winds may cause high mortality, especially in earlier instars. Cannibalism of eggs by adults can be considerable, averaging 19% in one study (Schrod et al., 1996). Cannibalism during the first instar is particularly common at high temperatures with poor humidity but negligible under moderate conditions when suitable foliage is present. Larvae from the same egg mass hatch synchronously and tend to remain grouped on the lower leaf surface until the first moult, after which they migrate to immature foliage on the plant. Larvae are voracious foliage feeders. Although total consumption depends on host plant, first instars consume ca 3% of the total foliage consumed during development and second, third and fourth instars consume ca 5, 15, and 77% of the total (Ferro et al., 1985). By the final or fourth instar, they may feed on petioles and stems, if the plants have become severely defoliated. The fully developed larvae descend to the ground and bury themselves in the soil at varying depths depending to soil situation. Pupation occurs in smoothly lined cells and the pupal period lasts 1020 days, after which the first generation adults emerge. These adults may walk or fly to the nearest host plant to feed. After feeding for several days, adults of the first and subsequent generations may mate and reproduce, or cease feeding and enter diapause, depending on temperature, photoperiod, and host plant condition. A portion of the population may produce some offspring before entering diapause (Tauber et al., 1988; Voss et al., 1988; Voss and Ferro, 1990a, b; Nault and Kennedy, 1999). The number of complete generations varies between one, near the colder extremes, to about four, in the warmest areas where development from egg to adult is completed in 30 days. The minimum requirements for completion of one full generation are at least 60 days during summer when the temperatures exceed 15°C and winter temperatures that remain above 8° C. There are some cold areas in which only a partial age band is produced and the beetles cannot permanently become established. In general, sunny weather with a mean daily temperature of 1720° C results in growth and spread of populations, but if temperatures do not exceed 1114° C and humidity is high, the populations may actually decrease (Svikle, 1976). In the northern part of the beetles' range in Europe, newly emerged first generation beetles feed and then burrow 2540 cm into the soil, where they enter diapause and hibernate over winter. Mortality during hibernation averaged 30% in the Ukraine, but may be as high as 83%, mainly because of fungal and bacterial infections (Koval, 1984). A considerable portion of pre-diapause adults migrate to field limitations near tree lines before burrowing into the soil, although large numbers of beetles also enter the soil and overwinter within potato fields (Weber and Ferro, 1993; Weber et al., 1994; Nault et al., 1997). In temperate regions, photoperiod is the most important factor inducing 'hibernal diapause' in teneral adults of L. decem-lineata, but ambient temperatures and food quality may have modifying effects (Hsiao, 1988). This species is a typical "long day" insect entering diapause after exposure to a critically short photoperiod, which varies with latitude. In general, populations from the south require a shorter photoperiod for diapause induction than those from the north (de Wilde and Hsiao, 1981). Stable photoperiods approach 16 h for northern populations with latitude 45°N (Tauber et al., 1988) and decline to about 12 h for southern populations with latitude 32°N.

The physiological condition of potato foliage on diapause induction and the activity of the corpora allata were demonstrated by De Wilde et al. (1969). The corpora allata of diapausing beetles were small and inactive, as expected, but were 26% smaller in beetles fed mature foliage than in beetles fed young potato foliage. In adults fed mature foliage, the difference in size of the corpora allata was correlated with a >50% reduction in mean daily egg production. High temperatures reduce the beetle's sensitivity to photo-period; a shorter photoperiod will induce diapause at higher temperatures (de Wilde and Hsiao, 1981). This allows the utilization of high-latitude areas that have relatively mild conditions. Response to temperature varies among populations. For instance, the critical photoperiod for diapause induction in an upstate New York population was longer and the induction of diapause by low temperature was greater, than for populations from the warmer coastal areas of Long Island (Tauber et al., *1988*).

9.8   MANAGEMENT

9.8.1   CULTURAL PRACTICES

Tardy immigration of the crop reduces the time available for second and subsequent generations to develop (Voss et al., 1988). Late colo-nization could be achieved by various means, such as crop rotation, manipulation by planting date, setting up different barriers such as plastic line trenches, portable trench barriers, mulching, and mechan-ical collection of beetles on overwintering sites or immediately after colonization. Crop rotation delays migration of the crop by overwin-tered adults and reduces the size of the population that subsequently develops within the crop. Population reductions of 90% or more have been reported in potato (Lashomb and Ng, 1984; Wright, 1984). The effect of crop rotation is increased with increased distance from the source of overwintering beetles. Follett et al. (1996) suggested 0.5 km as the minimum distance necessary to fully benefit from field rotation. Because rotated potato fields require fewer insecticides to control *L. decemlineata* (Speese and Sterrett, 1998), crop rotation is an important tool in delaying the development of resistance to insecticides (Roush et al., 1990). Saxon and Wyman (2005), in their suggestion of developing of an area wide *L. decemlineata* pest management strategy, reported that long distance rotations of more than 400 m were an effective cultural control management strategy to limit adult beetle infestations in the spring. This strategy can be optimized when collaborating growers are able to maximize their rotational distances by coordinating their rotational schemes over large areas. Deploying long distance rotations over a large area over many years would limit *L. decemlineata* populations and could result in significantly reduced *L. decemlineata* populations entering fields in the spring. The planting date of a potato crop can be manipulated to reduce the population of second generation larvae produced in the crop. Early planting of short season varieties allows the crop to mature before the second larval generation is produced. In contrast, colonization of late planted, rotated potato plantings occurs later in the season causing most summer generation adults to emerge after the critical photoperiod for diapause induction has been reached. Consequently, these adults do not produce a second generation of larvae (Weber and Ferro, 1993).

9.8.2 PHYSICAL PRACTICES

Plastic lined trenches, which serve as pitfall traps, and trap crops of early planted potato have both been shown to effectively intercept overwintered adults in the spring, before they colonize the potato crop. Plastic lined trenches that were V-shaped with an average width at the top of 740 ± 7 mm and depth of 223 ± 2 mm retained 95% of the beetles they trapped in controlled field experiments (Misener et al., 1993; Boiteau and Osborn, 1999). Because 50–75% of overwintered beetles disperse into a nearby potato crop by walking, properly designed trenches, positioned to intercept the dispersing beetles can provide a significant level of crop protection. At one location in New York, USA, more than 100,000 overwintered beetles were trapped in 91 m of trench (Moyer, 1993). Overwintering beetles are frequently found in windbreaks and hedgerows adjacent to the crop at densities of hundreds per square meter, whereas densities within the fields average only 37 beetles per square meter (Weber and Ferro 1993, Hunt and Tan 2000). This distribution of overwintering beetles clearly has important implications for the management of *L. decemlineata*, particularly when another solanaceous crop is planned for the same or an adjacent field in the following year. For some growers, crop rotation is not a viable option due to a shortage of land. They designed an above ground trench composed of an extruded, UV retarded PVC plastic trough, designed to allow *L. decemlineata* and other pests to enter the device and become trapped and killed inside. The above ground trench can capture thousands of newly emerged *L. decemlineata* as they walk from their overwintering habitat into an adjacent crop, and also reduce crop damage. This trap is easily installed in the spring and removed in the fall, and is reusable for several years. The trench is black to raise the temperature, which results in an increase in beetle mortality, and is designed to release water and beneficial insects.

9.8.3   MECHANICAL PRACTICES

Mechanical collection, use of propane flamers, use of pneumatic thermal machines or bio-collectors can prevent adult colonization and reduce larval damage (Boiteau et al., 1992; Karalus, 1994; Pelletier et al., 1995; Lagud et al., 1999a, b; Derafshi, 2006). The bio-collector is a novel control method approved for use in organic potato cultivation in Germany. The collector, which is attached to a tractor, blows chrysomelids off the potato plants and collects them in trays. Collections are made 2–3 times per year depending on the level of infestation (Karalus, 1994). However, some authors stated that mechanical control causes undesirable damage and its efficacy should be improved (Sablon et al., 2013). Although Boiteau et al. (2012) set up in laboratory trials that wood ash is toxic to adult and larval stages of the *L. decemlineata*, the considerable control observed in the laboratory did not extend to field application.

9.8.4   BIOLOGICAL CONTROL

None of the naturally occuring beneficial organisms display the ability to regulate potato beetle populations consistently, though the tachinid, *Miopluirus dor'phorae,* and the stink bugs, *Perillus bioculaius,* and *Podiszis, naculiz'enlris,* have been cultured and released to suppress beetle damage (Tamaki and Butt, 1978; Tamaki et al., 1983a; Hough-Goldstein and McPherson, 1996). Stink bugs are effective at destroying eggs and larvae, if a favorable ratio of predatory bugs to beetle egg clusters can be established, a ratio of one bug per about 100 beetle eggs can reduce injury by 80%. Efforts have been made to identify new biotic agents, the most promising of which is the egg parasitoid, *Edovurn putticri* (Hymen optera: Fulophidae). This wasp can kill more than 80% of the eggs in an egg cluster, but the effectiveness of *E. puftieri* is limited by its require- ment for warm temperatures in order to be active, and its inability to over- winter in temperate climates. Repeated release of *E. pullleri* into eggplant fields in New Jersey was reported to be an important element in producing high-quality fruit with minimal pesticide use and high financial returns. This parasitoid was described by many authors, including Grissell (1981). Pathogens have been investigated extensively for Colorado potato beetle suppression and strains of *Bacillus thuriugiensis,* mentioned above under "insecticides" are most promising. However, the fungal pathogen *Azsveria bassiana* has been used in Europe with modest success (Roberts et al., 1981). It is limited mostly by the economics of potato production, and the incompatibility of the entomopathogenic fungus with fungicides that must be applied to control foliar plant diseases. Entomopathogenic nematodes can be applied to the foliage to suppress feeding larvae or to the soil to kill larvae as they pupate (Toba et al., 1983; Berry et al., 1997).

9.8.5   HOST PLANT RESISTANT

The incorporation of varietal resistance to Colorado potato beetle has emphasized the transfer of resistance traits to *S. tuberosum* from other *Solanum* species using a variety of techniques to obtain successful inter specific crosses (Tingey and Yencho, 1994). Emphasis has been placed on resistance derived from *Solanum berthaultii,* which is mediated in large part by glandular trichomes on the foliage (Yencho et al., 1996) and on resistance derived from *Solanum* chacoense, which is mediated by high concentrations of leptine glycoalkaloids in the foliage (Sinden et al., 1986; Sanford et al., 1997; Yencho et al., 2000). Potato breeding lines with resistance to potato beetle have been released (Lorenzen and Balbyshev, 1997). Other poten- tially valuable mechanisms of resistance have been identified as well

9.8.6   GENETIC CONTROL

Transgenic potatoes expressing the gene for *B. thuringiensis* subsp. *tenebrionis* Cry3A delta endotoxin were approved for commercial use in the USA in 1995. In 1998, transgenic potato varieties expressing the Cry3A toxin were planted on approximately 20,000 hectares in the USA. Although these transgenic potato varieties are highly toxic to Colorado potato beetle (Perlak et al., 1993; Wierenga et al., 1996) and provide excellent control of the beetle, planting of Cry3A toxin expressing transgenic potato varieties declined dramatically by 2000 due to concern over consumer resistance to purchasing transgenic potatoes and products made from them. An additional factor contributing to this decline was competition from new, conventional insecticides that controlled a broader spectrum of potato pests. The future role of transgenic potato varieties in the Colorado potato beetle management is currently uncertain, despite their effectiveness and considerable evidence that they have no significant effect on populations of natural enemies (Hoy et al., 1998; Riddick et al., 1998). Transgenic lines of aubergine expressing the Cry3A endotoxin (Hamilton et al., 1997b) and lines expressing the Cry3B endotoxin (Arpaia et al., 1997; Mennella et al., 1998), although not commercially available, have been shown to provide control of Colorado potato beetle. Because of the potential that Colorado potato beetle will develop resistance to the Cry3A endotoxin if these transgenic potato varieties are widely planted; significant effort has been directed towards the development of resistance management strategies for transgenic potatoes. Proposed resistance management strategies for potato focus on using varieties that express a high dose of the toxin in conjunction with plantings of refuge areas of non transgenic potato or other hosts in which beetles are not controlled. The level of toxin expressed should be high enough to kill any individuals that are heterozygous for resistance alleles. The size and location of refuge plantings must be such that any homozygous resistant beetles selected on the transgenic crop mate with homozygous susceptible beetles produced in the refuge. The progeny of such mating would be heterozygous for the resistance allele and would be killed if they fed on the transgenic crop. Strategies for managing resistance to transgenic potatoes are discussed by Gould et al. (1994) and Hoy (1999). Ochoa Campuzano et al. (2013) identified prohibitin, an essential protein for *L. decemlineata* larval viability. They explored the possibility for prohiobitin1 silencing in *L. decemlineata* larvae in order to reach higher efficacy of Cry3Aa toxin. Cooper et al. (2006) assessed the effectiveness of the protein avidin against the Colorado potato beetle neonates in a no choice detached leaf bioassay at 0,

17, 34, 51, 102, and 204 µg avidin/ml over 12 d. The combined effects of avidin (136 µg avidin/ml) with Bt.Cry3A or leptines were evaluated with neonates and third instars over 12 and 6 days, respectively. Survival of third instars on the Bt.Cry3A with avidin was significantly reduced after 3 days compared with survival on the Bt.Cry3A, suggesting the addition of avidin may increase susceptibility to Bt.Cry3A. The Colorado potato beetle spiroplasma (CPBS) appears to be host (genus) specific and is transmitted among larvae and adults during regurgitation and defecation. However, researchers have adopted a strategy to engineer spiroplasma with an insect lethal gene because the spiroplasma appears to be a commensal (Hackett et al., 1988; Gasparich et al., 1993a, b). The delta-endotoxin gene for *B. thuringiensis* subsp.tenebrionis (Btt) may be an appropriate gene for this purpose because the Colorado potato beetle is susceptible to the toxin and the CPBS adheres to the midgut microvilli (Hackett and Clark, 1989), which is the site of action of the delta-endotoxin. This system would allow multiplication and spread of the genetically engineered microorganism throughout the crop in contrast to direct treatment of beetles with the Btt beta-endotoxin.

9.8.7   CHEMICAL APPROACHES

Chemical insecticides have constituted the primary method of control for Colorado potato beetle for much of its history as a pest of potato. However, the extraordinary ability of the beetle to develop resistance to virtually all insecticides used to control it has led repeated control failures in many areas (Casagrande, 1987; Georghiou and Lagunes Tejeda, 1991; Bishop and Grafius, 1996). The commercialization of several new insecticides with novel modes of action, which are effective in controlling Colorado potato beetle populations resistant to organophosphate, carbamate, chlorinated hydrocarbon and pyrethroid insecticides has restored the ability to control the beetle with insecticides in areas most severely affected by resistance. In the past 60 years *L. decemlineata* has developed resistance to 54 different insecticides, including imidacloprid and eight other neonicotinoids (Whalon et al., 2013). Up to 155-fold resistance to imidacloprod was reported in adult *L. decemlineata* from selected fields on Long Island, New York state, USA, after three seasons of use (Zhao et al., 2000), followed by thiamethoxam and clothianidin in other regions of the USA (MotaSanchez et al., 2006; Alyokhin et al., 2007; 2008). Because of the importance of insecticides in managing Colorado potato beetle in commercial potato production, great emphasis has been placed on the development of procedures for managing resistance. Key processes involved in adaptation of Colorado potato beetle populations to insecticides have been explored using simulation models (Follett et al., 1993, 1995). Resistance management efforts for Colorado potato beetle focus primarily on using pest management procedures, including crop rotation, scouting and action thresholds to minimize the use of insecticides in conjunction with resistance monitoring and rotations of insecticides having different modes of action (Kennedy and French, 1994; Grafius, 1997; Midgarden et al., 1997; Dively et al., 1998). A computerized approach to insecticide management for the Colorado potato beetle has been used in commercial potato fields on the eastern shore of Virginia, USA (Vencill et al., 1995). The Potato Insect Expert System (PIES) uses insect life stages, potato growth stage, percentage defoliation and other factors to determine whether the application of insecticide is necessary to prevent tuber yield loss. Recommendations from PIES were compared with commercial spray thresholds based on the number of *L. decemlineata* per stem. Tuber yields were not significantly different between the two methods, although on average PIES recommended fewer insecticide applications than the conventional method. Visual defoliation ratings for PIES also required less time than conventional sampling for *L. decemlineata* on potato stems. In order to establish an integrated pest management system, a temperature driven decision support system (SIMLEP DSS) was designed for Europe, consisting of two modules (Jorg et al., 2007). SIMLEP 1 is a regional forecasting model for the first occurrence of hibernating beetles and the start of egg laying. SIMLEP 3 is a field specific model which forecasts the occurrence of the developmental stages of *L. decemlineata*. From 1999 to 2004 SIMLEP 3 was validated in Germany, Austria, Italy, and Poland. In about 90% of cases SIMLEP 3 correctly predicted the periods of maximum egg laying and young larval occurrence, which are the optimal periods for field assessments and treatments with conventional and biological insecticides. SIMLEP 3 has since been validated in Slovenia (Kos et al., 2009). SIMLEP 3 is used in practice in Germany and Austria on a large scale and in the western part of Poland.

9.8.8 BIOTECHNOLOGICAL AND MOLECULAR APPROACHES

Zhu et al. (2011) reported the results of study in which the potential of feeding dsRNA expressed in bacteria or synthesized *in vitro* to manage populations of *L. decemlineata* was investigated. Feeding RNA interference (RNAi) successfully triggered the silencing of all five target genes tested and caused significant mortality and reduced body weight gain in the treated beetles. This study provides the first example of an effective RNAi response in insects after feeding dsRNA produced in bacteria. The obtained results suggest that the efficient induction of RNAi using bacteria to deliver dsRNA is a possible method for management of *L. decemlineata*. This method is still in the experimental stage.

9.8.9 LEGISLATIVE CONTROL

Countries at risk should require that consignments of any plants or plant products be found free from *L. decemlineata* after having been subjected to sorting and packaging techniques in suitable premises. In addition, potatoes and certain vegetable crops should be grown in fields which have been inspected during the growing season and found to be free from the pest, and/or in areas where either the pest does not occur or is under intensive official control.

KEYWORDS

- **integrated pest management**
- **tuber crops**
- ***Leptinotarsa decemlineata***

REFERENCES

Alyokhin, A. V.; Ferro D. N. Electrophoretic Confirmation of Sperm Mixing in Mated Colorado Potato Beetles (Coleoptera: Chrysomelidae). *Ann. Entomol. Soc. Am.* **1999,** *92* (2), 230235; 27.

Anonymous, Potato Bugs. The Colorado Potato Bug, its Past History and Future Progress. *Am. Entomol.*, **1868**, *1*, 44–49.

Arpaia, S.; Mennella, G.; Onofaro, V.; Perri, E.; Sunseri, F.; Rotino, G. L.Production of Transgenic Eggplant (*Solanum melongena* L.) Resistant to Colorado Potato Beetle (Leptinotarsa decemlineata Say). *Theor. Appl. Genet.* **1997**, *95* (3), 329–334.

Baker, A. M.; Lambdin, P. L.Fecundity, Fertility, and Longevity of Mated and Unmated Spined Soldier Bug Females. *J. Agric. Entomol.* **1985**, *2* (4), 378–382.

Berry, R. E.; Liii, J.; Reed, G.Comparison of Endemic and Exotic Entomopathogenic Nematode Species for Control of Colorado Potato Beetle (Coleoptera; Chrysomelidae). *J. Earn. Entonwl.***1997**, *90*, 1528–1533.

Bjegovic, P. Meigenia Mutabilis Fall. As a Parasite of Oulema melanopa L. and its Relation to *Leptinotarsa decemlineata* Say (SE). *Zastita Bilja*, **1967**, *18*,93–100.

Boiteau, G.; Osborn, W. P. L. Comparison of Plasticlined Trenches and Extruded Plastic Traps for Controlling *Leptinotarsa decemlineata* (Coleoptera: Chrysomelidae). *Canadian Entomol.* **1999**, *131* (4), 567–572.

Boiteau, G.; Singh, R. P.; McCarthy, P. C.; MacKinley, P. D. Wood Ash Potential for Colorado Potato Beetle Control. *Am. J. Potato. Res.* **2012**, *89* (2), 129–135.

Bolter, C. J.; Dicke, M.; Loon, J. J.; AvanVisser, J. H.; Posthumus, M. A. Attraction of Colorado Potato Beetle to Herbivore Damaged Plants During Herbivory and After its Termination. *J. Chem. Ecol.* **1997**, *23* (4), 1003–1023.

Boiteau, G.; Misener, G. C.; Singh, R. P.; Bernard, G. Evaluation of Vacuum Collector for Insect Pestcontrolin Potato. *Am. Potato. J.* **1992**, *69*, 157–166.

Bishop, B. A.; Grafius, E. An Onfarminsecticide Resistance Test Kit for Colorado Potato Beetle (Coleoptera: Chrysomelidae). *Am. Potato J.* **1991**, *68* (1), 53–64.

Brooks, W. M.; Becnel, J. J; Kennedy, G. G. Establishment of Endoreticulatus n.g. for Pleistophora fidelis (Hostounsky & Weiser, 1975) (Microsporida: Pleistophoridae) Based on the Ultrastructure of a Microsporidium in the Colorado Potato Beetle, *Leptinotarsa decemlineata* (Say) (Coleoptera: Chrysomelidae). *J. Protozool.* **1988**, *35* (4), 481–488.

Brown, J.; Jenny, J. T.; Butt, B. A. The Influence of an Alternate Host Plant on the Fecundity of the Colorado Potato Beetle, *Leptinolarsa decemlineata* (Coleoptera: Chrysomelidae). *Ann. Entomol. Soc. Am.* **1980**,*73*, 197–199.

CABI/EPPO, *Leptinotarsa decemlineata*. Distribution Maps of Plant Pests, No. 139. Wallingford, UK: CAB International, 2003.

Cappaert, D. L.; Drummond, F. A.; Logan, P. A. Incidence of Natural Enemies of the Colorado Potato Beetle, *Leptinotarsa decemlineata* (Coleoptera: Chrysomelidae) on a Native Host in Mexico. *Entomophaga.***1991**, *36* (3), 369–378.

Chittenden, F. H. On the Natural Enemies of the Colorado Potato Beetle. Bulletin United States Department of Agriculture, *Bur. Entomol.*, **1911**, *82*, 85–88.

Coll, M.; Mendoza, L. G.; de, Roderick, G. K, Population Structure of a Predatory Beetle: the Importance of Gene Flow for Intertrophic Level Interactions. *Heredity* **1994**, *72* (3), 228–236.

Cox, M. L.Egg Bursters in the Chrysomelidae, with a Review of Their Occurrence in the Chrysomeloidea and Curculionoidea (Coleoptera). *Syst. Entomol.* **1988**, *13* (4), 393–432

Derafshi, M. H. Design and Construction of a Pneumatic Thermal Machine for Controlling Colorado Potato Beetle (*Leptinotarsadecemlineata*). *J. Appl. Sci.* **2006**, *6* (4), 919–925.

de Wilde, J.; Bongers, W.; Schooneveld, H. Effects of Hostplant Age On Phytophagous Insects. *Entomologia Experimentalis et Applicata* **1969**, *12*, 714–720.

de Wilde, J.; Hsiao, T. H. Geographic Diversity of the Colorado Potato Beetle and its Infestation in Eurasia. In Advances in Potato Pest Management. Lashomb, J. H., Casagrande, R. A., Eds.; Hutchinson Ross Publishing Company: Stroudsberg, Pennsylvania, USA 47–68.

Dickens, J. C. Predatorprey Interactions: Olfactory Adaptations of Generalist and Specialist Predators. *Agric. Forest. Entomol.* **1999**, *1* (1), 47–54.

Drummond, F. A.; James, R. L.; Casagrande, R. A.; Faubert, H. Development and Survival of Podisus maculiventris (Say) (Hemiptera:Pentatomidae), a Predator of the Colorado Potato Beetle (Coleoptera: Chrysomelidae).*Environ. Entomol.* **1984**, *13* (5), 1283–1286.

Drummond, F. A.; Casagrande, R. A.; Groden, E. Biology of Oplomus dichrous (Heteroptera: Pentatomidae) and its Potential to Control Colorado Potato Beetle (Coleoptera: Chrysomelidae). *Environ. Entomol.* **1987**, *16* (3), 633–638.

Eisenschmidt, H. Ein starkes Auftreten der Buckelfliege Megaselia rufipes Meigen (Diptera, Phoridae) als Parasit das KartoffelkSfers(*Leptinotarsa decemlineata* Say) im Jahre. Zeitschrift für Angewandte Zoologie **1958**, *45*, 11–19.

Ferro, D. N.; Morzuch B. J.; Margolies D. Crop Less Assessment of the Colorado Potato Beetle (Coleoptera: Chrysomelidae) on Potatoes in Western Massachusetts. *I. Earn. Entomol.* **1983**, *76*, 349–356.

Ferro, D. N.; Tuttle A. F.; Weher. D. C. Ovipositional and Flightbehavior of Overwintered Colorado Potato Beetle (Coleopt era:(hrysomclidae). *Environ. Entomol.* **1991**, *20*, 1309–1314.

Ferro, D.N.; Logan, A.; Voss, R. H.; Elkington J. S. Colorado Potato Beetle (Coleoptera; Chrysomelidae) Temperature-Depend Ent Growth and Feeding Rates. *Environ. Enlomol.* **1985**, *14*, 343–348.

Ferro, D. N.; Alyokhin.A. V.; Tobin, D. B. Reproductive Status and Flight Activity of the Overwintered Colorado potato beetle. *Entomologia Experimentalis etApplicata* **1999**, *91* (3), 443–448.

Follett, P. A.; Cantelo, W. W.; Rderick, G. K. Local Dispersal of Overwintered Colorado Potato Beetle (Chrysomelidae:Coleoptera) Determined by Mark and Recapture. *Environ. Entomol.* **1996**, *25* (6), 1304–1311.

Forgash, A. J. Insecticide Resistance in the Colorado Potato Beetle. Research Bulletin, Massachusetts Agricultural Experiment Station, No.704, 33–52.

Franz, J. Observations on the Natural Mortality of the Colorado Potato Beetle, *Leptinotarsa decemlineata* (Say) in Canada. *Entomophaga* **1957**, *2*, 197–212.

Franz, J.; Szmidt, A. Observations Concerning Perillus bioculatus (Fabr.) (Heteropt., Pentatomidae) an Imported Predator of the Colorado Potato Beetle from North America. *Entomophaga* **1960**, *5*, 87–110.

Gauthier, N. L.; Hofmaster, R. N.; Semel, M. History of Colorado Potato Beetle Control. In Advances in Potato Pest Management. Lashomb J.H., Casagrande R., Eds.; Hutchinson Ross Stroudsburg, Penn., USA, 13–33.

Geismer, L. M. The Tachina Fly (*Phorocera doryphorae*), an Interesting Parasite on Potato Beetles. Potato Magazine, 3–8.

Glez, V. M. A Method of Determining the Abundance of the Colorado Beetle in the field. Zashchita Rastenii, 6–37.

Grafius, E. Economic Impact of Insecticide Resistance in the Colorado Potato Beetle (Coleoptera: Chrysomelidae) on the Michigan potato industry. *J. Econ. Entomol.* **1997**, *90* (5), 1144–1151;44.

Grissell, E. E.; Edovum puttleri, n.g., n.sp.(Hymenoptera: Eulophidae), an Egg Parasite of the Colorado Potato Beetle (Chrysomelidae). Proceedings of the Entomological Society of Washington, **1981**, *83* (4), 790–796.

Groden, E.; Drummond, F. A.; Casagrande, R. A.; Haynes, D. L. Coleomegilla maculata (Coleoptera: Coccinellidae): its Predation Upon the Colorado Potato Beetle (Coleoptera: Chrysomelidae) and its Incidence in Potatoes and Surrounding Crops. *J. Econ. Entomol.* **1990**, *83* (4), 1306–1315.

Hamilton, G. C.; Jelenkovic, G. L.; Lashomb, J. H.; Ghidiu, G.; Billings, S.; Patt, J. M..Effectiveness of Transgenic Eggplant (Solanum melongena L.) Against the Colorado Potato Beetle. *Adv. Hortic. Sci.* **1997**, *11* (4), 189–192.

Harcourt, D. G. Population Dynamics of *Leptinotarsa decemlineata* (Say) in Eastern Ontario: III. Major Population Processes. *Canadian Entomol.* **1971**, *103*, 1049–1061.

Hare, J. D. Ecology and Management of the Colorado Potato Beetle. *Annu. Rev. Entomology.* **1990**, *35*, 81–100.

Hazzard, R. V.; Ferro, D. N.; van Driesche, R. G.; Tuttle, A. F. Mortality of Eggs of Colorado Potato Beetle (Coleoptera: Chrysomelidae) from Predation by Coleomegilla maculata (Coleoptera: Coccinellidae). *Environ. Entomol.* **1991**, *20* (3), 841–848.

Heim, D. C.; Kennedy, G. G.;van Duyn J. W. Survey of Insecticide Resistance Among North Carolina Colorado Potato Beetle (Coleoptera:Chrysomelidae) Populations. *J. Econ. Entomol.* **1990**, *83* (4), 1229–1235.

Heimpel, G. E.; Hough Goldstein, J. A. A Survey of Arthropod Predators of *Leptinotarsa decemlineata* (Say) in Delaware potato fields. *J. Agric. Entomol.* **1992**, *9* (2), 137–142.

Hilbeck, A.; Kennedy, G. G. Predators Feeding on the Colorado Potato Beetle in Insecticide Free Plots and Insecticide Treated Commercial Potato Fields in Eastern North Carolina. *Biol. Control* **1996**, *6* (2), 273–282.

Hostounsky, Z.; Weiser, J. Nosema Polygrammae sp.n.and Plistophora fidelis sp.n. (Microsporidia, Nosematidae) Infecting Polygramma undecimlineata (Coleoptera: Chrysomelidae) in Cuba. *Vestnik Ceskoslovenske Spolecnosti Zoologicke* **1975**, *39* (2), 104–110.

Hostounsky, Z. Production of Microsporidia Pathogenic to the Colorado Potato Beetle (Leptinotarsa decemlineata) in Alternate Hosts. *J. Invertebr. Pathol.* **1984**, *44* (2), 166–171.

Hostounsky, Z.; Weiser, J. Nosema Gastroideae sp.n.(Nosematidae, Microsporidia) Infecting Gastroidea polygoni and *Leptinotarsa decemlineata* (Coleoptera, Chrysomelidae).*Acta Entomologica Bohemoslovaca* **1973**, *70* (5), 345–350.

Hough Goldstein, J.; McPherson, D.Comparison of Perillus bioculatus and Podisus maculiventris (Hemiptera: Pentatomidae) as Potential Control Agents of the Colorado Potato Beetle (Coleoptera: Chrysomelidae). *J. Econ. Entomol.* **1996**, *89* (5), 1116–1123.

Hoy, C. W.; Feldman, J.; Kennedy, G. G.; Reed, G.; Wyman, J. A. Naturally Occurring Biological Controls in Genetically Engineered Crop. In *Conservation Biological Control.* Barbosa P., Ed.; Academic Press: New York, USA, 185–202.

Hsiao, T. Ecophysiological Adaptions Among Geogragraphic Populations of the Colorado Beetle in North America. In *Advances in Potato Pest Management*. Lashomb J. H., Casagrande R, Eds.; Hutchinson Ross: Stroudsburg, Penn., USA, 1981, 69–85.

Hsiao, T. H. Host Specificity, Seasonality and Bionomics of Leptinotarsa beetles. p 581–599.

Hu, J. S., Gelman, D. B., Bell, R. A. Effects of Selected Physical And Chemical Treatments Of Colorado Potato Beetle Eggs On Host Acceptance And Development Of The Parasitic Wasp, Edovum puttleri. *Entomologia Experimentalis et Applicata* **1999**, *90* (3), 237–245.

Hunt, D. W. A. Tan, C. S. Overwintering Densities and Survival of the Colorado Potato Beetle (Coleoptera: Chrysomelidae) in and Around Tomato (Solanaceae) Fields. *Canadian Entomol.* **2000**, *132* (1), 103–105.

Humber, R. A.; Collection of Entomopathogenic Fungal Cultures: Catalog of Strains. Collection of Entomopathogenic Fungal Cultures, United States Department of Agriculture: Agricultural Research Service. Catalog of Strains, 1992, viii, 177p.

Jacques, R. L. The Potato Beetles. The Genus Leptinotarsa in North America (Coleoptera: Chrysomelidae). Flora and Fauna Handbook No.3., Brill E. J.: New York, USA, 1988.

Jansson, R. K.; Lashomb, J.; Groden, E.; Bullock, R. Parasitism of *Leptinotarsa decemlineata* (Coleoptera: Chrysomelidae) by Edovum puttleri (Hymenoptera: Eulophidae) in Different Cultivars of Eggplant. *Entomophaga* **1987**, *32* (5), 503–510.

Karalus, W. Control of Potato Beetles in Organic Farming. *Kartoffelbau* **1994**, *145* (4), 170–173.

Koval yuv, Characteristic of the Colorado beetle. Zashchita Rasteni, No. 5:34

Lagud, C.; Khelifi, M.; Gill, J.; Lacasse, B. Pneumatic and Thermal Control of Colorado Potato Beetle. *Canadian Agric. Eng.* **1999**, *41* (1), 53–57.

Lagud, C.; Khelifi, M.; Lacasse, B.; Evaluation of a Furrow Prototype Machine for Pneumatic Control of Colorado Potato Beetle. *Canadian Agric. Eng.* **1999**, *41* (1), 47–52.

Landolt, P. J.; Tumlinson, J. H.; Alborn, D. H. Attraction of Colorado Potato Beetle (Coleoptera: Chrysomelidae) to Damaged and Chemically Induced Potato Plants. *Environ. Entomol.* **1999**, *28* (6), 973–978.

Lashomb, J. H.; Ng, Y. S.; Ghidiu, G.; Green, E. Description of Spring Emergence by the Colorado Potato Beetle, *Leptinotarsa decemlineata* (Say) (Coleoptera: Chrysomelidae) in New Jersey. *Environ. Entomol.* **1984**, *13* (3), 907–910.

Lashomb, J.; Ng, Y. S.; Jansson, R. K.; Bullock, R. Edovum puttleri (Hymenoptera: Eulophidae), an Egg Parasitoid of Colorado Potato Beetle (Coleoptera: Chrysomelidae): Development and Parasitism on Eggplant. *J. Econ. Entomol.* **1987**, *80* (1), 65–68.

Lindroth, C. H. The Ground Beetles (Carabidae, excl. Cicindelinae) of Canada and Alaska, part 6. *Opuscula Entomologica Supplement* **1969**, *34*, 945–1192.

Logan, P. A.; Casagrande, R. A.; Hsiao, T. H.; Drummond, F. A. Collections of Natural Enemies of *Leptinotarsa decemlineata* (Coleoptera: Chrysomelidae) in Mexico, 1980–1985. *Entomophaga.* **1987**, *32* (3), 249–254.

Long, D. W.; Drummond, F. A.; Groden, E. Susceptibility of Colorado Potato Beetle (*Leptinotarsa decemlineata*) Eggs to *Beauveria bassiana*. *J. Invertebr. Pathol.* **1998**, *71* (2), 182–183.

Lorenzen, J. H.; Balbyshev, N. F. A Usual Source of Resistance to the Colarado Potato Beetle. *Am. Potato J.* **1997**, *74* (5), 331–335.

Mietkiewski, R.; Sapieha, A.; Tkaczuk, C. The Effect of Soilborne Entomogenous Fungi on the Mycoses of the Colorado Potato Beetle During Hibernation Period. *Bulletin OILB/SROP* **1996,** *19* (9), 162–165.

Misener, G. C.; Boiteau, G.; McMillan, L. P. A Plasticlining Trenching Device for the Control of Colorado Potato Beetle: Beetle Excluder. *Am. Potato J.* **1993,** *70* (12), 903–908.

JianChu, M.; Jia'an, C. 2003. Advances on the Study of Pest Resistance To Chloronicotinyl Insecticides. *Acta. Phytophylacica Sinica* **2003,** *30* (1), 91–96.

Moyer, D. D Trapping Beetles. Spudman, April 93,1993, 12–16.

Nault, B. A.; Hanzlik, M. W.; Kennedy, G. G. Location and Abundance of Adult Colorado Potato Beetles (Coleoptera: Chrysomelidae) Following Potato Harvest. *Crop Protection* **1997,** *16* (6), 511–518.

Nault, B. A.; Kennedy, G. G. Influence of Foliar Applied *Bacillus thuringiensis* subsp. Tenebrionis and an Early Potato Harvest on Abundance and Overwinter Survival of Colorado Potato Beetles (Coleoptera: Chrysomelidae) in North Carolina. *J. Econ. Entomol.* **1999,** *92* (5), 1165–1171.

Nault, B. A. Kennedy, G. G. Seasonal Changes in Habitat Preference by Coleomegilla maculata: Implications for Colorado Potato Beetle Management in Potato. *Biol. Control* **2000,** *17* (2), 164–173.

Nordlund, D. A.; Vacek, D. C.; Ferro, D. N. Predation of Colorado Potato Beetle (Coleoptera: Chrysomelidae) Eggs and Larvae by Chrysoperla rufilabris (Neuroptera: Chrysopidae) Larvae in the Laboratory and Field Cages. *J. Entomol. Sci.* **1991,** *26* (4), 443–449.

Ning, L.; YingChao, L.; RunZhi Z, Invasion of Colorado Potato Beetle, Leptinotarsa decemlineata, in China: Dispersal, Occurrence, and Economic Impact. *Entomologia Experimentalis et Applicata* **2012,** *143* (3), 207–217.

Obrycki, J. J.; Tauber, M. J.; Tauber, C. A.; Gollands, B. Edovum puttleri (Hymenoptera: Eulophidae), an Exotic Egg Parasitoid of the Colorado Potato Beetle (Coleoptera: Chrysomelidae): Responses to Temperate Zone Conditions and Resistant Potato Plants. *Environ. Entomol.* **1985,** *14* (1), 48–54.

Oetting, R. D.; Yonke, T. R. Immature Stages and Notes on the Biology of Euthyrhynchus floridanus (L.) (Hemiptera: Pentatomidae). *Ann. Entomol. Soc. Am.* **1975,** *68* (4), 659–662.

O'Neil, R. J. Functional Response and Search Strategy of Podisus maculiventris (Heteroptera: Pentatomidae) Attacking Colorado Potato Beetle (Coleoptera: Chrysomelidae). *Environ. Entomol.* **1997,** *26* (6), 1183–1190; 40.

Pelletier, Y.; McLeod, C. D.; Bernard, G. Description of Sublethals Injuries Caused to the Colorado Potato Beetle (Coleoptera, Chrysomelidae) by Propane Flamer Treatment. *J. Econ. Entomol.* **1995,** *88* (5), 1203–1205.

Perlak, F. J.; Stone, T. B.; Muskopf, Y. M.; Petersen, L. J.; Parker, G. B. McPherson, S. A.; Wyman, J.; Love, S.; Reed, G.; Biever, D.; Fischhoff, D. A. Genetically Improved Potatoes: Protection from Damage by Colarado Potato Beetles. *Plant Mol. Biol.* **1993,** *22* (2), 313–321.

Pope, R. D.; Madge, R. B. The 'When' and 'Why' of the Colorado Potato Beetle. *Antenna* **1984,** *8,* 175–177.

Reinhard, H. J. North American Two Winged Flies of the Genus Doryphorophaga (Tachinidae, Diptera). *J. New York Entomol. Soc.* **1935,** *43,* 387–394.

Riley, C. V. First Annual Report of the Noxious, Beneficial and Other Insects of the State of Missouri. Ellwood Kirby: Jefferson City, Missouri, USA,1869.

Roberts, D. W.; LeBrun R. A.; Semel, M. Control of the Colorado Potato Beetle with Fungi. Lashombe, J., Casagrande R. Eds.; Advances in Potato Pest Management. Ilutchinson Ross: Stroudsburg, Pennsylvania, pp 119–137.

Roush, R. T.; Hoy, C. W.; Ferro, D. N.; Tingey, W. M. Insecticide Resistance in the Colorado Potato Beetle (Coleoptera: Chrysomelidae): Influence of Crop Rotation and Insecticide Use. *J. Econ. Entomol.* **1990,** *83* (2), 315–319.

Ruberson, J. R.; Tauber, M. J.; Tauber, C. A. Biotypes of Edovum puttleri (Hymenoptera: Eulophidae): Responses to Developing Eggs of the Colorado Potato Beetle (Coleoptera: Chrysomelidae). *Ann. Entomol. Soc. Am.,* **1987,** *80* (4), 451–455.

Sablon, L.; Dickens, J. C.; Haubruge, É.; Verheggen, F. J. Chemical Ecology of the Colorado Potato Beetle, *Leptinotarsa decemlineata* (Say)(Coleoptera: Chrysomelidae), and Potential for Alternative Control Methods. *Insects* **2013,** *4* (1), 31–54.

Sablon, L.; Haubruge, E.; Verheggen, F. J. Consumption of Immature Stages of Colorado Potato Beetle by Chrysoperla carnea (Neuroptera:Chrysopidae) Larvae in the Laboratory. *Am. J. Potato. Res.* **2013,** *90* (1), 51–57.

Sanford, L. L., Kobayashi, R. S., Deahl, K. L. Sinden, S. L. Diploid and Tetraploid Solanum chacoense Genotypes That Synthesize Leptine Glytcoalkaloids and Deter Feeding by Colarado Potato Beetle. *Am. Potato J.* **1997,** *74* (1), 15–21.

Samsinakova, A.; Kalakova, S. Paecilomyces farinosus Br. and Smith (Deuteromycetes) a Study on Conditions of its Use for the Control of Colorado Beetles. *Zeitschrift für Angewandte Entomologie* **1978,** *87,* 68–75.

Schalk, J. M.; A. K. Stoner. Tomato Production in Maryland,1979.

Schrod, J.; Basedow, T.; Langenbruch, G. A. Studies on Bionomics and Biological Control of the Colorado Potato Beetle (*Leptinotarsa decemlineata* Say, Col., Chrysomelidae) at Two Sites in Southern Hesse (FRG). *J. Appl. Entomol.* **1996,** *120* (10), 619–626; 26.

Schroder, R. F. W.; Athanas, M. M. Potential for the Biological Control of *Leptinotarsa decemlineata* (Coleoptera:Chrysomelidae) by the Egg Parasite, Edovum puttleri (Hym.: Eulophidae) in Maryland, 198–184. *Entomophaga.* **1989,** *34* (1), 135–141.

Senanayake, D. G.; Holliday, N. J. Comparison of Visual, Sweepnet, and Whole Plant Bag Sampling Methods for Estimating Insect Pest Populations on Potato. *J. Econ. Entomol.* **1988,** *81* (4), 1113–1119.

Sexson, D. L.; Wyman, J. A. Effect of Crop Rotation Distance on Populations of Colorado Potato Beetle (Coleoptera: Chrysomelidae) Development of Area Wide Colorado Potato Beetle Pest Management Strategies. *J. Econ. Entomol.* **2005,** *98* (3), 716–724.Shapiro, I. D.; Turuleva, L. M.;. Fasulati, S. R.; Ivashchenko, L. S. Immunogenetic Barrier and Sources of Resistance in Potato to Colorado Beetle. Nauchno Tekhnicheskii Bulletin Vsesoyuznogo Ordena Lenina Ordena,1991.

Sinden, S. L.; Sanford, L. L.; Deahle Segregation of Reptine Glycoalkaloid in Solanum Chacoense Bitter. *J. Agric. Food Chem.* **1986,** *34,* 372–377.

Svikle MYa, Control of the Colorado beetle. Zashchita Rastenii,1976, No. 61011

Speese, J. III; Starrett, S. B. Crop Rotation Reduces the Cost of Colorado Potato Beetle Control in Potatoes. *Hort. Technol. 8,* 229–234.

Tamaki, G.; Butt, B. A. Impact of Perillus bioculatus on the Colorado Potato Beetle and Plant Damage. Technical Bulletin, United States Department of Agriculture, 1978, No. 1581: 11.

Tamaki, G.; Chauvin, R. L.; Burditt, A. K. Jr, Field Evaluation of Doryphorophaga doryphorp (Diptera: Tachinidae), a Parasite, and Its Host the Colorado Potato Beetle (Coleoptera: Chrysomelidae). *Environ. Entomol.* **1983,** *12* (2), 386–389.

Tamaki, G.; Chauvin, R. L.; Burditt, A. K. Jr. Laboratory Evaluation of Doryphorophaga doryphorp (Diptera: Tachinidae), a Parasite of the Colorado Potato Beetle (Coleoptera: Chrysomelidae). *Environ. Entomol.* **1983,** *12* (2), 390–392.

Tauber, M. J.; Tauber, C. A.; Obrycki, J. J.; Gollands, B.; Wright, R. J. Voltinism and the Induction of Pstival Diapause in the Colorado Potato Beetle, *Leptinotarsa decemlineata* (Coleoptera: Chrysomelidae). *Ann. Entomol. Soc. Am.* **1988,** *81* (5), 748–754.

Thomas, G.; Wood, F. Colorado Beetle in the Channel Islands. Bulletin, *Organisation Europe Mediterraneenne pour la Protection des Plantes,* **1980,** *10* (4), 491–498.

Tingey,W. M.; Yencho,G. C. Insect Resistance in Potato: a Decade of Progress. In *Advance in Potato Pest Biology and Management.* Zehnder G. W., Jansson R. K., Powelson, Raman K. V., Eds.; APS Press: St. Paul, USA,1994.

Trouvelot, B.Research on the Parasites and Predators Attacking the Colorado Potato Beetle in North America. *Annales des Epiphyties* (et de Phytogénétique) **1932,** *17* (1931), 408–445.

Tuba, H. H.; Turner, E. Evaluation of Baiting Techniques for Sampling Wireworms (Coleoptera: Elateridae) Infesting Wheat in Washington. *Econ. Entomol.* **1983,** *76,* 850–855.

Voss, R. H.; Ferro, D. N. Phenology of Flight and Walking by Colorado Potato Beetle (Coleoptera: Chrysomelidae) Adults in Western Massachusetts. *Environ. Entomol.* **1990,** *19* (1), 117–122.

Voss, R. H.; Ferro, D. N.; Logan, J. A. Role of Reproductive Diapause in the Population Dynamics of the Colorado Potato Beetle (Coleoptera:Chrysomelidae) in Western Massachusetts. *Environ. Entomol.* **1988,** *17* (5), 863–871.

Weber, D. C.; Ferro, D. N. Distribution of Overwintering Colorado Potato Beetle In and Near Massachusetts Potato Fields. *Entomologia\Experimentalis et Applicata* **1993,** *66* (2), 191–196.

Weber, D. C.; Ferro, D. N.; Buonaccorsi, J.; Hazzard, R. V. Disrupting Spring Colonization of Colorado Potato Beetle to Nonrelated Potato Fields. *Entomologia Experimentalis et Applicata* **1994,** *73* (1), 39–50.

WenChao, G.; Tuerxun; JianJun, X; Jian, L.; Jiang, H.; Jing, L; DeCheng, M.; Jun, W. Research on the Identification of Colorado Potato Beetle and Its Distribution Dispersal and Damage in Xinjiang. *Xinjiang Agric. Sci.* **2010,** *47* (5), 906–909.

Williams, C. E. Exploitation of Eggs of the Colorado Potato Beetle, *Leptinotarsa decemlineata* (Coleoptera: Chrysomelidae), by the Exotic Egg Parasitoid Edovum puttleri (Hymenoptera: Eulophidae) in Eggplant. *Great Lakes Entomol.* **1987,** *20* (4), 181–186.

Wierenga, J. M.; Norris, D. L.; Whalon, M. E.Stage Specific Mortality of Colarado Potato Beetle (Coleoptera:Chrysomelodae) Feeding on Transgenic Potatoes. *J. Econ. Entomol.* **1996,** *89* (5), 1047–1052.

Wright, R. J. Evaluation of Crop Rotation for Control of Colorado Potato Beetles (Coleoptera: Chrysomelidae) in Commercial Potato Fields on Long Island. *J. Econ. Entomol.* **1984,** *77* (5), 1254–1259.

Yencho, G. C.; Bonier, M. W.; Tingey, W. M.; Plaisted, R. L.; Tankley, S. D. Molecular Markers Locate Genes for Resistance to the Colorado Potato Beetle, Leptinotarsa decemlineata, in Hybrid Solanum tuberosum x S. berthaultii Potato Progenies. *Entomologia Experimentalis et Applicata* **1996,** *81* (2), 141–154.

Yooncho, G. C.; Kowalski, S. P.; Kennedy, G. C.; Sanford, L. L. Segregation of Leptine Glycoalkaloid and Resistance to Colarado Potato Beetle (*Leptinotarsa desemlineata* (Say)) in F Solanu tuberosum (4x) x S.chacoense (4x) Potato Progenies. *Am. J. Potato. Res.* **2000,** *77* (3), 167–178.

Zehnder, G. W.; Powelson, M. L.; Jansson, R. K.; Raman, K. V. Eds.; *Advances in Potato Pest Biology and Management.* American Phytopathological Society (APS): St Paul, USA, 1994, xiii + 655 pp.

–

CHAPTER 10

PROBLEM OF SNAILS AND SLUGS IN VEGETABLE CROPS AND THEIR MANAGEMENT

RANJEET KUMAR

P.G. Department of Entomology, Bihar Agricultural University, Sabour, Bhagalpur, India

E-mail: rkipm06@gmail.com

ABSTRACT

Besides the damage to agricultural and horticultural crops some terrestrial snails and slugs play a significant role in the forest ecosystem they are important feeders of leaf litter, and fallen leaf broken by these organism, they also enhance colonization of microorganisms for better decomposition and mineralization of leaves. The snail and slugs also accumulate the metal ions in the soil of forest ecosystem . Some species of this groups are used as indicator for monitoring of heavy metal pollution in fresh water eco system even though the abnormally high environmental concentrations of heavy metal affects numerous biological processes involved in the development and maintenance of snail and slug populations such as feeding, growth, reproduction, general physiological activities and maturity. The application of snails as bioindicators of heavy metals pollution reported. The nutritive value of snail meat is more as compare to other animal proteins and used by some people of the world. The snails and slug is ecologically and economically important pest of vegetables. Snails and slug are important invertebrate pests of the world and belongs to second largest phylum of the animal kingdom after arthropods. The snails belong to Phylum Mollusca and Class Gastropoda. Molluscs divided into six classes: Amphineura (Aplacophora, Polyacophora), Monoplacophora,

Gastropoda, Scaphopoda, Pelecypoda and Cephalopoda. The members of classes, such as Gastropoda, Pelecypoda and Cephalopoda are widely consumed all over the world. Land snails belong to the class Gastropoda, which can be divided into 3 subclasses, Prosobranchia, Opisthobranchia and Pulmonata. Pulmonate land snails are very large group with more than 60 families . The second largest animal phyla, members of Mollusca are found in the sea, fresh water and on land . There are approximately 50,000 living species and 60,000 known fossil records of molluscs, including snails, slugs, scallops, oysters, clams, squids, octopi and nautili. There are eight living classes and two extinct classes. Taxa are particularly difficult to identify, because shell morphology is very similar of several species. All over the world about 35,000 species of land snails have been described and yet to be described. The Giant African Snail is one of the world's largest and most damaging land snail pests. The Global Invasive Species Database has included this snail among the "100 World's Worst" invaders. It is non host specific and can consume at least 500 different types of plants, including several vegetables, cereals, pulses and fruits crops. The snail is native to coastal East Africa (Kenya and Tanzania), but is now widespread on all continents except Antarctica. It is highly adaptive to a wide range of environmental conditions and is capable of modifying its life cycle to suit local conditions. Giant African Snail is a threat to the sustainability of crop systems and others ecosystems, has a negative impact on native fauna, and acts as a vector of human diseases. The introduction of Giant African Snail outside its native range dates back to the early 1800s, when it spread to Ethiopia, Somalia, Mozambique and Madagascar. The first occurrence outside Africa was in West Bengal (India) through Mauritius in 1847. In the Asia-Pacific region, the snail is recorded from Bangladesh, China, Fiji, India, Indonesia, Japan, Kiribati, Malaysia, New Zealand, Palau, Papua New Guinea, Philippines, Samoa, Solomon Islands, Sri Lanka, Vanuatu and Vietnam and its range is still expanding. From India 1488 species belonging to 26 families and 140 genera have been recorded.

10.1   INTRODUCTION

The snails and slug are ecologically and economically important pests of vegetables. Snails and slug are important invertebrate pests of the world and belong to the second-largest phylum of the animal kingdom after arthropods. The snails belong to Phylum Mollusca and Class Gastropoda. Molluscs are divided into six classes: Amphineura (Aplacophora, Poly-acophora), Monoplacophora, Gastropoda, Scaphopoda, Pelecypoda, and Cephalopoda. The members of classes, such as Gastropoda, Pelecypoda, and Cephalopoda are widely consumed all over the world. Land snails belong to the class Gastropoda, which can be divided into three subclasses, Prosobranchia, Opisthobranchia, and Pulmonata. Pulmonate land snails are very large group with more than 60 families (Tompa, 1984). The second-largest animal phyla, members of Mollusca are found in the sea, freshwater, and on land (Barnes, 1980). There are approximately 50,000 living species and 60,000 known fossil records of molluscs (Brusca and Brusca, 1990), including snails, slugs, scallops, oysters, clams, squids, octopi, and nautili. There are eight living classes and two extinct classes. Taxa are particularly difficult to identify because shell morphology is very similar of several species. All over the world about 35,000 species of land snails have been described and yet to be described (Lydeard et al., 2004). The Giant African Snail is one of the world's largest and most damaging land snail pests. The Global Invasive Species Database has included this snail among the "100 World's Worst" invaders. It is nonhost specific and can consume at least 500 different types of plants, including several vegetables, cereals, pulses, and fruit crops. The snail is native to coastal East Africa (Kenya and Tanzania) but is now widespread on all continents except Antarctica. It is highly adaptive to a wide range of environmental conditions and is capable of modifying its life cycle to suit local conditions. Giant African Snail, a threat to the sustainability of crop systems and other ecosystems, has a negative impact on native fauna and acts as a vector of human diseases. The introduction of Giant African Snail outside its native range dates back to the early 1800s when it spread to Ethiopia, Somalia, Mozambique and Madagascar. The first occurrence outside Africa was in West Bengal (India) through Mauritius in 1847. In the Asia-Pacific region, the snail is recorded from Bangladesh, China, Fiji, India, Indonesia, Japan, Kiribati, Malaysia, New Zealand, Palau, Papua New Guinea, Philippines, Samoa, Solomon Islands, Sri Lanka, Vanuatu, and Vietnam and its range is still expanding. From India 1488 species belonging to 26 families and 140 genera have been recorded (Ramakrishna and Mitra, 2002; Madhyastha et al., 2004). Recent analysis of Indian land and freshwater molluscan literature has confirmed that there are hardly some studies found on the ecology and conservation of Indian land snails compared with the wide range of historical literature available on taxonomy (Aravind et al., 2010). In the period of intensive study, there was a drastic decline in studies on Indian land snails. There are no studies on the population status, phylogeny, and taxonomic revision of different families or genera of Indian land snails (Sen et al., 2012).

Besides the damage to agricultural and horticultural crops, some terrestrial snails and slugs play a significant role in the forest ecosystem; they are important feeders of leaf litter, and fallen leaf broken by these organisms, they also enhance colonization of micro-organisms for better decomposition and mineralization of leaves (Mason, 1970; Seifert and Shutov, 1979). the snail and slugs also accumulate the metal ions in the soil of forest ecosystem (Beeby and Eaves, 1983). Some species of this groups are used as indicator for monitoring of heavy metal pollution in freshwater ecosystem (Nuenberg, 1984; Bryan and Langston, 1992; Kiffiney and Clement, 1993) even though the abnormally high environmental concentrations of heavy metal affects numerous biological processes involved in the development and maintenance of snail and slug populations, such as feeding, growth, reproduction, general physiological activities, and maturity. The application of snails as bioindicators of heavy metals pollution reported by (Viard et al., 2004; Notten et al., 2005). The nutritive value of snail meat is more as compared with other animal proteins and used by some people of the world (Imevbore and Ademosun, 1988).

10.1.1 DAMAGE CAUSED BY MOLLUSCS

The Giant African Snail is a macrophytophagous herbivore but it also eats sand, very small stones, bones from carcasses, and concrete as sources of calcium. It is a threat to native snails and affects native ecosystems by altering the food chain by providing alternative food sources for predators. This voracious snail feeds on a variety of vegetables and is considered to be a major agricultural pest; it also attacks plantations of teak, rubber, coffee, and tea. It causes severe damage to horticultural and medicinal plants. In most parts of the world, the damage is greatest when the species is first established. Giant African Snail also acts as an intermediate vector of the Rat Lungworm *Angiostrongylus cantonensis*, causing eosinophilic meningoencephalitis in humans, as well as a Gram-negative bacterium, *Aeromonas hydrophila*, causing a wide variety of symptoms, especially in persons with compromised immune systems. The parasites carried by the snail are usually passed to humans through the consumption of raw or improperly cooked snails. It also enhances the spread of plant diseases like black pod disease of cacao caused by *Phytophthora palmivora*, which it spreads through its feces. It is a general nuisance in human habitations since their decaying bodies release a bad stench and the calcium carbonate in the shell neutralizes the acid soils, altering soil properties and also the types of plants that can grow in the soil. In Florida, the cost of eradicating Giant African Snail after 5 years of its establishment was over US $1 million. It also affects international tourism because the snail thrives in warm, tropical conditions which are tourist destinations.

10.2   IMPORTANCE OF MOLLUSCS

10.2.1   HABIT AND HABITAT OF MOLLUSCS

Molluscs occur in various habitats and are divided into freshwater, marine, and terrestrial forms. The freshwater molluscs constitute an important fraction of the ecosystem. Freshwater molluscs are common in rivers, ponds, lakes, water pools, flowing water lakes, irrigation canals, etc. Gastropods are generally found attached to submerged vegetation, rocks, sticks, etc. Representatives of the Limacidae and Milacidae are found in Europe, Asia, Africa, and America. *Deroceras reticulatum* has spread to South America and *D. caruanae* to California and South Africa and later to India. Amongst the stylommatophoran snails, the Mediterranean *Helix aspersa* has spread worldwide and is today a pest throughout Europe including England, Sweden and even Asia. *Theba pisana* from the Mediterranean region has become a pest of Western Europe, South Africa, Saudi Arabia, and America. Apart from human activity, the spread of molluscs can result from winds (hurricanes) and birds (Newell, 1966). Food and feeding behavior of different species of land molluscs have been studied by Ghose et al. (1969), Raut and Ghose (1983), Chang (1991), Oli (1996), and Gupta and Oli (1997). Normally snails hide under damp and dense vegetation during daytime and only after dusk they emerge in search of food. The pest snails have spread in recent times by travel and trade to many countries and continents, even over oceans, so that many species are now widely distributed and no longer limited to their region of origin. Thus, the field slugs *Deroceras reticulatum* and *D. agreste*, as well as *Arion circumscriptus* and *Lehmannia marginata*, have spread from Europe to Australia and then Asia while the Mediterranean species *Cochlicella ventricosa* and *C. acuta* have spread northwards (Godan, 1983). Snails generally are nocturnal. Thus, feeding, reproduction, and locomotory activities are at the peak in the night (Yoloye, 1994). Dietary Calcium is an essential nutrient for growth, reproduction, and construction of shells in terrestrial gastropods (Boycott, 1934; Wagge, 1952; McKillop and Harrison, 1972; Crowell, 1973; Tompa and Wilbur, 1977). The plant matter upon which the land snails feed is not a sufficient source of calcium and land snails fed on lettuce only result in weak, thin, and fragile shells (Crowell, 1973). Various attempts have been made to breed molluscs on artificial diet and have been shown that different growth rates in molluscs in laboratory (Williamson and Cameron, 1976). The snails spend most of the daytime under stones, soil, or litter of decaying matter and consume and convert many households and farm wastes into body nutrients (Ajayi et al., 1978). Snail hatchlings need to build a strong shell soon after emergence and the parental effort extends in some species to providing calcium-rich feces or coating the eggs with a layer of calcium-rich soil (Tompa, 1980). These studies have shown that food and feeding habits also determine choice of a particular habitat (Crawley, 1983). Some snails select sites for oviposition which provide supply of calcium after emergence (Beeby and Richmond, 1988). Diets based on green vegetables do not ensure a good growth rate of terrestrial snails (Gomot et al., 1989). Some species have been very selective in their habitat preference (Chang and Emlen, 1993) and choice of food (Linhart and Thompson, 1995; Dallinger et al., 2001). Provision of calcium carbonate and nutrients to the shell is the most common form of parental investment (Baur, 1994). Odiete (1999) reasoned that the nocturnal behavior is a form of adaptation that assists the snails in reducing body moisture lost common during the day.

10.2.2   DAMAGE CAUSED BY MOLLUSCS

Snails may directly transmit disease, or they may serve as intermediate hosts for parasites of humans and animals. Since *Deroceras* sp. and *Arion* sp. feed readily on human and animal feces, they ingest eggs of worm and disperse them with their own excreta. Alternatively, these sources of infection may adhere to the slime present on the body of the snail and as they crawl around, the worm eggs may be deposited on vegetables and fruits (Dainton, 1954). When slugs were fed on leaves infected with false mildew (*Phytophthora phaseoli*) and thereafter were offered healthy lima bean sprouts, the latter were infected by the fungus after 5 days (Wester et al., 1964). Snails are carriers of plant pathogens and thus spread diseases of cultivated plants. Experiments on *D. reticulatum*, *L. marginata*, *Discus rotundatus*, and *Oxychilus draparnaudi* have shown that the tobacco mosaic virus, introduced into the mouth, was detected in the digestive tract after two days and thus it could be transferable to plants. Fungal spores of *Alternaria* sp., *Fusarium* sp., and *Phytophthora* sp. have been found in the faeces and slime of the bodies of *Arion* sp., *Limax* sp., *Cepaea neemoralis*, *Helicigona arbustorum*, and *Helicella obcoia* (Hassan and Vago, 1966). *Deroceras reticulatum* spread black root rot on cabbage and other crucifers in Brazil. Spores of *Phytophthora palmivora*, the causal organism of black pod disease in cocoa plants were dispersed by the giant African snail *A. fulica* through its feces. *Achatina fulica* and other snails are also responsible for the spread of foot rot in black pepper (*Piper nigrum*) *Achatina* sp. can spread the fungi *Phytophthora parasitica* and *P. colocasiae*. Spores of rust were also traced in the faeces of *H. arbustorum*, *Bradybaena fruit-icum*, *Helicella obvia*, *Succinea putris*, and *Arion rufus* (Turner, 1967). Snails are of great concern in agriculture, medical, and veterinary practices due to their damage in agriculture, horticulture, and forestry as well as their main role as intermediate hosts for the trematodes causing schistosomiasis and fascioliasis in humans and domestic animals (Godan, 1983). Snail infection rates are usually part of the epidemiological data required for the evaluation of the control of transmission (Sturrock, 1986). Therefore, alternative eco-friendly control measures need to be developed. Most of the snails serve as intermediate hosts for certain parasitic worms of man and his domestic animals. The predilection of snails for fungal foods increases the attractiveness of diseased plant and possibility of spreading of disease by these snails. In the tropics and subtropics, schistosomiasis is the second most important parasitic disease after malaria in terms of prevalence, public health, and socioeconomic importance (WHO, 1993; Pointier and Giboda, 1999; Chitsulo et al., 2000). Thus, human health counters risk when these eatables are consumed fresh or cooked without washing thoroughly. Human schistosomiasis is a parasitic disease caused by digenetic trematode species of the genus *Schistosoma* which cohabitate the venous plexuses of the mammalian viscera and transmitted by freshwater gastropods which serve as intermediate hosts (Lockyer et al., 2003). *Achatina fulica* is one of the most destructive pests affecting subtropical

and tropical areas, causing large damages to farms, commercial plantations, and domestic gardens. It can also be found on trees, decaying material in decomposition, and next to garbage deposits. Furthermore, *A. fulica*
could be an intermediate host of *Angiostrongylus costaricencis*, the etio-
logical agent of abdominal angiostrongylosis and its dispersion could
imply a possible risk of transmission of this disease (Mead, 1995). The
Giant African Snail (*Achatina fulica*; Bowdich, 1822) promotes substantial ecological and economic impacts in areas where it has been introduced (Raut and Barker, 2002). Herbivorous molluscs are significant
pests of cultivated plant species in many regions of the world, affecting
ornamental, horticultural, arable, pastoral, and silvicultural crop species.
Reflecting their economic status, there has been substantial research
and developmental effort on terrestrial mollusc control. Consequently,
molluscs represent some of the most thoroughly studied pest species,
with a substantive body of literature relating to population and behavioral
ecology and control. Yet molluscs are also among the most intractable
of pests (Barker and Watts, 2002). The pest gastropods not only directly
damage the agricultural crops in the field but also lower the quality by
soiling with slime and feces. The snail affected portions of agricultural
product are contaminated by rotting agents such as bacteria and fungi,
which lead to further damage of fruits and vegetables in storage. *Achatina
fulica* feeds on the stems, leaves, flowers, and/or fruits of a broad range of
agriculturally-important plants, including the following:

- Banana (*Musa* spp.)
- Bean (*Phaseolus* spp.)
- Breadfruit (*Artocarpus altilis*)
- Cabbage (*Brassica oleracea*, Capitata group)
- Cacao (*Theobroma cacao*)
- Carrot (*Daucus carota* subsp. *sativus*)
- Cauliflower (*Brassica oleracea*, Botrytis group)
- Cassava (*Manihot esculenta*)
- Cotton (*Gossypium hirsutum*)
- Cucumber (*Cucumis sativus*)
- Eggplant (*Solanum melongena*)
- Marigold (*Tagetes patula*)
- Melons (*Cucumis* spp.)
- Noni (*Morinda citrifolia*)
- Okra (*Abelmoschus esculentus*)

- Papaya (*Carica papaya*)
- Peas (*Pisum sativum*)
- Pumpkin (*Cucurbita pepo*)
- Sponge gourd (*Luffa cylindrica*)
- Taro (*Colocasia esculenta*)

10.2.3 BIOLOGY AND STRUCTURE OF MOLLUSCS

Snails have conspicuous spiral shell, made up of Calcium carbonate. Slugs and snails are closely related molluscs, snails bearing the obvious distinction of an external shell large enough to enclose the entire animal. The ability to withdraw the soft body parts into a rigid shell is an advantage that allows the snail to survive under rather severe conditions of drought and heat. Higher levels of moisture tend to correlate with increased snail densities, growth rate, and larger adult size (Brown, 1913). Lack of a protecting shell, however, leaves them vulnerable to desiccation if trapped in the open during the day. They prefer an undisturbed habitat with adequate moisture and good food supply. The large requirement for lime ($CaCO_3$) for construction of the shell, on the other hand, tends to restrict snails to areas where the soil is rich in Calcium. The land molluscs, which refer to as slugs, are usually without an external shell. Besides economy in Calcium, slugs have the advantage of being able to alter their form so as to slip through narrow places and to burrow into the ground in search of food or shelter (Morton, 1960). Molluscs are bilaterally symmetrical protostomes whose coelom functions as a hydrostatic skeleton. All are soft-bodied individuals, most with a shell of Calcium carbonate for protection; for some, this is an internal remnant (e.g., the squid), whereas others, like the octopus, lack a shell completely. Snails are best known for their shell, which can appear in various forms but normally is coiled (helical). Unlike most animals, it is not obvious that snail display bilateral symmetry (the left and right halves of the animal are mirror images). In fact, the bodies of snails are mostly symmetrical, but their shells tend to be asymmetrical. This is due to the helical nature of the shell, which winds to the right (the shell opening is to the right when held spire upwards) most often, but to the left occasionally. The shape of the shell varies considerably. It may range from being quite conical, resulting from an elevated spire to globose, which is almost spherical in form, to depressed or discoidal, which is nearly flat. Snails secrete an acidic material from the sole of their foot that dissolves Calcium in the soil and allows uptake so the shell can be secreted. Calcium carbonate is also deposited in the shell of their eggs. Many marine snails have retractable covering on the dorsal end of the foot that serves to close the shell opening which is called an operculum. However, it is absent from nearly all terrestrial snails. Some terrestrial snails have a temporary operculum, which is called the epiphragm. The epiphragm is basically a mucous secretion but sometimes contains Calcium carbonate for reinforcement, making it hard and durable. The purpose of this secretion is to seal the shell and prevent dehydration during period of inactivity, including the winter or dry season. A snail's shell and the secretion of mucus epiphragm over the shell aperture are its primary structural means of protection from water loss (Machin, 1967). Retraction of the entire body inside the shell or avoiding daytime heat and dryness are behavioral means to reduce water-loss (Machin, 1967; Cameron, 1978). Terrestrial snails survive long-lasting unfavorable situations (hot and dry weather conditions or lack of food) by withdrawing into their shells and becoming inactive called dormancy or aestivation (Schmidt-Nielsen et al., 1971; Machin, 1975). Water loss in land snails occurs mainly from the mantle cavity via respiration, through the shell aperture and through the shell itself depending on age and thickness. In general, moisture conditions appear to be the best documented environmental correlate of shell size across a wide range of terrestrial land snail species. Shell size and shape can be highly variable, often reflecting age or adaptations to particular habitats. Reproduction output is often positively correlated with shell size. This is not found at faunal level as both large and small snails occur in dry and wet conditions (Goodfriend, 1984). Physiological means include aestivation and the retention of pallial fluid within the mental cavity (Blinn, 1964; Riddle, 1981; Smith, 1981). High snail densities probably correlate with both increased snail fecundity and more efficient snail mating. These will be due in part to more frequent or extended periods of activity and higher food intake rates. However, there are many possible confounding variables, such as Calcium levels, population density, and food availability (Sacchi, 1965; Gould, 1984). Within species, larger snails have greater survivorship compared with smaller snails in low-humidity environments (Heatwole and Heatwole, 1978; Knights, 1979). The only environmental stimulus to which the inactive animals can respond is increased humidity, which activates the animal. In gastropods, both the activation process and the increased behavioral responsiveness to six external stimuli are a consequence of the increased level of central arousal, which is mediated by serotonin (Weiss et al., 1981; Palovcik et al., 1982; Chase, 2002; Elliott and Susswein, 2002). Hodasi (1982) reported that perpetual light (both day and night) promoted rapid growth as continuous light at night have activatory effect on the snails, thus increasing their activity and rate of food consumption. Because surface area to volume ratios decrease with increasing size, one would predict that conditions of greater desiccation stress would favor larger snails (Nevo et al., 1983; Goodfriend, 1984). Temporal variation in population density, growth, and reproductive activity in land snails is largely controlled by environmental conditions (Albuquerque de Matos, 1989, 1990; Elmslie, 1989). Pulmonates minimize water loss via three primary means: Structure, behavior, and physiology (Asami, 1993). Species in this phylum are also of particular interest to scientists because they are great models of how nervous system complexity correlates with habitat and evolution. For example, clams are sessile organisms with simple sensory organs whose central nervous system is comprised of chain of ganglia circling the body; meanwhile, an octopus has the most sophisticated nervous system of all invertebrates. Their large brain along with their large eyes and rapid conduction along giant axons correlate well with the active predatory life (Campbell, 1993). Most land snails are nocturnal but following a rain may come out of their hiding places during the day. They move with a gliding motion by means of a long flat muscular organ called as foot. Mucus, constantly secreted by glands in the foot, facilitates movement and left a silver color like slimy trail. The reproductive organs of both sexes occur in the same individuals and each is capable of self-fertilization, although cross-fertilization is normal. Adults deposit eggs. The pulmonates have two sets of tentacles, the eyes are at the tip of the upper pair, there is no operculum, they breathe by lungs and both sexes are present in each individual (hermaphrodites). Among the more unusual features of snail, biology is the mode of reproduction. Terrestrial snails are hermaphrodites, copulate, and inseminate each other simultaneously and even self-fertilization may occur. Cross-fertilization is thought to be more common; however, in many snails, the male reproductive system matures earlier than the females called protandric condition. In some snails, there is only a single act of copulation, whereas, in others, mating can occur repeatedly. Mating requires high humidity and often occurs following precipitation.

10.3 SLUG

Slugs and snails can be troublesome pests in the garden. There are many species of these molluscs, but only a few present a serious problem. Some of these are the gray garden slug, *Deroceras reticulatum*, the tawny garden slug, *Limax flavus*, the spotted garden slug, *Limax maximus*, and the brown garden snail, *Helix aspersa*. Slugs and snails travel by means of a large foot that glides over a trail of mucus or slime, secreted from glands located under their heads. The slime provides a cushion over rough areas. This slime also serves as a trail marker. They return to their favorite hiding and feeding places night after night unless disturbed or the site becomes too dry. Slugs and snails tend to avoid crossing dusty or dry materials. The shiny dried slime trails left behind on plants or along the ground are good indicators of their activity. Both eat large, ragged holes in the leaves of plants and may also completely consume young seedlings. They begin feeding early in the spring and continue through the growing season until frost. Slugs and snails feed on a wide variety of plant material and can be especially troublesome on hostas, violets, ageratum, strawberries, lettuce, and cabbage. Slugs and snails are hermaphroditic animals. Each individual has both male and female sex organs, but they usually require another individual to fertilize the eggs that each carry in its body. Each one is capable of laying a total of about 300 eggs. These are laid any time conditions are optimal but peak times are the spring and fall. The gel-like eggs are laid in clusters of 25, about an inch below loose soil in damp locations and are covered with a layer of mucus. Eggs hatch in about 30 days. However, under dry conditions, the eggs may remain unhatched until adequate moisture has been absorbed. Slugs and snails reach adult size in 3–12 months and can live for several years. Temperature and moisture are the prime factors that influence their activity. In times of unfavorable environmental conditions, slugs can survive by burrowing as deep as 3 feet into the soil and snails will close up within their shells.

10.4 MANAGEMENT STRATEGIES

Indirect control strategies for these pests involve the removal of their favored habitat and moisture reduction. Direct control strategies focus on handpicking, trapping, and barriers.

The use of salt, vinegar solutions, or boiling water is not recommended as it can cause plant damage.

Natural Enemies: Toads, turtles, predacious beetles, and their larvae, and wild birds all contribute to reducing high populations of snails and slugs.

10.4.1 HABITAT MODIFICATION

Eliminate shelter and favorable conditions for reproduction by plant thinning, planting ground covers less favored by slugs, and removal of bricks, boards and excessive mulch. Steps that increase air circulation and sun penetration will promote drying of the area, discouraging slugs and snails.

10.4.2 PHYSICAL REMOVAL

Picking by hand, or with tweezers or tongs although not an enjoyable task, is a very effective method of immediate control and can last a reasonably long time if combined with other techniques. The best time for handpicking is two hours after sunset. Use a flashlight and place captured slugs and snails in a container of hot soapy water for later disposal.

10.4.3 TRAPS

Trapping is another successful tactic for control. Overturned flower pots, boards, shingles, inverted grapefruit halves, commercially available slug, and snail traps, or just about anything that will provide daytime shelter can be an effective trap. These traps can also be baited with pieces of potato, apple, or lettuce leaves to increase their effectiveness. Most people are familiar with using shallow containers filled with beer as a method of baiting. Apparently, the yeast is what attracts them, and a simple mixture of any yeast and water will also work. Slugs attracted to beer become tipsy and fall in to drown. Drowned or trapped slugs and snails should be disposed of in hot soapy water. The drawback to this type of trapping is that it requires frequent monitoring and bait replacement.

10.4.4 DIATOMACEOUS EARTH

DE is an organic product mined from fossilized algae known as diatoms. It is ground into a fine dust for horticultural use. Apply an inch-wide band of DE around plants needing protection. When slugs and snails crawl over the DE dust, it adheres to their bodies, causing desiccation (drying out). Reapply DE after heavy rains. DE dust can cause irritation, so use goggles and a dust mask during application.

10.4.5 BARRIERS

The best commercial barriers against slugs and snails are strips of copper that can be fastened around beds, flower boxes, or greenhouse benches. Homemade strips can be made from copper sheeting. Copper screening also works but is not as durable. The green oxidation of the copper does not seem to decrease its effectiveness. Apparently, they have a reaction to the copper similar to an electric shock. Now available is a coconut oil repellant called "Slug Stop." Sold in a tube, this paste is applied in a barrier circle around plants to be protected. An important point to remember is that barriers will not be effective against slugs and snails already present in the area to be protected. Barriers are also ineffective if plant parts bridge over the barrier material or if they are less than an inch wide.

10.4.6 PICK THEM OFF BY HAND

This is one of the most effective methods. Go out after dark with a torch—damp nights are good—and wear rubber gloves if you do not fancy their slimy skin. You will be amazed how many slugs and snails you find. Dispose of them as you will (but not over the garden wall).

10.4.7 NATURAL PREDATORS

One of the best ways to control slugs and snails in the garden is to encourage their natural predators. Slugs are eaten by beetles, slow worms, hedgehogs, toads, birds, and especially frogs. Snails are eaten by song thrushes, mistle thrushes, blackbirds, hedgehogs, toads, mice, centipedes, and the larvae of glow worms. In fact, snails are the exclusive diet of glow-worm larvae, so without snails, they would not exist. Song thrushes prefer large snails, cracking them open on stones, but mistle thrushes and blackbirds have not yet mastered this trick, so they eat smaller snails with softer shells. To encourage these natural predators in the garden, ensure there is a place for them to breed, shelter and hibernate. A pond is good for frogs and a log pile will encourage hedgehogs and lots of other useful creatures to stay. Parasitic nematode worms are an effective and safe biological control. They burrow into the mollusc and eat it from the inside. The packs of powder contain millions of microscopic nematodes that you simply water into the earth. A ring of sharp grit, crushed eggshells, pine needles or wood ash around a plant can help keep slugs and snails away. This will need to be replenished after rain. To protect young plants from late frosts as well as slugs and snails, remove the cap from a plastic drink bottle and cut off the bottom half. Place over the plant and remove it during the day if you wish.

10.4.8 PREVENTION

Poor quarantine regulations and the animal's high reproductive capacity are the main reasons for the rapid dispersal of this snail. Preventing its introduction is the most cost-effective option. Because of the huge risk that GAS poses and also its multiple methods of dispersal, strict quarantine, and surveillance activities are necessary to control its spread. Creating awareness about the various negative impacts of the snail can help stop the illegal import of GAS for trade and its international spread.

10.4.9 CONTROL

Control methods like collecting the snails by hand or incinerating them with flame throwers have been used to control infestations of GAS but neither of these methods is very effective. In China, the snail is used as a food item, which controls the population to a large extent, but may encourage further deliberate spread. Chemical control involves using Metaldehyde, Methiocarb or a combination of these chemicals with other molluscicides as bait formulations or foliar sprays. But these chemicals can also harm nontarget snails and endemic forms. Other methods like creating frigid temperatures or saturating the snail in ethanol are also used. In India, cuttings of *Anona glabra* are used as a snail repellant to protect nursery beds. Kerosene and common salt are used in some countries to control the snail. The predatory rosy wolf snail, *Euglandina rosea*, which is native to the southeastern United States, has been introduced to islands in the Caribbean, Pacific, and Indian Oceans as a biological control agent for the GAS. However, in Hawaii and French Polynesia, this predatory snail has caused the extinction of numerous endemic tree snails. *Platydemus manokwari*, a turbellarian flatworm, has been reported to be successful in controlling GAS in Guam, Philippines, and the Maldives. However, this worm has also been implicated in the decline of native snails.

10.4.10   COMMONLY USED PESTICIDES

The chemical control of snail host is done with the use of compounds commonly known as molluscicide. The molluscicidal activity of carbamates relates to their disruption of the neurotransmitter cholinesterase (Frain, 1982). Terrestrial snails are destructive agricultural pests causing damages to a wide variety of plants including vegetables, forage crops, tree fruits, shrubs, flowers, ground green cover, and newly sown lawn grasses. Moreover, they play an important role in transmitting and spreading disease to cultivated plants. The molluscicidal activity is affected both through ingestion and dermal contact. In molluscs, the toxicant causes rapid paralysis and loss of muscle tone (Godan, 1983). Molluscicides immobilize terrestrial gastropods and interfere with neural control of feeding behavior (Bailey et al., 1989). There are three major classes of compounds used in control of terrestrial 12 mollusc pests, namely metaldehyde, carbamates, and metal chelates. The molluscicidal effect of metaldehyde is based primarily on its effect on the mucous cells; it causes major disruption of the water balance physiology of the mollusc, resulting in their desiccation (Triebskorn and Ebert, 1989). Ingestion of iron chelate does not cause paralysis in molluscs but arrests feeding and kills within 24

h. Chelates offer advantages over metaldehyde and carbamate products in that the level of mortality is independent of the eater relations and thus not dependent on prevailing environmental moisture conditions (Henderson et al., 1989). The principal carbamates used to control terrestrial mollusc pests are carbaryl, isolan, mexacarbate, cloethocarb, methiocarb, and thiodicarb. The mode of toxicity of metal-containing chelates in molluscs is poorly known at present but that of iron EDTA principally involves ferric ion interference with the oxygen uptake by hemocyanin, the respiratory pigment of hemolymph in molluscs. The molluscicides currently in use are either synthetic or from natural origin. Several of them possess the nitro group functionality since the only compound that has been widely and effectively used as molluscicides in the control of schistosomiasis is the synthetic compounds niclosamide. The high cost of synthetic molluscicides used in the control of the intermediate host of schistosomiasis has resulted in renewed interest in plant molluscicides (Clark et al., 1995). Thus, the use of specific molluscicides is considered one of the most effective measures for control of terrestrial molluscs. The use of molluscicides is one of the procedures recognized by the World Health Organization (WHO, 1998) against schistosomiasis. Metaldehyde has a secondary neurotoxic effect, contributing to loss of motor activity (Coloso et al., 1998). Arehen et al. (2001) reported the synthesis of electrochemistry and molluscicidal activity of nitro-aromatic compounds. Application of pesticides (molluscicides) is today regarded as the most pragmatic approach to control of terrestrial mollusc pests and there has been extensive research and development of these, although there is renewed interest in nonchemical approaches in response to concerns over adverse environmental effects (Barker and Watts, 2002). Delivery of molluscicidal chemicals to target pest populations has primarily focused on bait formulations and there has been substantial investment in bait technologies by government and commercial agencies in many parts of the world. In general, other molluscicide formulations, such as sprays and dust, have proved ineffective in control of field infestations. Pesticides are often broad-spectrum, impacting negatively on 13 nontarget species, such as arthropods, annelids, and molluscs, in agroecosystems (Don-Pedro, 2010).

The destructiveness of molluscan pests is far greater today than in former times, since limits for their spread not only from one country to another but also from one province to another as a result of ever denser and faster transport and traffic, if both climate and available food are suitable, large pest populations may be built up, hand collection with subsequent destruction of animals is the oldest method of control of pest molluscs (Godan, 1983) and has been used effectively in conjunction with chemical methods for management of infestations in agricultural areas and in eradication of incipient infestations of invasive species. The large numbers of natural enemies of molluscs are known, recorded enemies include pathogenic bacteria, fungi, viruses, parasitic protozoa, microsporidia, predatory molluscs, flatworms, and arthropods and predatory vertebrates, including amphibians, reptiles, birds, and mammals. There have been relatively few classical biological control programs for pest mollusc species. Carbamate molluscicides are known to act as nerve toxins by inhibition of cholinesterase (Wilkinson, 1976; Young and Wilkins, 1989). Bailey et al. (1989) claimed that methiocarb shortens meal by interfering with the neural control feeding. Cytological effects induced by carbamates have been recorded (Triebskorn and Ebert, 1989; Triebskorn and Kunast, 1990; Triebskorn et al., 1996). Carbamate molluscicides are known to act as nerve toxins by inhibition of cholinesterase. Metaldehyde molluscicides caused an excessive increase of fluid excretion in soft snail body, leading to snail death (Kassem et al., 1993). The 20th century witnessed the emergence of land snails as important crop pests in temperate and tropical regions. Furthermore, in some crops the significance of land snails is only now becoming apparent with the decline in the importance of other pest groups such as insects (Barker, 2002). Most of our knowledge of carbamate is restricted either to vertebrates or to insects and there is little information on terrestrial gastropods (Henderson and Triebskorn, 2002). Control of land snails on different crops is heavily dependent on the use of molluscicides that limits the effect of these pests below damaging level. Hence, the synthetic molluscicides are the most effective measures available at present for the control of terrestrial gastropods (Heiba et al., 2002; Abd-El-Ali, 2004; Zedan et al., 2006; Genena and Mostafa, 2008). Bait formulation of molluscicides was the most effective application method in the field for controlling terrestrial gastropods rather any other technique (Kassem, 2004). Transaminases enzymes and acetylcholine esterase, as well as total proteins and total lipids, are important in the biological processes in the land snails (Abd-El-Ali, 2004). The importance of terrestrial gastropods as crop pests has greatly increased and in the process demands for effective control have outstripped the development of chemical control measures. Recently much effort has been directed toward snail screening and evaluation of chemicals with different pharmacological and physiological modes of action to control land snails, especially, against their mucous and shell. Also, there is very little information available regarding the use of additive, capable of enhancing the action of molluscicides and increase its toxicity. Few pesticides are available to combat terrestrial snails. Most Asian snails with polished, flattened, semitransparent and closely coiled shells were initially grouped within *Macrochlamys sensu stricto* (Benson, 1832; Godwin-Austen, 1883). Although investigations of genital morphology within *Macrochlamys sensu lato* revealed a large number of distinct groups (Godwin-Austen, 1908), most species within *Macrochlamys sensu stricto* are known only from their shells and relatively few generic placements have been critically determined by examination of the reproductive organs (Godwin- Austen, 1907; Blanford and Godwin-Austen, 1908; Solem, 1966; Schileyko, 2002). The land snail *Macrochlamys indica* belongs to the family Zonitidae, subfamily Ariophantinae, genus *Macrochlamys*. Family Ariophantidae is tropical family, widespread and has a medium to large shell which is usually narrowly umbilicate or perforate. The genus *Macrochlamys* has one medium size depressed species. This species was for a long time identified with *Helix vitrinoides*, a shell of unknown origin and described as imperforate. *M. indica* occurred from Calcutta to Cawnpore, India. The first complete description of this species was given by Godwin-Austen and the name *M. indica* is accepted (Blandford and Godwin-Austen, 1908). *Macrochlamys indica* is considered to represent a potentially serious threat as a pest, an invasive species which could negatively affect agriculture, natural ecosystems, human health, or commerce. The determination of chemical pesticide application for the control of terrestrial molluscs is an important area that needs to be given attention in Maharashtra state. As compared with molluscan pest control there is more attention in controlling the other crop pests as insect pest control because of lack of study of biology and behavior of molluscan pests. The study presented here aims to investigate control of terrestrial crop pest *Macrochlamys indica* (Benson, 1832) by the application of two chemical pesticides Thaimethoxam (Arrow 25% WG) and Diafenthiuron (Pegasus 50% WP).

KEYWORDS

- snail
- slug
- pest management
- vegetable crops
- damaging symptoms of snail and slug

REFERENCES

Abd-El-Ali, S. M. Toxicity and Biochemical Response of *Eobania vermiculata* Land Snail to Niclosamide Molluscicide Under Laboratory and Field Conditions. *J. Agric. Sci. Mansoura Univ.* **2004,** *29*, 4751–4756.

Ajayi, S. S.; Tewe, O. O.; Morianty, C.; Awesu, M.O. Observation on the Biology and Nutritive Value of the African Giant Land Snail, *Archachatina marginata. J. of East Afri. Wildlife*. **1978,** *16*, 85–95.

Akinnusi, O. *Introduction to Snail Farming*; Triolas Publishing Company: Abeokuta, 2002. p 70.

Albuquerque de Matos, R. M. Contributions of Genetics to Snail Farming and Conservation. In *Slugs and Snails in World Agriculture*; Henderson, I. F., Ed.; British Crop Protection Council Monograph 41. Thornton Health: Surrey, UK, 1989; pp 11–18.

Albuquerque de Matos, R. M. Genetic and Adaptive Characteristics in *Helix aspersa* of Direct Interest in Snail Farming. *Snail Farm Res* **1990,** *3*, 33–43.

Alvarez, J.; Willig, M. R. Effects of Tree Fall Gaps on the Density of Land Snails in the Luquillo Experimental Forest. *Biotropica* **1993,** *25*, 100–110.

Amusan, J. A.; Omidiji, M. O. Edible Land Snails. *A Technical Guide to Snail Farming in the Tropics*; Verity Printers: Ibadan, 1999; pp 1–16.

Annandale, N. The Fauna of Certain Small Streams in Bombay Presidency. Records of the Indian Museum, 1919, Vol. 16, pp 117–120.

Annandale, N.; Prashad, B. Some Freshwater Mollusca from the Bombay Presidency. Records of the Indian Museum, 1919, Vol. 16, pp 139–152+4, 5pls.

Aravind, N. A.; Naggs, F. Snailing up the Canopies of Western Ghats. In *Canopies of South Asia- A Glompse*; Devy, M. S. T., Ganesh, A.T, Eds: ATREE: Bangalore, 2012; pp 43–46.

Aravind, N. A.; Rajashekhar, K. P.; Madhyastha, N. A. A Review of Ecological Studies on Patterns and Processes of Distribution of Land Snails of Western Ghats, India. Proc. World Congress Malacol. **2010,** 222.

Aravind, N. A.; Rajashekhar, K. P. and Madhyastha, N. A. Species Diversity, Endemism and Distribution of Land Snails of Western Ghats, India. Records of Western Museum Supplement, **2005,** *68*, 31–38.

Aravind, N. A.; Rajashekhar, K. P.; Madhyastha, N. A. Micromolluscs of Western Ghats, India: Diversity, Distribution and Threats. *Zoo Symposia*. **2008,** *1*, 281–294.

Arebeu, de F. C.; Paula, de F. S.; Santos, A. F. D.; Sant Ana, A. E. G.; Almeida, M. V. D.; Cesar, E. T.; Trindade, M. N.; Goulart, M. O. F. Synthesis of Electrochemistry and Molluscicidal Activity of Nitro-aromatic Compounds. *Bioorg. Med. Chem.* **2001,** *9,* 659–664.

Asami, T. Interspecific Differences in Desiccation Tolerance of Juvenile Land Snails. *Fun. Ecol.* **1993,** *7,* 371–577.

Bailey, S. E. R. Foraging Behaviour of Terrestrial Gastropods: Integrating Field and Laboratory Studies. *J. Molluscan Stud.* **1989,** *55,* 263–272.

Barker, G. M. *Molluscs as Crop Pests*; CAB International Publishing, 2002; p 468.

Barker, G. M.; Watts, C. *Management of the Invasive Alien Snail Cantareus aspersus on Conservation Land*; Doc Science Internal Series 31, Published by Department of Conservation: Wellington, New Zealand, 2002; p 8.

Barnes, R. D. *Invertebrate Zoology*; 5th ed; Saunders College Publishing: New York, New York, 1980; pp 342–464.

Baur, B. Parental Care in Terrestrial Gastropods. *Experientia* **1994,** *50,* 5–14.

Beeby, A.; Eaves, S. L. Short Term Changes in Pb, Zn and Cd Concentrations of Garden Snail *Helix aspersa* (Muller) from a Central London Car Park. *Environ. Pollut.* **1983,** *30,* 233–244.

Beeby, A.; Richmond, L. Calcium Metabolism in Two Populations of *Helix aspersa* on a High Lead Diet. *Arch. Environ. Contam. Toxicol.* **1988,** *17,* 507–511.

Bergy, B.; Dallinger, R. Terrestrial Snails as Quantitative Indicators of Environmental Metal Pollution. *Environ. Monit. Assess.* **1993,** *25,* 60–84.

Blanford, W. R. Descriptions of *Cremnocochus syhadrensis* and *Lithotis rupicola* Two New Generic Forms of Mollusca Inhabiting Cliffs in the Western Ghats of India. *Annal. Mag. Nat. Hist.* **1863,** *12* (3), 184–187.

Blanford, W.T. Contributions to Indian Malacology. No. XI, Descriptions of Helicidae from various parts of India. *J. Asia. Soc. Bengal* **1870,** *39* (2), 9–25.

Blanford, W. T. Contributions to Indian Malacology, No XII, Descriptions of New Land and Freshwater Shells from Southern and Western India, Burmah, the Andaman Islands. *J. Asia. Soc. Bengal* **1880,** *49* (2), 181–222+1, I1pls.

Blanford, W. T.; Godwin-Austen, H. H. Mollusca: Testacellidae and Zonitidae. In *The Fauna of British India including Ceylon and Burma*; Bingham, C. T., Ed; Taylor and Francis: London, 1908; pp 1–311.

Blinn, W. C. Water in the Mantle Cavity of Land Snails. *Physiol. Zoo.* **1964,** *37,* 329–337.

Bloch, C. P.; Willig, M. R. Context-dependence of Long-term Responses of Terrestrial Gastropod Populations to Large-scale Disturbance. *J. Trop. Ecol.* **2006,** *22,* 111–122.

Bodhankar, D. S. *Effects of Pesticides on Some Ecophysiological Aspects of the Land Slugs, Laevicaulis alte.* Ph.D. Thesis, Dr. B.A.M. University, Aurangabad, India, 1984.

Bonnelly de Calventi, I. Copper Poisoning in the Snail *Helix pomania* and its Effect on Mucous Secretion. *Annal, New York Acad. Sci.* **1965,** *118,* 1015–1020.

Borkakati, R. N.; Gogoi, R.; Borah, B. K. Snail: From Present Perspective to the History of Assam. *Asian Agri-Hist.* **2009,** *13* (3), 227–234.

Boyce, C. B. C.; Tyssul Jones, T. W.; Van Tongeren, W. A. The Molluscicidal Activity of N-Tritylmorpholine. *Bull. Org. Mond. Sante./Bull. Wld. Hlth. Org.* **1967,** *37,* 1–11.

Boycott, A. E. The Habitats of Land Mollusca in Britain. *J. Ecol.* **1934,** *22,* 1–38.

Brown, A. P. Variation in Two Species of *Lucidella* from Jamaica. *Proc. Acad. Nat. Sci. Philadelphia.* **1913,** *63,* 3–21.

Brusca, R. C.; Brusca, G. J. *Invertebrates*; Sinauer Associates, Inc.: Sunderland, Massachusetts, 1990.

Bryan, G. W.; Langston, W. J. Bioavailability, Accumulation and Effects of Heavy Metals in Sediments with Special Reference to United Kingdom Estuaries: A Review. *Environ. Pollut.* **1990,** *76* (2), 89–131.

Calabrese, A.; Thurberg, F. P.; Gould, E. Effects of Cadmium, Mercury and Silver on Marine Animals. *Mar. Fish. Rev.* **1977,** *39,* 5–11.

Cameron, R. A. D. Differences in the Sites of Activity of Coexisting Species of Land Mollusc. *J. Conchol* **1978,** *29,* 273–278.

Campbell, N. A. *Biology*; 3rd ed; The Benjamin/Cummings Publishing Company, Inc.: Riverside, California, 1993; pp 695–765.

Carlos, E. D. Biological Control for Freshwater Snails. *Parasitol. Today* **1991,** *7,* 124.

Chang, H. Food Preference of the Land Snail, *Cepaea nemoralis* in a North American Population. *Malacol. Rev.* **1991,** *24,* 107–114.

Chang, H. W.; Emlen, J. M. Seasonal Variation of Microhabitat Distribution of the Polymorphic Land Snail *Cepaea nemoralis.* *Oecologia* **1993,** *93,* 501–507.

Chase, R. *Behavior and its Neural Control in Gastropod Mollusks*; Oxford University Press: Oxford-New York, 2002.

Chaudhari, R. D.; Kulkarni, A. B. X Effect of Monocrotophos on Protein Metabolism of Terrestrial Snail, *Zootecus insularis.* *J. Ecotoxicol. Environ. Mounit.* **1993,** *3* (2), 157–159.

Chaudhari, T. R.; Patil, P. N.; Rao, K. R.; Deshmukh, S. B.; Diwate, S. G. Effect of Pesticide Rogor on Some Biochemical Constituents in Freshwater Snail, *Thiara lineate.* *Environ. Ecol.* **1999,** *17* (1), 146–148.

Chitsulo, L.; Engels, D.; Montresor, A.; Savioli, L. The Global Status of Schistosomiasis and its Control. *Acta Trop.* **2000,** *77,* 41–51.

Clark, S. J.; Coward, N. P.; Dawson, G. W.; Henderson, I. F.; Martin, A. P. Metal Chelate Molluscicides: The Redistribution of Iron Diazaalkanolates from the Gut Lumen of the Slug, *Deroceras reticulatum* (Muller) (Pulmonata: Limacidae). *Pest Sci.* **1995,** *44,* 381–388.

Coloso, R. M.; Borlongan, I. G.; Blum, R. A. Use of Metaldehyde as a Molluscicide in Semi-commercial and Commercial Milkfish Ponds. *Crop Prot.* **1998,** *17,* 669–674.

Cortet, J.; Gomot-de Vaufleury, A.; Poinsot-Balaguer, N.; Gomot, L.; Cluzeau, D. The Use of Invertebrate Soil Fauna in Monitoring Pollutants Effects. *Eur. J. Soil Biol* **1999,** *35,* 115–134.

Coupland, J. B. The Efficacy of Metaldehyde Formulations Against Helicid Snails: The Effects of Concentration, Formulation and Species. *Slug and Snail Pests in Agriculture*; Henderson, I. F., Ed.; British Crop Protection Council Symposium Proceedings, 1996, Vol. 66, pp 151–156.

Cowie, R. H.; Dillon, R. T. Jr.; Robinson, D. G.; Smith, J. W. Alien Non-marine Snails and Slugs of Priority Quarantine Importance in the United States: A Preliminary Risk Assessment Amer. *Malac. Bull.* **2009,** *27,* 113–132.

Crawley, M. J. *Herbivory: The Dynamics of Animal-plant Interactions*; Blackwell: Oxford, 1983; p 420.

Crowell, H. H. Laboratory Study of Calcium Requirements of the Brown Garden Snail, *Helix aspersa* Muller. *Proc. Malacol. Soc. London.* **1973**, *40*, 491–503.

Dainton, B. H. The Activity of Slugs: The Induction of Activity by Changing Temperature. *J. Exp. Biol.* **1954**, *31*, 165–187.

Dallinger, R.; Berger, B.; Triebskorn-Kohler, R.; Kohler, H. Soil Ecology and Ecotoxicology. In *The Biology of Terrestrial Mollusks*; Barker, G. M. Ed.; Cab International: Oxford, England, 2001; pp 489–525.

Don-Pedro, K. N. Pesticide Science and Toxicology. University of Lagos-Nigeria, 2010, p 270.

El-Khodary, A. S.; Sharshir, F. A.; Helal R. M.; Shahawy Wafaa A. Evaluation of Some Control Method Against the Land Snail *Monacha catiana* (Montagu) at Kafr El-Sheikh Governorate, Egypt. *J. Agric. Res Tanta Univ.* **2002**, *27* (2), 290–300.

Elliott, C. J.; Susswein, A. J. Comparative Neuroethology of Feeding Control in Molluscs. *J. Exp. Biol.* **2002**, *205*, 877–896.

Elmslie, L. J. Snail Farming in Field Pens in Italy. In *Slugs and Snails in World Agriculture*; Henderson, I. F., Ed.; British Crop Protection Council Monograph 41. Thornton Heath: Surrey, UK, 1989; pp 19–25.

EL-Okda, M. M.; Emara, M. M.; Selim, A. M. The Response of the Harmful and Useful Terrestrial Mollusca Towards Several Toxicants: I. Efficacy of Six Toxicants Under Laboratory Conditions. *Alex. Sci. Exch.* **1989**, *10* (3), 375– 384.

El-Saadany, M. M.; Mohamed, A. A.; Hanna, M. Y.; Mersal, H. Effect of Photoperiod on Glycogen Content, Total Protein and Gametogenesis of the Land Snail *Eobania vermiculata. J. Egypt. Ger. Soc. Zool.* **1993a**, *12*, 269–289.

El-Saadany, M. M.; Mohamed, A. A.; Hanna, M. Y. and Mersal, H. Effect of Temperature on The Glycogen Content, Total Protein and on the Gametogenesis of Land Snail *Eobania vermiculata. J. Egypt. Ger. Soc. Zool.* **1993b**, 12, 351–373.

El-Wakil, H. B.; Radwan, M. A. Biochemical Studies on the Terrestrial Snail *Eobania vermiculata* (Muller) Treated with Some Pesticide. *J. Environ. Sci. Health.* **1991**, *26* (596), 479–489.

Frain, J. Chemical Control of Molluscs Using Metaldehyde. *Int. Pest Cont.* November/ December, **1982**, 150–151.

Genena M. A. M. Studies on the Gastropods at Dakhlia Governorate, M.Sc., Thesis, Faculty of Agriculture, Mansoura University, 2003, p 136.

Genena, M. A. M.; Mostafa, F. A. M. Efficacy of Four Pesticides Applied Against the Land Snail, *Monacha cantiana* (Montagu) (Gastropoda: Helicidae) at Three Exposure Periods. *J. Agric. Sci. Mansoura Univ.* **2008**, *27*, 7767–7775.

Ghose, K. C.; Biswas, A.K. and Halder, D. P. Application of Cellulose. I. Degradation of Vegetable Foodstuffs with Bacterial Enzymes. *J. Food Sci. Technol.* **1969**, *6*, 29–32.

Godan, D. Untersuchungen Uber Die Molluskizide Wirkung der Carbamate. II Teil. Abhangigkeit von Art, Grosse und Ernahrungder Schnecke. *Z. Angew. Zool.* **1966**, 53, 417–430.

Godan, D. *Pest Slugs and Snails: Biology and Control*; Springer Verlag: Berlin, Heidelberg, 1983; p 445.

Godwin-Austen, H. H. *Land and Freshwater Mollusca of India, Including South Arabia, Baluchistan, Afghanistan, Kashmir, Nepal, Burmah, Pegu, Tenasserim, Malay Peninsula,*

Ceylon and other Islands of the Indian Ocean; Taylor and Francis: London., 1882–1914; Vol. 1, (1882–1888): pp 1–257, Pls. 1–62; Vol. 2 (1897-1914): pp 1–442, Pls. pp 63–158.

Godwin-Austen, H. H. The Importance of the Animal in the Land Mollusca, Shown by Certain Evolutionary Stages in Some Genera of the Zonitidae. *J. Malacol.* **1910,** *13,* 33–41.

Gomot, L.; Gomot, S.; Boukras, A.; Bruckert, S. Influence of Soil on the Growth of the Land Snail *Helix aspersa*: An Experimental Study of the Absorption Route for the Stimulating factors. *J. Moll. Stud.* **1989,** *55,* 1–7.

Goodfriend, G. A. Variation in Land-snail Shell Form and Size and its Causes: A Review. *Syst. Zoo.* **1984,** *35,* 204–223.

Gould, S. J. Covariance Sets and Ordered Geographic Variation in Cerion from Aruba, Bonaire and Curacao: A Way of Studying Non-adaptation. *Syst. Zoo.* **1984,** *33,* 217–137.

Gupta, P. K.; Oli, B. P. Food Preference of Some Species of Terrestrial Mollusks in Kumaon Himalayan Forests. *Vasundhara* **1997,** *2,* 27–39.

Hanafy, A. H. A.; Youssef H. M.; El-Shahat S. M. Preparation of Methomyl Baits and Efficacy of Against Certain Land Mollusca in Different Vegetation. *Adv. Agric. Res.* **1998,** *3* (3), 435–441.

Hassan, S.; Vago, C. Transmission of *Alternaria brassicola* by Slugs. *Plant Dis. Rep.* **1966,** *50,* 764–767.

Hata, T. Y.; Hara, A. H.; Hu, B. K. S. Molluscicides and Mechanical Barriers Against Slugs, *Vaginula plebeia* Fischer and *Veronicella cubensis* (Pfeiffer) (Stylommatophora Veronicellidae). *Crop Prot.* **1997,** *16,* 501–506.

Heatwole, H.; Heatwole, A. Ecology of the Puerto Rican Comaenid Tree Snails. *Malacologia* **1978,** *17,* 241–315.

Heiba, F. N.; Al-Sharkawy, I. M. and Al-Batal, A. A. Effect of the Insecticide, Lannate, on the Land Snails, *Ebonia vermiculata* and *Monacha contiana*, Under Laboratory Conditions. *J. Biol. Sci.* **2002,** *2,* 8–13.

Henderson, I. F.; Triebskorn, R. Chemical Control in Terrestrial Gastropods. In *Molluscs as Crop Pests*;Barker, G. M., Ed.; CAB International: London, 2002; pp 1–13.

Henderson, I. F.; Briggs, G. G.; Coward, N. P.; Dawson, G. W.; Pickett, J. A. A New Group of Molluscicidal Compounds. *Slugs and Snails in World Agriculture;* Henderson, I. F., Ed.; British Crop Protection Council Monograph, 1989; Vol. 41, pp 289–294.

Hendrix, P. F.; Langner, C. H. L.; Odum, E. P. Cadmium in Aquatic Microcosm. Implications for Screening the Ecological Effects of Toxic Substance. *Environ. Manage.* **1982,** *6,* 543–553.

Hodasi, J. K. N. The Effect of Different Light Regimes on the Behavior and Biology of *Achatina achatina. J. Molluscan Stud.* **1982,** *48* (3), 1–7.

Hora, S. L. On Some Inter-acting Features of the Western Ghats. J. *Bombay Nat Hist Soc.* **1925,** *31,* 447–449.

Hussein, H. I.; Al-Rajhy, D.; El-Shahawi, F. I.; Hashem, S. M. Molluscicidal Activity of *Pergularia tomentosa* (L.), Methomyl and Methiocarb Against Land Snails. *Int. J. Pest Manag.* **1999,** *45* (3), 211–213.

Imevbore, E. A.; Ademosun, A. A. The Nutritive Value of African Giant Land Snail *Archachatina marginata. J. Anim. Prod. Res.* **1988,** *8* (2), 76-87.

Ismail, S. A.; Abd-Allah, S. A.; El-Massry, S. A.; Hegab, A. M. Evaluation of Certain Chemicals and Insecticides Against *Monacha cartusiana* Snails Infesting Some

Vegetable Crops at Sharkia, Governorate. *J. Agric. Sci. Mansoura Univ.* **2005**, *30* (10), 6283–6292.

Kassem, F. A. Metaldehyde Inducing Histological Alterations of Brown and White Garden Snail's Digestive Glands. *J. Agric. Sci. Mansoura Univ.* **2004**, *29*, 925–933.

Kassem, F. A.; Sabra, F. S.; Koudsieh, S. S.; Abdallah, E. A. M. Molluscicidal Efficacy of Plant Extracts Against Molluscs Species. *Natl. Cong. Pests Fruits Egypt Ismailia.* **1993**, *1*, 998–108.

Kelly, J. R.; Martin, T. J. Twenty-One Year Experience with Methicarb Bait. Monograph, British Crop Protection Council, 1989, Vol. 41, pp 131–145.

Ketiku, A. O.; Adeleke, A. O. Nutrient Composition of Black Fleshed and White Fleshed African Giant Land Snails (*Archachatina marginata*). *West Afr. J. Biol. Appl. Chem.* **1998**, *44*, 21–23.

Kiffiney, P. M.; Clement, W. H. Bioaccumulation of Heavy Metals by Benthic Invertebrates at the Arkansas River, Colorado. *Environ. Toxicol. Chem.* **1993**, *12* (8), 1507–1517.

Knights, R. W. Experimental Evidences for Selection on Shell Size in *Cepaea hortensis* (Mull). *Genetica.* **1979**, *50*, 51–60.

Linhart, Y. B.; Thompson, J. D. Terpene-based Selective Herbivory by *Helix aspersa* (Mollusca) on *Thynus vulgaris* (Labiatae). *Oecologia.* **1995**, *102*, 126–132.

Lockyer, A. E.; Olson, P. D.; Ostergaard, P.; Rollinson, D.; Johnston, D. A.; Attwood, S. W.; Southgate, V. R.; Horak. P.; Snyder, S. D; Le T. H.; Agatsuma, T.; McManus, D. P.; Carmichael, A. C.; Naem, S.; Littlewood, D. T. J. The Phylogeny of the Schistosomatidae Based on Three Genes with Emphasis on the Interrelationships of *Schistosoma*. *Parasitology* **2003**, *126*, 203–224.

Lydeard, C.; Cowie, R. H.; Ponder, W. F.; Bogan, A. E.; Bouchet, P.; Clark, S. A.; Cummings, K. S.; Frest, T. J.; Gargominy, O.; Herbert, D. G.; Hershler, R.; Perez, K. E.; Roth, B.; Seddon, M; Strong, E. E.; Thompson, F. G. The Global Decline of Non-marine Molluscs. *Bioscience* **2004**, *54*, 321–330.

Machin, J. Structural Adaptation for Reducing Water-loss in Three Species of Terrestrial Snail. *J. Zoo. London* **1967**, *152*, 55–65.

Machin, J. Water Relationships. In *Pulmonates;* Fretter, V., Peake, J., Eds.; Academic Press, London, 1975; Vol. 1, pp 105–164.

Madhyastha, N. A.; Mavinkurve, R. G.; Shanbhag, S. P. Land Snails of Western Ghats. In: Gupta, A. K.; Kumar, A. and Ramakantha, V. (eds). Wildlife and Protected Areas, Conservation of Rain Forest in India, ENVIS Bulletin-4, 2004, Vol. 4, pp 143–151

Madsen, H. Biological Method for the Control of Freshwater Snails. *Parasitol. Today* **1990**, 6, 237–341.

Malek, E. A.; Cheng, T. C. Control of Economically and Medically Important Snails. In *Medical and Economic Malacology*; Academic Press: New York and London, 1974.

Mason, C.F. Food, Feeding Rates and Assimilation in Woodland Snails. *Oecologia* **1970**, *4*, 358–373.

McCullough, F. S.; Malik, E. A. Note on the Molluscan host of *Parasgonimus Spp.* and Their Possible Role in Biological Control. *Ann. Trop. Med. Parasitol.* **1984**, *78*, 339.

McKillop, W. B.; Harrison, A. D. Distribution of Aquatic Gastropods Across an Interface Between the Canadian Shield and Limestone Formations. *Can. J. Zool.* **1972**, *50*, 1433–1445.

Mead, A. R. Anatomy, Phylogeny and Zoogeography in African Land Snail Family Achatinidae. In *Proceedings of the 12th International Malacological Congress*, Vigo, Spain, 1995, pp 422–423.

Medina, A. B.; Griffond, B.; Gomot, P. Influence of Photoperiod on Differentiation of Male Cell in *Helix aspersa*, an Autoradiographic Study. *Reprod. Nut. Develop.* **1988**, *28*, 617–623.

Meirreles, L. M. O.; Silva, L. C.; Junqueira, F. O. and Bessa, E. C. A. The Influence of Diet and Isolation on Growth and Survival in the Land Snail *Bulimulus tenuissimus* (Mollusca: Bulimulidae) in Laboratory. *Rev. Bras. Zool.* **2008**, *25*, 224–227.

Morton, J. E. *Molluscs. Harper Torchbooks*, Harper & Bros.: New York, 1960; p 232.

Murphy, S. D. Toxic Effects of Pesticides. In *Toxicology: The Basic Science of Poisons*; Klaassen, C. D., Amdur, M. V., Doull, J., Eds.; 3rd ed;. McMillan: New York, 1986; pp 519–581.

Naggs, F. William Benson and the Study of Land Snails of British India and Ceylon. *Arch.. Nat. Hist.* **1997**, *24*; 37–88.

Nevo, E.; Bar-El, C.; Bar, Z. Genetic Diversity, Climatic Selection and Speciation of *Sphincterochila* Land Snails in Israel. *Biol. J. Linn. Soc.* **1983**, *19*, 339–373.

Newell, P. F. The Nocturnal Behaviour of Slugs. Med. Biol. Illust. **1966**, *16*, 146–159.

Notten, M. J. M.; Oosthoek, A. J. P. and Rozema, J. Heavy Metal Concentration in Soil-plant-Snail Food Chain Along a Terrestrial Soil Pollution Gradient. Environ. Pollut. **2005**, *138*, 178–190.

Nuenberg, H. W. Bio-accumulation of Heavy Metals by Bivalves from Limfford (North Adriatic Sea). *J. Mar. Biol. Sci.* **1984**, *81* (2), 177–180.

Nuenberg, K. J. Fresh Water Mollusks as Sentinel Organisms of Heavy Metal Accumulation in Fresh Water Ecosystem, *J. Environ. Pollut.* **1984**, *16*, 26–33.

Odiete, W. O. Environmental Physiology of Animals and Pollution. Diversified Resources LTD, 1999.

Ohayo, G. J. A.; Heederik, D. J. J.; Kromhout, H.; Omondi, B. E. O.; Boleij, J. S. M. Acetylcholinesterase Inhibition as an Indicator of Organophosphate and Carbamate Poisoning in Kenyan Agricultural Workers. *Int. J. Occup. Environ. Health* **1997**, *3*, 210–220.

Okka, M. A.; Ahmad, F. A. M. and Sharshar, F. A. Efficacy of Certain Pesticides Against the Land Snail, *Monacha contiana* (Muller) Found on Some Orchards Under Laboratory Conditions. Fourth Arabic Conf. for Horticulture Crops, El-Minia, Egypt, 1996, pp 903–910.

Oli, B. P. Ecology of Terrestrial Mollusks in Kumaun Himalayan Forest, Ph.D. Thesis, Kumaun University, Nainital, India, 1996.

Orisawuyi, Y. A. Practice Guide to Snails Rearing. Gratitude Enterprises, Lagos, 1989, p 27.

Palovcik, R. A.; Basberg, B. A.; Ram, J. L. Behavioral State Changes Induced in *Pleurobranchaea* and *Aplysia* by serotonin. *Behav. Neural Biol.* **1982**, *35*, 383–394.

Patil, H. Z.; Mahale, P. N. Toxicity Effect of Mercury Chloride on the Freshwater Gastropod, *Indoplanorbis exustus*. *J. Ecobiot.* **2010**, *2/8*, 21–24.

Patil, P. N.; Rao, K. R.; Sasane, S. R.; Vedpathak, A. N.; Shejule, K. S.; Chudhari, T. R. (Impact of Zolone on the Organic Constituents of the Freshwater Gastropod, *Thiara lineata* from Panzara River at Dhule. *Proc. Acad. Environ. Biol.* **1993**, *2* (2), 193–197.

Patil, S. S.; Talmale, S. S., A Checklist of Land and Freshwater Mollusca of Maharashtra State. *Zoos' Print J.* **2005**, *20* (6), 1912–1913.

Patil, S. G. Occurrence of Freshwater Bivalves (Bivalvia: Unionidae) in Pusad, Yavatmal District, Maharashtra. *Zoos' Print J.* **2003**, *18* (9), 1195.

Patil, S. G.; Talmale, S. S. Occurrence of Pestiferous Land Mollusca from Maharashtra on New Host Plants. *Bionotes.* 5 (3), 71.

Patil, V. T.; Kulkarni, A. B.; Shinde, S. V. Seasonal LC50 Variations in Monochrotophos Exposed Snail, *Indoplanorbis exustus. J. Ecotoxicol. Environ. Mount.* **1991**, *1* (4), 321–324.

Pointier, J. P.; Giboda, M. The Case for Biological Control of Snail Intermediate Hosts of *Schistosoma mansoni. Parasitol. Today* **1999**, *15*, 395–397.

Preston, H. B. Fauna of British India including Ceylon and Burma. Mollusca: Freshwater Gastropoda and Pelecypoda, London, 1915, p xix +244.

Radwan, M. A.; El-Wakil, H. B.; Osman, K. A. Toxicity and Biochemical Impact of Certain Oxime Carbamates Pesticide Against Terrestrial Snail, *Theba pisana* (Muller). *J. Environ. Sci. Health* **1992**, *B27* (6), 759–773.

Ramakrishna; Mitra, S. C. Endemic Land Mollusks of India. Records of the Zoological Survey of India, Occasional Paper, 2002, Vol. 196, pp 1–65.

Raut, S. K.; Barker, G. M. *Achatina fulica* Bowdich and Other Achatinidae as Pest in Tropical Agriculture. In *Molluscs as Crop Pests*; Barker, G. M., Ed.; CABI Publishing: Wallingford, U.K., 2002; pp 55–114.

Raut, S. K.; Ghose, K. C. Food Preference and Feeding Behavior of Two Pestiferous Snails, *Achatina fulica* (Bowdich) and *Macrochlamys indica*. Record of Zoological Survey of India, 1983, Vol. 80, pp 421–440.

Riddle, W. A. Cold Hardiness in Wood Snail, *Anguispira alternate* (Say) (Endodontidae). *J. Ther. Biol.* **1981**, *6*, 117–120.

Ritchie, L. S. Chemical Control of Snails. In *Epidemiology and Control of Schistosomiasis*; Karger, Basel, Eds.; University Park Press: Baltimore, 1973; pp 248–532.

Rittschof, D.; McClellan-Green, P. Molluscs as Multidisciplinary Models in Environment Toxicology. *Mar. Pollut Bull.* **2005**, *50*, 369–373.

Rosenstock, L.; Daniell, W.; Barnhart, S.; Schwartz, D.; Demers, P. A. Chronic Neurophysiological Sequel of Occupational Exposure to Organophosphate Insecticides. *Am. J. Ind. Med.* **1990**, *18*, 321–325.

Sacchi, C. F. Ecological and Historical Bases for a Study of the Iberian Terrestrial Mollusca. Proceedings of the First European Malacologists Congress (1962), 1965, pp 243–257.

Salem, H.; Olajos, E. J. Review of Pesticides: Chemistry, Uses, and Toxicology. *Toxicol. Indus. Health.* **1988**, *4*, 291–321.

Schileyko, A. A. Treatise on Recent Terrestrial Pulmonate Molluscs: Helicarionidae, Gymnarionidae, Rhysotinidae, Ariophantidae. Ruthenica, Supplement 2, 2002, Part 9: pp 1167–1307.

Schmidt-Nielsen, K.; Taylor, K. C. R.; Shkolnik, A. Desert Snails: Problems of Heart, Water and Food. *J. Exp. Biol.* **1971**, *55*, 385–398.

Schulze-Baldes, M. Lead Uptake from Sea Water and Food, and Lead Loss in the Common Mussel *Mytilus edulis. Mar. Biol.* **1974**, *25*, 177–193.

Schuytema, G. S.; Nebeker, A. V.; Griffis, W. L. Effects of Dietary Exposure to Forest Pesticides on the Brown Garden Snail, *Helix aspersa* (Muller). *Arch. Environ. Contam. Toxicol.* **1994**, *26*, 23–28.

Seifert, D. V.; Shutov, S. V. Role of Certain Terrestrial Mollusks in the Transformation of Leaf Litter. *Ekologiya* **1979**, *5*, 58–61.

Sen, S.; Ravikanth, G.; Aravind, N. A. Land Snails (Mollusca: Gastropoda) of India: Status, Threats and Conservation Strategies. *J. Threa Taxa* **2012**, *4* (11), 3029–3037.

Shain, W.; Bush, B.; Seegal, R. Neurotoxicity of Polychlorinated Biphenyls: Structure-Activity Relationship of Individual Congeners. *Toxicol. Appl. Pharmacol.* **1991**, *111*, 33–42.

Silva, L. C.; Meirreles, L. M. O.; Junqueira, F. O.; Bessa, E. C. A. Development and Reproduction in *Bulimulus tenuissimus* (Mollusca: Bulimulidae) in Laboratory. *Rev. Bras. Zool.* **2008**, *25*, 220–223.

Smith, G. Copulation and Oviposition in *Lymnaea truncatula* (Muller). J. Molluscan Stud. **1981**, *47*, 108–111.

Solem, A. Some Non-marine Mollusks from Thailand, with Notes on Classification of the Helicarionidae. *Spolia Zoologica Musei Hauniensis* **1966**, *24*, 1–110.

Sturrock, R. F. Snail Collection to Detect Schistosome Transmission Sites. *Parasitol. Today* **1986**, *2*, 59–63.

Subba Rao, N. V. *Handbook of Freshwater Molluscs in India*; Zoological Survey of. India: Clacutta, 1989; p 289.

Subba Rao, N.V.; Dey, A. Freshwater Molluscs in Aquaculture, p225–232. In *Handbook of Freshwater Mollusca of India*; Zoological Survey of India: Clacutta, 1989; p 289.

Subba Rao, N. V.; Mitra, S. C. On Land Freshwater Molluscs of Pune District, Maharashtra. Records of the Zoological Survey of India, 1979, Vol. 75, pp 1–37.

Surya Rao, N. V.; Mitra, S. C.; Maitra, S. *Mollusca of Ujani Wetland*; Wetland Ecosystem Series 2: Fauna of Ujani, Zoological Survey of India: Kolkata, 2002; pp 110–115

Tompa, A. S. Studies on the Reproductive Biology of Gastropods: Part III: Calcium Provision and the Evolution of Terrestrial Eggs Among Gastropods. *J. Conchol.* **1980**, *30*, 145–154.

Tompa, A. S. Land Snails. In *The Mollusca*; *Reproduction*; Tompa, A. S., Verdonk, N. H.; van den Biggelaar, J. A. M., Eds.; Academic Press, New York, 1984; Vol. 7, p 486.

Tompa, A. S.; Wilbur, K. M. Calcium Mobilization During Reproduction in Snail *Helix aspersa*. *Nature* **1977**, *270*, 53–54.

Tonapi, G. T. Studies on the Freshwater and Amphibious Mollusca of Poona with Notes on Their Distribution- Part II. *J. Bombay Nat. Hist. Soc.* **1971**, *68* (1), 115–126.

Tonapi, G. T.; Mulherkar, L. On the Freshwater Mollusks of Poona. *J. Bombay Nat. Hist. Soc.* **1963**, *60* (1), 104–120.

Triebskorn, R.; Ebert, D. The Importance of Mucus Production in Slug's Reaction to Molluscicides and the Impact of Molluscicides on the Mucus Producing System. In *Slugs and Snails in World Agriculture*; Henderson, I. F., Ed.; Monograph on 41. British Crop Protection Council: Thornton Health, 1989; pp 373–378.

Triebskorn, R.; Kunast, C. Ultrastructural Changes in the Digestive System of *Deroceras reticulatum* Induced by Lethal and Sub-leathal Concentrations of the Carbamate Molluscicide *Cloethocarb*. *Malacologia*. **1990**, *32*, 87–104.

Triebskorn, R.; Henderson, I. E.; Martin, A.; Kohler, H. R. Slugs as Target or Non-target Organisms for Environmental Chemical. In *Slug and Snail Pest in Agriculture*;

Henderson, I. F., Ed.; British Crop Protection Council Monograph 66: Farnharm, 1996; pp 65–72.

Turner, G. I. Snail Transmission of the Species of *Phytopthora* with Special Reference to Foot Rot of *Piper nigrum*. Trans. Br. Mycol. Soc. **1967**, *50*, 251–258.

Viard, B.; Maul, A.; Pihan, J. C. Standard Use Conditions of Terrestrial Gastropods in Active Bio Monitoring of Soil Contamination. *J. Environ. Monitor.* **2004**, *6*, 103–107.

Wagge, L. E. Quantitative Studies of Calcium Metabolism in *Helix aspersa*. *J. Exp. Zool.* **1952**, *120*, 311–342.

Weiss, K. R.; Koch, U. T.; Koester, J.; Mandelbaum, D. E.; Kupfermann, I. Neural and Molecular Mechanisms of Food Induced Arousal in *Aplysia californica*. In *Advances in Physiological Sciences*; Salanki, J., Ed.; Neurobiology of Invertebrates Pergamon: Oxford, 1981; pp 305–344.

Wester, R. E.; Goth, R. W.; Wedd, R. E. Transmission of Downy Mildew (*Phytophthora phaseoli*) of Lima Beans by Slugs. *Phytopathology* **1964**, *54*, 749.

WHO The Control of Schistosomiasis, Second report of the WHO Expert committee, WHO Technical Report Series 830, WHO Geneva 05/01 (CG2005 -A01), 1993.

WHO *Report of the WHO Informal Consultation on Schistosomiasis Control*, Geneva 2-4 December, WHO/CDS/CPC/SIP/99.2, 1998.

Wilkinson, C. F. Insecticide Biochemistry and Physiology. Heyden Verlag, London, 1976.

Williamson, P.; Cameron, R. A. D. Natural Diet of the Land Snail *Cepaea Nemoralis*. *Oikos* **1976**, *27*, 493–500.

Willig, M. R.; Camilo, G. R. The Effect of Hurricane Hugo on Six Invertebrate Species in Luquillo Experimental Forest of Puerto Rico. *Biotropica* **1991**, *23*, 455–461.

Winberg, G. G. Experimental Application of Various Systems of biological indication of water pollution. *Proceedings of 1st and 2nd USA-USSR Symposium on Effect of Pollutants upon Aquatic Ecosystems*. Mount, D. I., Ed.;. *Environ. Res. Lap.* **1978**, *1*, 146–149.

Winner, R. W.; Boesel, M. W.; Farrell, M. P. Insect Community Structure as an Index of Metal Pollution in Licit Ecosystems. *Can. J. Fish. Aqua. Sci.* **1980**, *37*, 647–655.

Yoloye, V. I. *Basic Invertebrate Zoology*; Code and Quanta: Nigeria LTD, 1994; pp 140–145.

Young, A. G.; Wilkins, R. M. A New Technique for Accessing the Contact Toxicity of Molluscicides to Slugs. *J. Molluscan Stud.* **1989**, *53*, 533–536.

Zedan, H. A.; Mortada, M. M.; Shoeib Amera A. Assessment of Molluscicidal Activity of Certain Pesticide Against Two Land Snails Under Laboratory and Field Circumstances at Dakahlia Governorate. *J. Agric. Sci. Mansoura Univ.* **2006**, *31* (6), 3957–3962.

INDEX